T0350439

Applied AI and Multimedia Technologies for Smart Manufacturing and CPS Applications

Emmanuel Oyekanlu
Drexel University, USA

A volume in the Advances in Computational Intelligence and Robotics (ACIR) Book Series

Published in the United States of America by
IGI Global
Engineering Science Reference (an imprint of IGI Global)
701 E. Chocolate Avenue
Hershey PA, USA 17033
Tel: 717-533-8845
Fax: 717-533-8661
E-mail: cust@igi-global.com
Web site: http://www.igi-global.com

Copyright © 2023 by IGI Global. All rights reserved. No part of this publication may be reproduced, stored or distributed in any form or by any means, electronic or mechanical, including photocopying, without written permission from the publisher. Product or company names used in this set are for identification purposes only. Inclusion of the names of the products or companies does not indicate a claim of ownership by IGI Global of the trademark or registered trademark.

Library of Congress Cataloging-in-Publication Data

Names: Oyekanlu, Emmanuel, 1979- editor.
Title: Applied AI and multimedia technologies for smart manufacturing and CPS applications / Emmanuel Oyekanlu, editor.
Other titles: Applied artificial intelligence and multimedia technologies for smart manufacturing and cyber-physical systems applications
Description: Hershey, PA : Engineering Science Reference (an imprint of IGI Global), [2021] | Includes bibliographical references and index. | Summary: "This book provides a review of the state of the art technology regarding the integration of AI and multimedia technologies for smart manufacturing applications, offering detailed industrial integration examples for specific AI and multimedia technologies using both R&D and empirical analysis to examine the impacts of the integration of specific multimedia and AI technologies for stated industrial manufacturing applications"-- Provided by publisher.
Identifiers: LCCN 2021000177 (print) | LCCN 2021000178 (ebook) | ISBN 9781799878520 (hardcover) | ISBN 9781799878537 (paperback) | ISBN 9781799878544 (ebook)
Subjects: LCSH: Computer integrated manufacturing systems. | Cooperating objects (Computer systems) | Multimedia systems.
Classification: LCC TS155.63 .A66 2021 (print) | LCC TS155.63 (ebook) | DDC 670.285--dc23
LC record available at https://lccn.loc.gov/2021000177
LC ebook record available at https://lccn.loc.gov/2021000178

This book is published in the IGI Global book series Advances in Computational Intelligence and Robotics (ACIR) (ISSN: 2327-0411; eISSN: 2327-042X)

British Cataloguing in Publication Data
A Cataloguing in Publication record for this book is available from the British Library.

All work contributed to this book is new, previously-unpublished material. The views expressed in this book are those of the authors, but not necessarily of the publisher.

For electronic access to this publication, please contact: eresources@igi-global.com.

Advances in Computational Intelligence and Robotics (ACIR) Book Series

Ivan Giannoccaro
University of Salento, Italy

ISSN:2327-0411
EISSN:2327-042X

Mission

While intelligence is traditionally a term applied to humans and human cognition, technology has progressed in such a way to allow for the development of intelligent systems able to simulate many human traits. With this new era of simulated and artificial intelligence, much research is needed in order to continue to advance the field and also to evaluate the ethical and societal concerns of the existence of artificial life and machine learning.

The **Advances in Computational Intelligence and Robotics (ACIR) Book Series** encourages scholarly discourse on all topics pertaining to evolutionary computing, artificial life, computational intelligence, machine learning, and robotics. ACIR presents the latest research being conducted on diverse topics in intelligence technologies with the goal of advancing knowledge and applications in this rapidly evolving field.

Coverage

- Algorithmic Learning
- Evolutionary Computing
- Cognitive Informatics
- Robotics
- Natural Language Processing
- Computational Logic
- Cyborgs
- Fuzzy Systems
- Neural Networks
- Artificial Life

IGI Global is currently accepting manuscripts for publication within this series. To submit a proposal for a volume in this series, please contact our Acquisition Editors at Acquisitions@igi-global.com or visit: http://www.igi-global.com/publish/.

The Advances in Computational Intelligence and Robotics (ACIR) Book Series (ISSN 2327-0411) is published by IGI Global, 701 E. Chocolate Avenue, Hershey, PA 17033-1240, USA, www.igi-global.com. This series is composed of titles available for purchase individually; each title is edited to be contextually exclusive from any other title within the series. For pricing and ordering information please visit http://www.igi-global.com/book-series/advances-computational-intelligence-robotics/73674. Postmaster: Send all address changes to above address.
© © 2023 IGI Global. All rights, including translation in other languages reserved by the publisher. No part of this series may be reproduced or used in any form or by any means – graphics, electronic, or mechanical, including photocopying, recording, taping, or information and retrieval systems – without written permission from the publisher, except for non commercial, educational use, including classroom teaching purposes. The views expressed in this series are those of the authors, but not necessarily of IGI Global.

Titles in this Series

For a list of additional titles in this series, please visit: http://www.igi-global.com/book-series/advances-computational-intelligence-robotics/73674

Developments in Artificial Intelligence Creativity and Innovation
Ziska Fields (University of Johannesburg, South Africa)
Engineering Science Reference • © 2023 • 300pp • H/C (ISBN: 9781668462706) • US $270.00

Applying AI-Based IoT Systems to Simulation-Based Information Retrieval
Bhatia Madhulika (Amity University, India) Bhatia Surabhi (King Faisal University, Saudi Arabia) Poonam Tanwar (Manav Rachna International Institute of Research and Studies, India) and Kuljeet Kaur (Université du Québec, Canada)
Engineering Science Reference • © 2023 • 300pp • H/C (ISBN: 9781668452554) • US $270.00

Handbook of Research on Applications of AI, Digital Twin, and Internet of Things for Sustainable Development
Brojo Kishore Mishra (GIET University, India)
Engineering Science Reference • © 2023 • 538pp • H/C (ISBN: 9781668468210) • US $325.00

Constraint Decision-Making Systems in Engineering
Santosh Kumar Das (Sarala Birla University, Ranchi, India) and Nilanjan Dey (Techno International New Town, Kolkata, India)
Engineering Science Reference • © 2023 • 312pp • H/C (ISBN: 9781668473436) • US $270.00

Handbook of Research on Artificial Intelligence Applications in Literary Works and Social Media
Pantea Keikhosrokiani (School of Computer Sciences, Universiti Sains Malaysia, Malaysia) and Moussa Pourya Asl (School of Humanities, Universiti Sains Malaysia, Malaysia)
Engineering Science Reference • © 2023 • 376pp • H/C (ISBN: 9781668462423) • US $325.00

Convergence of Deep Learning and Internet of Things Computing and Technology
T. Kavitha (New Horizon College of Engineering (Autonomous), India & Visvesvaraya Technological University, India) G. Senbagavalli (AMC Engineering College, Visvesvaraya Technological University, India) Deepika Koundal (University of Petroleum and Energy Studies, Dehradun, India) Yanhui Guo (University of Illinois, USA) and Deepak Jain (Chongqing University of Posts and Telecommunications, China)
Engineering Science Reference • © 2023 • 349pp • H/C (ISBN: 9781668462751) • US $270.00

Deep Learning Research Applications for Natural Language Processing
L. Ashok Kumar (PSG College of Technology, India) Dhanaraj Karthika Renuka (PSG College of Technology, India) and S. Geetha (Vellore Institute of Technology, India)
Engineering Science Reference • © 2023 • 290pp • H/C (ISBN: 9781668460016) • US $270.00

701 East Chocolate Avenue, Hershey, PA 17033, USA
Tel: 717-533-8845 x100 • Fax: 717-533-8661
E-Mail: cust@igi-global.com • www.igi-global.com

Table of Contents

Preface xii

Chapter 1
Data Engineering for the Factory of the Future: From Factory Floor to the Cloud – Part 1: Performance Evaluation of State-of-the-Art Data Formats for Time Series Applications 1
Emmanuel Oyekanlu, Corning Incorporated, New York, USA
David Kuhn, Corning Incorporated, New York, USA
Grethel Mulroy, Corning Incorporated, New York, USA

Chapter 2
Data Engineering for the Factory of the Future, Multimedia Applications and Cyber-Physical Systems: Part 2 – Algorithms and Python-Based Software Development for Time-Series Data Format Conversion 28
Emmanuel Oyekanlu, Corning Incorporated, New York, USA
David Kuhn, Corning Incorporated, New York, USA
Grethel Mulroy, Corning Incorporated, New York, USA

Chapter 3
Deep Learning-Based Industrial Fault Diagnosis Using Induction Motor Bearing Signals 176
Saiful Islam, Ahsanullah University of Science and Technology, Bangladesh
Sovon Chakraborty, European University of Bangladesh, Bangladesh
Jannatun Naeem Muna, United International University, Bangladesh
Moumita Kabir, European University of Bangladesh, Bangladesh
Zurana Mehrin Ruhi, Brac University, Bangladesh
Jia Uddin, Woosong University, South Korea

Chapter 4
Next-Generation Industrial Robotics: An Overview 205
Khalid H. Tantawi, University of Tennessee at Chattanooga, USA
Victoria Martino, University of Tennessee at Chattanooga, USA
Dajiah Platt, University of Tennessee at Chattanooga, USA
Yasmin Musa, Motlow State Community College, USA
Omar Tantawi, Motlow State Community College, USA
Ahad Nasab, University of Tennessee at Chattanooga, USA

Detailed Table of Contents

In this chapter, the benefits that can be derived by using different existing data formats for industrial IoT (IIoT) and factory of the future (FoF) applications are analyzed. For factory floor automation, in-depth performance evaluation in terms of storage memory footprint and usage advantages and disadvantages are provided for various traditional and state-of-the-art data formats including: YAML, Feather, JSON, XML, Parquet, CSV, TXT, and Msgpack. Benefits or otherwise of using these data formats for cloud based FoF applications including for setting up robust Delta Lakes having very reactive bronze, silver, and gold data tables are also discussed. Based on extensive literature survey, this chapter provides the most comprehensive data storage performance evaluation of different data formats when IIoT and FoF applications are considered. The companion chapter, Part II, provides an extensive Pythonlibraries and examples that are useful for converting data from one format to another.

This chapter is the companion chapter to “Part 1: State-of-the-Art Time-Series Data Formats Performance Evaluation.” In this chapter, algorithms for converting data from one format to other formats are presented. To implement the algorithms, existing open-source Python libraries are used extensively, and where needed, new Python routines for converting data formats are developed. It is envisaged that the algorithms and Python libraries and routines that are freely provided in this chapter will be useful for data engineers, data scientists, and for industrial IoT, cyber-physical systems (CPS), multimedia, and big data practitioners who are on the quest to use different types of data formats that are compatible with memory-constrained factory floor IoT devices. It will also be useful for Delta Lake and big data engineers, who are on the quest for delivering robust bronze, silver, and gold data lakes in the cloud.

Chapter 3

Saiful Islam, Ahsanullah University of Science and Technology, Bangladesh
Sovon Chakraborty, European University of Bangladesh, Bangladesh
Jannatun Naeem Muna, United International University, Bangladesh
Moumita Kabir, European University of Bangladesh, Bangladesh
Zurana Mehrin Ruhi, Brac University, Bangladesh
Jia Uddin, Woosong University, South Korea

Earlier detection of faults in industrial types of machinery can reduce the cost of production. Observing these machines for humans is always a difficult task, for that purpose we need an automated process that can constantly monitor these machines. Without continuous monitoring, a huge downfall can happen that can cost enormous monitory value. In this research, we propose some transfer learning models along with LSTM for earlier detection of faults from vibration signals. Open source Case Western Reserve University (CWRU) dataset has been used to detect four types of signals using transfer learning models. The four classes are Normal, Inner, Ball, Outer. The dataset has divided into three parts namely set1, set2, and set3. VGG19, DenseNet-121, ResNet-50, InceptionV3, and LSTM are applied to that dataset for detecting faults in this signal. The earlier result shows VGG19, LSTM and InceptionV3 can predict the faults in signal with 100% accuracy in the validation set where DenseNet-121, Resnet-50 show an accuracy of 97% and 98% respectively.

Chapter 4

Khalid H. Tantawi, University of Tennessee at Chattanooga, USA
Victoria Martino, University of Tennessee at Chattanooga, USA
Dajiah Platt, University of Tennessee at Chattanooga, USA
Yasmin Musa, Motlow State Community College, USA
Omar Tantawi, Motlow State Community College, USA
Ahad Nasab, University of Tennessee at Chattanooga, USA

The Covid-19 pandemic resulted in a disruption across all industries; market data suggests that the demand for industrial robotics has steadily increased despite the pandemic, and possibly due to the impact of the pandemic. Particularly in industries that were not historically robotic markets, such as hospitals and distribution industries. In addition to that, non-traditional markets such as electronics and pharmaceuticals have become the dominant markets for industrial robots, taking over that position that has always been held by the automotive industry. The main challenge that faces increased artificial-intelligence (AI)-based technologies in next-generation collaborative robotics is the need for established preventive and corrective maintenance protocols on the AI-based technologies, as well as the need for established technician training programs on the new technologies and the wide availability of trained technicians. Another challenge that faces deploying mobile collaborative robotics in industry is the lack of safety standards for that technology.

Chapter 5
A Review of Big Data Analytics for the Internet of Things Applications in Supply Chain Management 221
Kamalendu Pal, City, University of London, UK

The internet of things (IoT) presents opportunities that enable communication between virtual and physical objects. It produces new digitized services that improve supply chain performance. Moreover, artificial intelligence (AI) techniques resolve unpredictable, dynamic, and complex global product development and supply chain-related problems. In this operating environment, heterogeneous enterprise applications, either manufacturing or supply chain management, either inside a single enterprise or among network enterprises, require sharing information. Thus, data management and its analytical interpretation have become a significant drivers for management and product development in networked enterprises. This chapter describes an information systems framework for the global product development purpose, and it also highlights how businesses can use business intelligence from gathered data from IoT applications. Finally, the chapter describes important categories of Big Data analytics applications for the supply chain operations reference (SCOR) model, and it also presents a data processing framework for supply chain management (SCM).

Chapter 6
Cognitive Load Measurement Based on EEG Signals 246
Tamanna Tasmi, Bangladesh University of Health Sciences, Bangladesh
Mohammad Parvez, Brac University, South Korea
Jia Uddin, Woosong University, South Korea

Measurement of the cognitive load should be advantageous in designing an intelligent navigation system for visually impaired people (VIPs) when navigating unfamiliar indoor environments. Electroencephalogram (EEG) can offer neurophysiological indicators of the perceptive process indicated by changes in brain rhythmic activity. To support the cognitive load measurement by means of EEG signals, the complexity of the tasks of the VIPs during navigating unfamiliar indoor environments is quantified considering diverse factors of well-established signal processing and machine learning methods. This chapter describes the measurement of cognitive load based on EEG signals analysis with its existing literatures, background, scopes, features, and machine learning techniques.

Chapter 7
A Study on Deep Learning Methods in the Concept of Industry 4.0 256
Mehmet Ali Şimşek, Tekirdag Namik Kemal University, Turkey
Zeynep Orman, Istanbul University-Cerrahpasa, Turkey

Nowadays, the main features of Industry 4.0 are interpreted to the ability of machines to communicate with each other and with a system, increasing the production efficiency, and development of the decision-making mechanisms of robots. In these cases, new analytical algorithms of Industry 4.0 are needed. By using deep learning technologies, various industrial challenging problems in Industry 4.0 can be solved. Deep learning provides algorithms that can give better results on datasets owing to hidden layers. In this chapter, deep learning methods used in Industry 4.0 are examined and explained. In addition, data sets, metrics, methods, and tools used in the previous studies are explained. This study can lead to artificial intelligence studies with high potential to accelerate the implementation of Industry 4.0. Therefore, the authors believe that it will be a handbook and very useful for researchers who want to do research on this topic.

Chapter 8
Smart Manufacturing: Post-Pandemic and Future Trends 278
Khalid H. Tantawi, University of Tennessee at Chattanooga, USA
Ismail Fidan, Tennessee Tech University, USA
Yasmin Musa, Motlow State Community College, USA
Anwar Tantawy, Smart Response Technologies, Canada

In this chapter, the current state and future trends in smart manufacturing (SM) and its technologies are presented with the perspective of economic growth and evolution of policies and strategies that steer its growth. The long-term effect of the COVID-19 pandemic on manufacturing is investigated. As a result of the COVID-19 pandemic, a long-lasting effect on manufacturing is foreseen, particularly in the supply chain dependency. To overcome future supply chain disruptions, attention is expected to shift towards incorporating industrial and service robotics, additive manufacturing, and augmented and virtual reality. Additive manufacturing will continue to play an increased role in customized product manufacturing. More demand is expected in the long term of additive manufacturing to counter future supply chain interruption.

Chapter 9
Social Perspective of Suspicious Activity Detection in Facial Analysis: An ML-Based Approach for Digital Transformation 301
Rohit Rastogi, ABES Engineering College, India
Parul Singhal, ABES Engineering College, India

Technology is demanded on to curb crimes, especially image recognition, which can be used to detect suspicious activities. Image, object, and face recognition along with speech identification can be used as great tools to achieve this target. The machine lerning algorithm gave immense capabilities to detect faces, objects, and speech to identify malicious activities, and with several epochs, the accuracy can be enhanced. The chapter applies the various ML algorithms on real-time video data to increase the accuracy and gets satisfactory results in this social cause of utmost importance.

Chapter 10
Nonlinear Filtering Methods in Conditions of Uncertainty 323
Rinat Galiautdinov, Independent Researcher, Italy

In the chapter, the author considers the possibility of applying modern IT technologies to implement information processing algorithms in UAV motion control system. Filtration of coordinates and motion parameters of objects under a priori uncertainty is carried out using nonlinear adaptive filters: Kalman and Bayesian filters. The author considers numerical methods for digital implementation of nonlinear filters based on the convolution of functions, the possibilities of neural networks and fuzzy logic for solving the problems of tracking UAV objects (or missiles), the math model of dynamics, the features of the practical implementation of state estimation algorithms in the frame of added additional degrees

of freedom. The considered algorithms are oriented on solving the problems in real time using parallel and cloud computing.

Preface

In the past decade, artificial intelligence (AI), data analytics, multimedia technologies and methods of integrating smart systems in manufacturing industries and in Industry 4.0 applications have been steadily growing in availability and diversity. However, for industrial leaders, finding cost effective AI and robust multimedia technologies that are easily implementable, remains a challenging endeavor. In addition, due to the lack of uniform standards for Industry 4.0 and cyber-physical systems (CPS), different types of data sets in different formats are always generated from differently standardized devices at different layers of manufacturing CPS. Thus, finding means and methods of reliable data governance, data integration, inter-operable data generation and data analytics for CPS and for general industrial applications is also quite challenging. These problems are even more pervasive since the ecosystem of industrial and CPS devices and applicable technologies keeps diverging.

Applied AI and Multimedia Technologies for Smart Manufacturing and CPS Applications provides an in-depth review of the challenges and applicable state-of-the art multimedia, deep learning, data conversion and data integration solutions for IIoT and industrial CPS. In the book, challenges relating to the integration of data, machine learning (ML) and multimedia technologies for smart manufacturing applications and CPS are discussed in detail. In-depth and wide-ranging hands-on examples and solutions for ML, data engineering, data conversion and data integration strategies for industrial CPS, and for academic and industrial captains are also provided in form of algorithms and Python software implementations.

The book is valuable and quite timely due to its substantive coverage of data engineering, schema and metadata generation strategies for common time series data types that are found in many manufacturing and medical CPS. From extensive literature searches, the book provides the most expansive coverage of Python-based time series data conversion strategies in literature. Algorithms and Python-based data conversion, data engineering, schema and metadata generation solutions that are provided in the book can be used to integrate or convert YAML, CSV, Tensors, HDF5, JSON, Avro, Feather, Parquet etc., data formats to other formats. These data conversion solutions can provide means through which different data formats and data types that are generated from many subsystems of CPS can be collectively useful for providing needed AI solutions for CPS. Methods of making different data formats that are generated from different parts of a CPS to be more inter-operable and readily convertible from one format to another are also extensively provided with the aid of hands-on Python implementations.

Moreover, being able to source data from different parts of CPS and being able to convert generated data sets from one format to another can provide avenues for implementing very reactive and scalable distributed deep learning (DL) solutions for CPS. Time series semantic data interpretation described in detail in one of the book chapters can also contribute to providing needed meaning and context for better understanding of the intrinsic interactions of different parts of CPS. The suite of algorithms and Python solutions provided in the book can also assist experienced data engineers, cloud engineers, IIoT and DevOps practitioners to provide needed data conversion solutions to most problems that are always encountered in industry 4.0 and CPS projects. Other likely beneficiaries of the algorithms and Python implementations presented in the book include applied AI and ML scholars, practitioners, researchers, educators, ML enthusiast and industrial captains.

Topics covered in the book include: Applied DL for industrial fault diagnosis, DL methods in the concept of Industry 4.0, in-depth study on data engineering for Factory of the Future (FoF), performance evaluation of state-of-the-art data formats for time series applications, algorithms and Python-based software development for time series data formats conversion, data engineering for multimedia applications and cyber-physical systems, an overview of the characteristics and importance of next generation industrial robotics, a review of Big Data analytics for the Internet of Things (IoT) applications in supply chain management, cognitive load measurement based on electroencephalogram (EEG) Signals, nonlinear filtering methods in conditions of uncertainty for mobile CPS such as UAVs and airplanes, ML based approach for suspicious activity detection using agents' facial analysis, etc.

CHAPTER 1

Data Engineering for Factory of the Future: From Factory Floor to the Cloud – Part 1: Performance Evaluation of State-of-the-Art Data Formats for Time Series Applications

Different types of data sets in different formats are always generated in industrial and medical CPS. In this chapter, the benefits that can be derived by using different existing data formats for industrial IoT (IIoT) and Factory of the Future (FoF) applications are discussed. For factory floor automation, in-depth performance evaluation in terms of storage memory footprint, usage advantages and disadvantages are provided for various traditional and state-of-the-art data formats including: YAML, Feather, JSON, XML, Parquet, CSV, TXT, and Msgpack. A stock market time-series dataset containing more than 6000 rows of datapoints with each row having fields containing float, int and string data types is used to show the storage memory saving benefits that could be obtained by using each type of data format. Benefits or otherwise of using these data formats for cloud based FoF applications including for setting up robust Delta Lakes having very reactive bronze, silver and gold data tables are also discussed. Based on extensive literature survey, this chapter provides the most comprehensive data storage performance evaluation of different data formats when IIoT and FoF applications are considered. While this chapter provides the overview and performance evaluation of the different data formats; the companion chapter "Part II: Python-based Software Development for Time-Series Data Formats Transformation" provides an extensive expose on Python-based software libraries and examples that are useful for converting big data from one format to another for IIoT and FoF applications.

CHAPTER 2

Data Engineering for Factory of the Future, Multimedia Applications, and Cyber-Physical Systems: Part 2 – Algorithms and Python-Based Software Development for Time-Series Data Formats Conversion

This chapter is the companion chapter to "Part 1: State-of-the-Art Time-Series Data Formats Performance Evaluation". In Part 1, a comprehensive performance evaluation of traditional and state-of-the-art data formats for both factory floor and cloud-based applications for industrial IoT (IIoT) and Factory of the Future (FoF) applications was provided. A time-series data set was the focus of study, and an extensive evaluation of the benefits of data formats conversion for the time-series data set including for factory floor analytics; and for setting up bronze, silver and gold data lakes in the cloud was conducted. Other methods useful for data engineers for setting up robust ETL pipelines from factory floor to the cloud are also discussed. In this chapter, algorithms for converting data from one format to other formats are presented. To implement the algorithms, existing open-source Python libraries are used extensively; and where needed, new Python routines for converting data formats are developed. It is envisaged that our algorithms and Python libraries and routines which are freely provided in this paper will be useful for data engineers, and for industrial IoT, Cyber-Physical Systems (CPS), Multimedia and Big Data practitioners who are on the quest to use different types of data formats that are compatible with memory-constrained factory floor IoT devices. It will also be useful for Delta Lake and Big Data engineers, who are on the quest for delivering robust bronze, silver and gold data lakes in the cloud.

CHAPTER 3

Deep Learning-Based Industrial Fault Diagnosis Using Induction Motor Bearing Signals

Deep learning applications are used extensively in industrial IoT (IIoT) and manufacturing CPS. This chapter discusses the application of DL towards the early detection of faults in industrial machines. Earlier detection of faults in industrial types of machinery can reduce the cost of production. Observing these machines for humans is always a difficult task, for that purpose, there is need for an automated process that can constantly monitor these machines. Without continuous monitoring, a huge downfall can happen that can cost enormous monetary value. In this research, authors proposed some transfer learning models along with LSTM for earlier detection of faults from vibration signals. Open-source Case Western Reserve University (CWRU) dataset was used to detect four types of signals using transfer learning models. The four classes are Normal, Inner, Ball, Outer. The dataset used was divided into three parts namely set1, set2, and set3. For training purposes, authors used set1 while set2 and set3 was used for validation purposes. VGG19, DenseNet-121, ResNet-50, InceptionV3, and LSTM are applied to test dataset for fault detection. Earlier result shows VGG19, LSTM and InceptionV3 can predict the faults in test signal with 100% accuracy in the validation set where DenseNet-121, Resnet-50 show an accuracy of 97% and 98% respectively.

CHAPTER 4

Next-Generation Industrial Robotics: An Overview

Robotic systems are vital to reactive and scalable industrial CPS. This chapter discusses the advantages of deploying next generation intelligent industrial robotics, in place of the current generation of robots. Current generation of industrial robots are always isolated in cages and can only operate in highly- controlled and deterministic environments for safety reasons. Benefits of next generation that are discussed includes: The ability of intelligent robotics to self-learn enhanced mobility, free movement, and ability to collaborate. The high flexibility achieved from intelligent robotics also improves productivity and safety significantly.

CHAPTER 5

A Review of Big Data Analytics for the Internet of Things Applications in Supply Chain Management

Supply chains are integral parts of manufacturing CPS. This chapter provides a systematic approach for using frameworks such as SCOR for supply chains, identifying the potential of Big Data within them, and identifying useful analytics applications for the manufacturing supply chain management. The current business environment pushes manufacturing enterprises to optimize business processes and hold costs. At the same time, the extensive availability of data makes customers increasingly knowledgeable about products, prices, and other essential information. At first, manufacturing supply chain networks appear to easily manage abstract operational frameworks such as SCOR's Plan, Source, Make, Deliver and Return. However, upon deeper examination, manufacturing chains need real-time business process integration, coordination, and collaboration to deliver a higher level of corporate performance.

CHAPTER 6

Cognitive Load Measurement Based on EEG Signals

Reliable navigation is very important in many CPS including for medical CPS. This chapter highlights the importance and role of intelligent navigation system for the visually impaired people (VIPs) whenever they are navigating unfamiliar indoor environments. Electroencephalogram (EEG) signals can offer neurophysiological indicators of perceptive process indicated by changes in brain rhythmic activity. To support the cognitive load measurement by means of EEG signals, the complexity of the tasks of the VIPs during navigating unfamiliar indoor environments is quantified considering diverse factors of well-established signal processing and machine learning methods. This chapter describes the measurement of cognitive load based on EEG signals analysis with its existing literatures, background, scopes, features, and machine learning techniques

CHAPTER 7

A Study on Deep Learning Methods in the Concept of Industry 4.0

In this chapter, a review of articles that uses DL approaches to solve Industry 4. 0 problems are is conducted. The review encompasses Industry 4.0 fields, applicable DL methods, metrics and databases used. The literature study conducted can be used as a guide into other studies in this area.

CHAPTER 8

Smart Manufacturing: Post-Pandemic and Future Trends

In this chapter, important aspects of Smart Manufacturing technologies and practices are introduced. The chapter starts with an overview of the major industrial policies in the world that govern the practices in Smart Manufacturing and direct research fundings. After that, the main technologies are presented, with an emphasis on the future trends in these technologies taking into account the challenges and opportunities that resulted from the CoVid-19 pandemic.

CHAPTER 9

Social Perspective of Suspicious Activity Detection in Facial Analysis: An ML-Based Approach for Digital Transformation

The chapter describes the basics of ML, data analytics, tools and techniques of face recognition technologies in crime detection. Brief review of data mining, regressions and other applicable AI-based detection technologies in use in CPS are also discussed. The work of notable authors and context of their work are reviewed. In the application section, latest classification techniques with reference to applications of support vector machines (SVM), CNN and RNN are also discussed. In the result section, the different agents' emotions with respect to recognized objects and their object have been recognized and their accuracy are discussed. Then in the later part the recommendations, novelty, applications, and limitations of facial analyses technologies are discussed.

CHAPTER 10

Nonlinear Filtering Methods in Conditions of Uncertainty

Unmanned aerial vehicle (UAV) networks are playing important roles in various areas of human endeavors due to their versatility and ease of deployment. In conventional UAVs, embedded systems are used with communication devices, computing, and control modules in the presence of a closed loop network for exchanging data and information, decisioning and instruction execution. These combination demands the integration of cyber processes into physical devices. Therefore, the UAV network could be considered a CPS.

In this chapter, authors considered the possibility of applying modern IT technologies to implement information processing algorithms in UAV motion control system. Filtration of coordinates and motion parameters of objects under a priori uncertainty is carried out using nonlinear adaptive filters: Kalman and Bayesian filters. The author considers numerical methods for digital implementation of nonlinear filters based on the convolution of functions, the possibilities of neural networks and fuzzy logic for solving the problems of tracking UAV objects (or missiles), the math model of dynamics, the features of the practical implementation of state estimation algorithms in the frame of added additional degrees of freedom. The considered algorithms are oriented on solving the problems in real time using parallel and cloud computing.

Emmanuel Oyekanlu
Drexel University, USA

Chapter 1
Data Engineering for the Factory of the Future:
From Factory Floor to the Cloud – Part 1: Performance Evaluation of State-of-the-Art Data Formats for Time Series Applications

Emmanuel Oyekanlu
Corning Incorporated, New York, USA

David Kuhn
Corning Incorporated, New York, USA

Grethel Mulroy
Corning Incorporated, New York, USA

ABSTRACT

In this chapter, the benefits that can be derived by using different existing data formats for industrial IoT (IIoT) and factory of the future (FoF) applications are analyzed. For factory floor automation, in-depth performance evaluation in terms of storage memory footprint and usage advantages and disadvantages are provided for various traditional and state-of-the-art data formats including: YAML, Feather, JSON, XML, Parquet, CSV, TXT, and Msgpack. Benefits or otherwise of using these data formats for cloud based FoF applications including for setting up robust Delta Lakes having very reactive bronze, silver, and gold data tables are also discussed. Based on extensive literature survey, this chapter provides the most comprehensive data storage performance evaluation of different data formats when IIoT and FoF applications are considered. The companion chapter, Part II, provides an extensive Pythonlibraries and examples that are useful for converting data from one format to another.

DOI: 10.4018/978-1-7998-7852-0.ch001

Copyright © 2023, IGI Global. Copying or distributing in print or electronic forms without written permission of IGI Global is prohibited.

INTRODUCTION

In many vertical industries all over the world, numerous legacy machines generate huge amount of operational data that the legacy machines have no means of directly funneling into Industry 4.0 applications. These trapped operational data would have otherwise been valuable for generating insights that may have served to improve manufacturing processes (Accenture, n.d.; Fogg, 2020; Oyekanlu, 2018a). Also, across many industries, data scientists spend over 80 percent of their time just scrubbing data to make it fit for analytical purposes (Snyder, 2019); and by extension, to make it fit for inclusion in Gold Tables in the cloud. In some other cases, companies often find themselves unable to control and manage data at scale. This implies that due to excessive data volume, speed, and lack of transparent methods of ensuring data veracity, companies key decision makers may not be able to control, derive insight from, or operationalize generated dataset. These issues always prevent key decision makers from leveraging data for strategic smart manufacturing initiatives. In addition, most factory floors are still populated with legacy machines; and this often hamper efforts to capture, and process varied data types and deliver analytics insights with high agility.

Being able to easily integrate legacy machines into industrial IoT (IIoT) initiatives will enable manufacturers to begin to have deeper insight on factory shop floor performance from newly generated data as soon as the dataset are generated from those legacy machines. Connecting equipment and systems to FoF and IIoT systems will allow for greater visibility, remote machine monitoring, and accurate performance reporting (Fogg, 2020). In many instances, due to difficulties resulting from lack of easy means of integrating legacy machine situated at factory floors with data analytics repositories at the factory's edge computing platform, or in the cloud; analytics data lakes are often built without sufficiently factoring-in the factory floors' entire analytics needs. Such integration issue oftentimes leads to compromising the factory floor's data preparation, structure, veracity, and lineage. Sometimes, reactive data analytics and useful insights will need data inputs from both the internal factory floor and from external factory customers. However, data from customers may not be available. At some other times however, the data may be obtainable, but may not be readily available in formats that are compatible with the data formats being used on factory floor analytics system.

Naturally, people lacked trust in the data (Accenture, n.d.). This issue makes the challenges of data security and verification difficult. It also makes obtaining strategic values from data almost impossible (Accenture, n.d.). Additionally, employees, due to lack of required training skills, are often limited to using data in some specific legacy formats such as comma separated values (CSV), text (TXT) and JavaScript Object Notation (JSON). For such employees, working with data that are generated in recently available Big Data formats such as Feather, Avro, Optimized Row Columnar (ORC) and Parquet formats is a difficult proposition. Employees' lack of comfortability with these new data formats may sometimes make generating insights and deriving values from data, right from the factory shop, and floor across several FoF tools, such as the factory edge computing platform, Hadoop Big Data platform, and the company's cloud system to be quite burdensome.

In the IIoT ecosystem, enormous amounts of industrial data are generated through several different devices and systems throughout the supply chain. This includes machines, assembly lines, mobile devices, utility meters, smart sensors, automated appliances, routers, robots, and others (Databerg, 2019). The data generated from these industrial devices can be in structured, semi-structured and unstructured formats. As shown in Figure 1, generating data in structured formats allows for greater efficiency in terms of storage and data usage performance, especially during computation. For trustful analytics and adequate data governance, metadata (Kamiya et al., 2021) are sorely needed to provide context at the lower layer since most data are still processed in-situ near the operational machine instead of the cloud. Unlike structured formats such as Parquet and Avro that can readily integrate metadata into their data structures (Yavus et al., 2017), unstructured and semi-structured data formats such as JSON, CSV, images, and TXT do not have associated means by which machine and users' metadata can be readily integrated.

Figure 1. Structured data formats generally performs better in terms of storage and in-memory computation

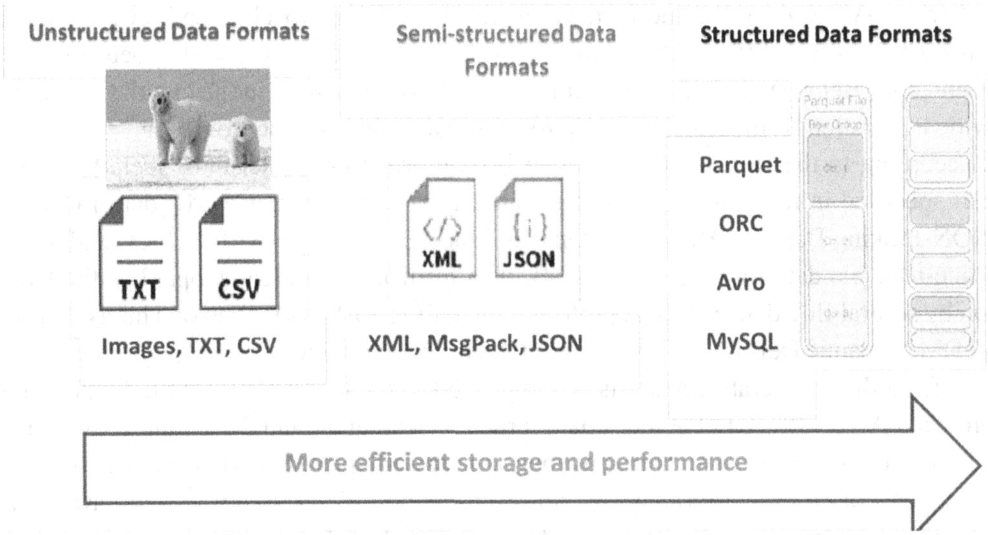

As a result of the lack of metadata, unstructured dataset generally require context around the data to be parse able. This means that due to lack of metadata, to be able to effectively work with the data, users always need to have additional context or detail about the data. Due to extensive availability of dated machines, most sources of IIoT data are unstructured. The cost of having unstructured formats is that it becomes cumbersome to extract value out of these industrial data sources since numerous additional transformations and feature extraction techniques are required to interpret these datasets (Yavus et al., 2017). In semi-structured data formats, each data record is always augmented with its schema information. They always lack well-defined global schema that span all records. As a result, while parsing semi-structured data sets in IIoT computations, they always incur extra parsing overheads. These extra parsing overheads always need further processing; and this may impact time obtain useful analytics that are crucial to the success of many FoF applications.

In addition, semi-structured data formats are not particularly suitable for ad-hoc querying (Yavus et al., 2017).

There have been numerous cost-effective approaches regarding how legacy machines can be enabled to participate and contributes to the successful enablement of Industry 4.0 (Kancharla, 2021; Lima et al., 2019; Oyekanlu, 2017; Oyekanlu, 2018b; Oyekanlu, 2018c; Oyekanlu & Scoles, 2018a; Oyekanlu & Scoles, 2018b; Silva et al., 2021). However, many of these methods focuses on hardware approaches without sufficient digital transformation roadmap by which associated softwarization and toolchain development can be achieved. This issue is even more so in the context of finding approaches of easily integrate new types of reactive and highly useful data formats currently being used in the Big Data industry in the factory floor environment.

Solutioning methods to data formats integration challenges is our aim in both Part I and its companion chapter, Part II. Our contribution in Part I include comparative analysis of the performance of several IIoT data exchange formats that are not included in (Belov et al., 2021).). We have also provided a performance evaluation of several available data formats in terms of storage memory needs for most of the use cases emphasized as being important by the Industrial Internet Consortium (IIC) in (Kamiya et al., 2021). In (Belov et al., 2021).), researchers described the five most popular formats for storing big data. Avro, CSV, JSON, ORC and Parquet are considered. An Apache Spark evaluation and comparative analysis for each of these data formats was accomplished. However, the focus of the research work presented in (Belov et al., 2021).) was for Apache Hadoop system in the cloud. Consideration for system-wide integration from factory floor up to the cloud layer was not provided.

Also, according to the IIC, a comprehensive IIoT information model must include ontology and metamodel specified in the Unified Modeling Language (UML) format. UML data formats include JSON, JSON-Linking Data (JSON-LD), Extensible Markup Language (XML), TXT, and other human and machine-readable data formats. These UML data formats are widely adopted as the language of interaction by several IIoT data exchange protocols, including, OPC-UA, Web of Things, SensorThingsAPI, OneDM/Semantic Definition Format, Asset Administration Shell etc. Many of these data formats are missing from the comparative analysis accomplished in (Belov et al., 2021).). Authors in (Naidu, 2022) compared Avro, ORC and Parquet's data formats advantages, and disadvantage, with regards to their use in the Hadoop ecosystem. Our contribution in this chapter (Part I) also discusses advantages and disadvantages of each data formats. We however extend the discussion to more available legacy and state-of-the-art data formats. We also considered the benefits of using Big Data state-of-the-art data formats for data analytics and other Industry 4.0 applications on the factory floor.

In (Ahmed et al., 2017), authors explore data formats useful for different bioinformatics applications, algorithms, and big data analytics such as CSV, Parquet, Avro, JSON and LibSVM. However, other formats such as MsgPack and Feather are omitted. In (Plase et al., 2017), researchers compared the Avro to the Parquet data format with regards to their data queries performance. In contrast, our storage performance comparison in this chapter considers a wider array of file and data models and formats.

In the companion chapter (Part II: Python-based Software Development for Time-Series Data Formats Transformation), we have provided open source methods by which most existing legacy machines that are generating data in common JSON, TXT, Tab-Separated Values (TSV), Yet Another Markup Language/ YAML Ain't Markup Language (YAML), CSV and XML formats can have those data transformed into formats that are compatible with most current Big Data, edge computing and cloud computing formats such as Avro, Hierarchical Data Format (HDF5), MessagePack (MsgPack) and Parquet data formats.

In section II of this chapter, benefits of using Big Data file formats for data sets that are generated in the IIoT factory floors are discussed. Section III presents a discussion and comparison of advantages and disadvantages of different types of data formats and model. In section IV and V, results of data storage performances of each data format types are presented. The conclusion of the chapter is presented in section VI.

BENEFITS OF DATA FORMAT TRANSFORMATION FOR FACTORY OF THE FUTURE APPLICATIONS

New tools that can enable dated machines to fully participate in IIoT and Big Data revolution that is currently driving the industry 4.0 era are needed in many industries. Many factories and companies often have proprietary data and trade secrets that they always want to keep on premise. In most cases, companies are always reluctant, and they always refrain from send such proprietary data onto cloud repositories since industrial key decision makers can sometimes believe that cloud repositories may be subjected to data breaches. How may data engineers enjoy the benefit offered by Big Data file formats such as Avro and Parquet formats while working with those proprietary, top-secret datasets on the factory floor?

A direct solution to this issue may be that data engineers may have to design methods by which this natively Big Data friendly formats can be made available for analytics applications at the factory floor or at the lower layer of the IIoT analytics pipeline. However, for this solution to work as intended, proprietary data generated by IIoT devices at factory floor level must be stored in data formats that support organized storage and quick retrieval.

Impact of Data Formats on Metadata Generation and Storage

Most legacy data formats such as TXT and CSV does not naturally store metadata along with stored datasets. For companies that does not want to send their proprietary data to the cloud, the use of data formats that do not use data archiving or metadata can lead to an increase in data retrieval time (Belov et al., 2021).). If such companies want to implement real-time smart manufacturing applications, storing data in formats that does not encourage the use of metadata may defeat their Industry 4.0 initiative. Hence a data format conversion from legacy formats such as TXT to state-of-the-art formats such as Avro or Parquet may serve their needs in terms of metadata storage and lower storage memory requirements.

Impact of Data Formats on Data Types and Structures

Legacy data format types such as CSV formats always present data in form of a table. A CSV file does not support different data types and structures. All data types in CSV data formats are generally in string formats (Belov et al., 2021).). If IIoT devices at the factory floor generates data in the form of such legacy file formats, for more reactive and reliable Industry 4.0 analytics, it may be highly useful to transform data from such legacy formats to other relatively new formats such as Avro, ORC or Parquet.

Impact of Data Formats on Time to Read Data

According to report in (Belov et al., 2021).), a most important parameter in data processing and analytics is the time required to read the data. Also important are the times required to filter, sort, group, and search for unique strings in the entire data. According to studies in (Belov et al., 2021).), when data reading time is considered, data analytics machines use more time when reading data in legacy data formats such as JSON and CSV. For the same set of data, Avro, ORC and Parquet, which are state-of-the-art data formats requires less than 50% data reading time when compared to the JSON format (Belov et al., 2021)).

Changing Data Format at Factory Floor Level Can Lead to Storage Cost Reduction in Cloud Delta Lake

In recent times, cloud-based data engineering paradigm has shifted from the concept of a Data Lake to a Delta Lake. In the conventional Data Lake, data can be dumped into the cloud repository in any format without means of data governance enforced by using metadata tracking and schema persistence. Without the metadata, Data Lake, can, with time, become a Data Swamp.

Figure 2. Current Delta Lake Architecture Supports Storing Any Data Format in the Bronze Table, Leading to High Data Storage Cost Due to Large Sizes of Legacy Data Formats Such as CSV, TXT and JSON

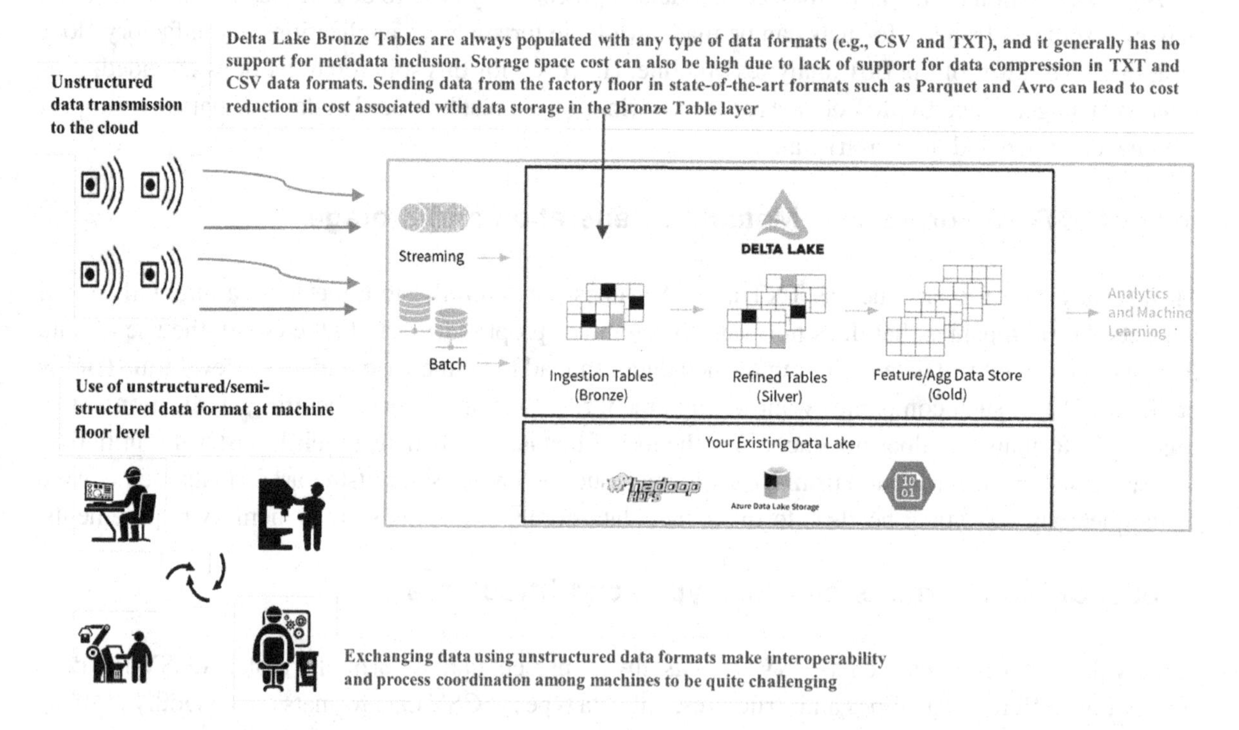

Storing some types of data formats, such as CSV and JSON in the Data Lake can also result to huge storage footprint. Data storage cost analysis accomplished in (Timescale, 2020) reveals that recently, cost of cloud-based data storage has increased. A 10TB disk volume in the cloud is more than $12,000 per annum at $0.10/GB/month for AWS EBS storage subscription. An additional high availability (HA) replicas and backups can grow this number by another 2-3 times relative to the initial storage cost. However, storing data in formats that can achieve a 95% storage compression can save the subscriber $10,000 to $25,000 per annum in storage costs alone. The cost saving can multiply to about $22,800 if the replica HA storage cost are considered (Lee & Heintz, 2019).

The Delta Lake is an improvement over the Data Lake. In the Delta Lake, metadata and metadata changes tracking is always provided in the Silver Table while raw data in any format are always stored in the Bronze Table (Lee & Heintz, 2019). The Gold Table always contain the most refined dataset that can be used by any branch of the company for analytics. However, in this Chapter, we hypothesize that due to storage cost, there must be a paradigm change in the Delta Lake architecture with regards to storing any type of data format in the Delta Lake Bronze Table. The suggested paradigm is as shown in Fig. 3.

Figure 3. Data Format Transformation from the Factory Floor to the Cloud Can Lead to Significant Savings in Storage Cost in the Delta Lake Cloud. It Can Also Lead to Better Analytics and Process Coordination at the Factory Floor.

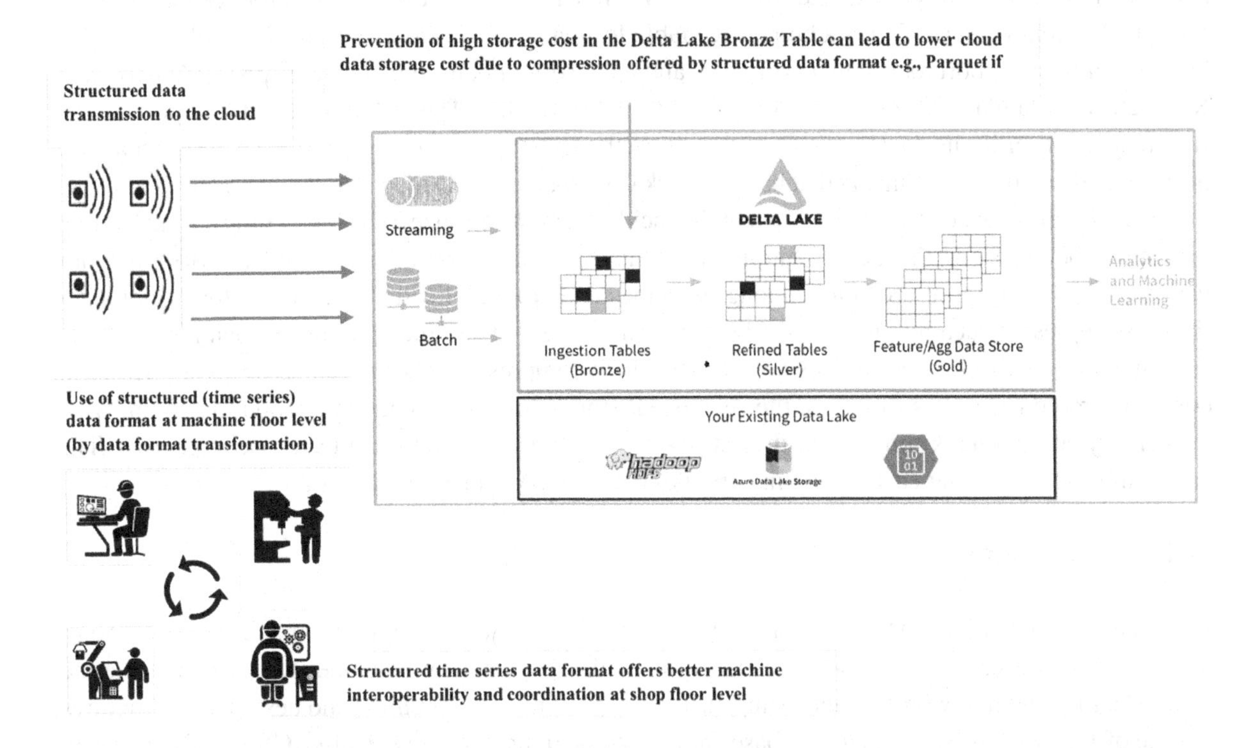

As explained in Figure 3, data format transformation from legacy data formats such as JSON or TXT to state-of-the-art format such as Parquet or Avro from the factory floor to the Bronze Table in the Delta Lake cloud can lead to significant savings in cost of data storage in the cloud. It can also lead to better process coordination, metadata generation and improved process analytics at the factory floor level. The new paradigm change in the usage of data formats can lead to significant benefits if the transformation is possible in a two-way format. The uplink transformation can include data transformation from legacy format (e.g., TXT) to new formats (e.g., Avro) from the factory floor to the Bronze Table in the Delta Lake. While the downlink transformation can, if need be, occur in the reverse format.

BRIEF OVERVIEW OF ADVANTAGES AND DISADVANTAGES OF DIFFERENT DATA FORMATS

As mentioned earlier, using human-readable data formats such as CSV, JSON and TXT have numerous advantages and demerits when FoF, Industry 4.0 and IIoT applications at the factory floor are considered.

Binary formats such as Avro, Parquet, Pickle and Msgpack have numerous advantages and shortcomings as well. This section presents a brief review of these advantages and demerits.

CSV Data Format

In most modern industrial environments, CSV data format is still used extensively (Idreos et al., 2011) since many existing machines or industrial systems may not be able to ingest or produce data in other formats different from CSV. It is a human-readable format, and it is easy to parse (Data Flair, 2018). It has very limited support for schema evolution; and it also does not have metadata support (Ackerman & King, 2019; Data Flair, 2018) as a natural part of its data structure. The lack of metadata support in CSV files may however limits the flow of CSV data from the factory floor to the cloud layer. There may be a need for dedicated data engineers that have to track CSV data in the Delta Lake Bronze Table and enrich the data with the needed metadata to prevent the incidence of the Bronze Table becoming a Data Swamp.

On the factory floor, the lack of automatic methods of integrating metadata on CSV files may prevent seamless sharing of CSV data among machines and IIoT applications on the factory floor. CSV files also represent each row in a CSV file as an ordered record and the location of each record on the file cannot be easily changed. It does not support block compression (Data Flair, 2018). CSV files also consumes more memory space for either persistent storage or for in-memory computing. Also, when in-memory computing is considered, the entire CSV file must be compressed and decompressed. This exerts significant reading cost computing engines during CSV file computation.

JSON Data Format

JSON data format is used extensively in legacy and modern industrial environments. It is a lightweight data format that is easy for humans to read. It is also relatively easy for machines to parse and generate. Due to factory and real-world objects being easier to represent as hierarchical and nested data structures, the use of JSON-based storages in databases have become quite pervasive. Unlike CSV, JSON naturally stores meta data with the actual data; and as such, it fully supports schema evolution. Each record on each row is flexible and the data order can be rearranged. It is also a human-readable data format. However, due to its basically being a text file, it does not support block data compression. Based on needs and applications on the factory floor, users can split files stored in Parquet, ORC and Avro formats across multiple disks for processing. Splitting data enables scalability and parallel processing. JSON data formats cannot be split. This makes processing JSON files in parallel to be impossible (Ackerman & King, 2019).

Table 1. Comparison of different data formats in terms of schema and data compression (Ramm, 2021) (McKinney & Richardson, 2020), (IBM, 2021)

Data Format	Schema Integration	Compression Support
JSON	Yes	No
Parquet	Yes	Yes
CSV	No	No
TXT	No	No
TSV	No	No
ORC	No	Yes
Sequence File	No	No
Avro	Yes	Yes
Feather	Yes	Yes (Feather v2)
XML	Yes	No

Text (TXT) Data Format

Like JSON and Tab-Separated Values (TSV), TXT data format is a simple human readable data format. TXT files uses control characters to separate values; thus, increasing the total TXT file size. Extra computation is also required to adequately parse TXT data. TXT files does not naturally have schema or metadata. It is also complex to handle since it can embrace numerous coding schemes such as ASCII or UFT-8 (Xavier, 2021). TXT files are generally slow to process. Like CSV and TSV file formats, queries engines always must scan entire TXT files to obtain results since the row or string of interest could be anywhere in the file. At Big Data level in the cloud, representing data in TXT formats have significant drawbacks due to the time it always takes to obtain query results. Generally, as shown in Table 1 (Ahmed et al., 2017), (Ramm, 2021),

(McKinney & Richardson, 2020), (IBM, 2021), human-readable data formats, including CSV, TXT, XML and JSON generally, except for XML does not support schema integration. Also, they always take huge amount of storage space due to their lack of support for data compression schemes.

XML Data Format

The Extensible Markup Language (XML) is a human-readable data format that make use of custom tags to define elements in a document. The custom tags can be utilized to support a wide range of data elements. Like JSON, it performs well for storing structured data. Its syntax is as shown in Fig. 4. From Fig. 4, it can be observed that XML's structure is like JSON and YAML structure. XML files are saved in TXT format. It is platform independent (Singh, 2018) and it also support Unicode encoding. Being platform and program independent, it is generally easy to parse and read XML data. Thus, it is a common choice for sharing data between applications and programs. XML support schema based on document type definition (DTD) elements and attributes definition.

However, XML syntax does not support element arraying, it is verbose and less readable when compared to other human-readable formats such as JSON. The verbosity of XML data formats results to higher storage and transportation cost when XML formatting is used to transmit data (Singh, 2018).

Figure 4. Relationship between XML, YAML & JSON

XML

```
<Servers>
  <Server>
    <name>Server1</name>
    <owner>John</owner>
    <created>123456</created>
    <status>active</status>
  </Server>
</Servers>
```

JSON

```
{
  Servers: [
    {
      name: Server1,
      owner: John,
      created: 123456,
      status: active
    }
  ]
}
```

YAML

```
Servers:
  -     name: Server1
        owner: John
        created: 123456
        status: active
```

YAML Data Format

The YAML data format is a superset of the JSON data format as it can accomplish all that JSON does; and much more. It is human-readable. However, while JSON uses brackets and braces as shown in Fig. 4, YAML uses indentations and newlines. It is also different from XML data formats that uses opening and closing tags. It instead uses colon and the dash as delimiters. Thus, it is generally less verbose and more readable that either JSON or XML. YAML data format is good for creating log and configuration files. It is also a good choice for data transmission, one-pass processing, for object persistence across different programming languages (Sharma, 2022), inter-process calls and messaging; and for representing complex data structures using a less verbose format.

HDF5 Data Format

Usage of human-readable data formats such JSON, YAML, TSV, TXT and CSV is very widespread for representing general purpose data. However, their simplicity in most cases makes their use inconvenient for industrial and scientific applications. A common challenge with using human-readable data formats is that reading these type of file formats can be inefficient for large data files. This issue has led to the creation of more appropriate data formats that can adequately address large industrial or scientific requirements.

An example of such data format is the HDF5. It is the current version of the HDF data format. It is very useful in storing complex and large volume of experimental, scientific, and industrial datasets (MPHY0021, 2022). The HDF5 is a binary file data format, and it can be used to store different kinds of heterogenous data of arbitrary sizes and types. HDF5 is also self-describing since it has support for metadata embedding. It also supports file compression. However, even when compressed, HDF5 sizes can still be quite large (Wasser, 2020). Another powerful attribute of HDF5 is data slicing. With data slicing, a subset of a dataset can be onboarded onto memory for in-memory processing in lieu of the entire dataset.

Figure 5. Avro Data Format Structure

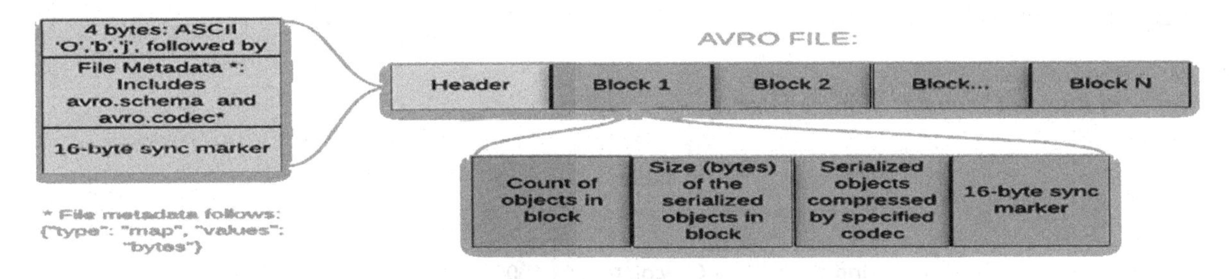

Avro Data Format

The use of binary data formats such as Avro, ORC and Parquet data formats is very predominant for handling Big Data in the industry (Khandelwal, 2020). They are basically machine-readable and not directly human-readable like JSON, TXT and CSV. With these binary data formats, a huge dataset can be split across many disks for parallel data processing. Parallel processing can increase the speed of data processing. Binary data formats are also on-the-wire formats since computing clusters can swap specific datasets even during processing. IIoT and FoF applications at the factory floor level can potentially benefit from increased speed (Ackerman & King, 2019) offered by these Big Data formats.

Specifically, the Avro data format is a row-optimized, schema-based, cross-language data format. By supporting compression, it has huge advantages in terms of space savings; and its basic structure (Bhatia, 2021) is as shown in Figure 5. It is easier to pass and faster to load onto memory since it has no encoding issues. It can be highly useful for real-time IIoT applications since it supports streaming analytics due to its having schema persistence. It can also be very useful for analytics that have need for low processing latency (Dremio, n.d.). Avro schema can be transferred alongside the dataset (Belov et al., 2021).) which makes Avro data sets to be self-describing. For real time IIoT and FoF applications, row-based data formats because of having fast data write times can be quite useful.

Parquet Data Format

Similar to Avro, Parquet is a binary, machine-readable data format. As opposed to Avro however, Parquet is column optimized. This means that specific columnar chunks of Parquet data can be selected for analysis. Figure 6 (Staubli, 2017) is a graphical illustration of how row arrangement of data differs from columnar arrangement; and Figure 7 (Chehaibi, 2017) is an illustration (Chehaibi, 2017) of how Parquet's columnar data layout is implemented along with a nested schema layout. Parquet support data compression, and additional Parquet columns can always be appended as a new column at the end of all the previous Parquet columns in a file. In addition to supporting compression, Parquet data is always small on disk due to its internal use of dictionary and run-length encoding (Levy, 2022). Parquet also supports schema evolution, and it is very efficient in terms of processing speed even though it supports many data types including string, integer, float and most date and time representation. It is also the data format of choice when On-Line Analytics Processing (OLAP) applications are being considered. When compared to Avro, Parquet is also a self-describing data format. For FoF applications, being self-describing could be very useful on the factory floor and in the cloud. Self-describing generally means that inter-machine transmission of Parquet files is possible without any of the transmitted files losing interpretability (Chehaibi, 2017).

Figure 6. Row vs Columnar Storage Format

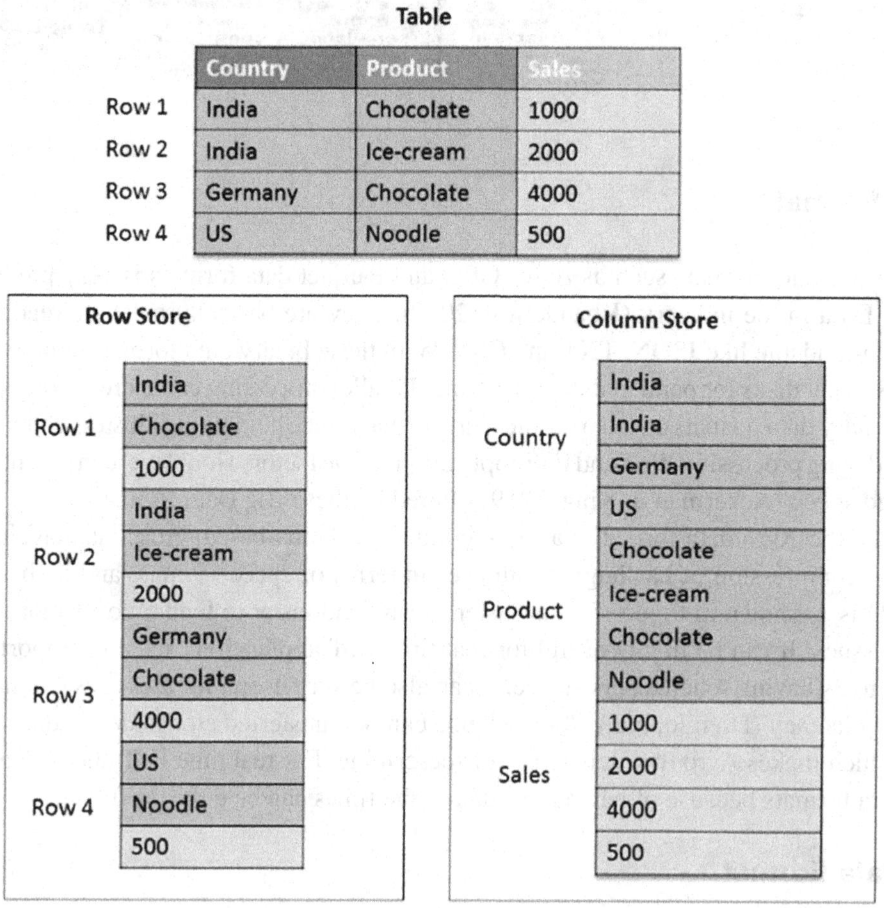

Table

	Country	Product	Sales
Row 1	India	Chocolate	1000
Row 2	India	Ice-cream	2000
Row 3	Germany	Chocolate	4000
Row 4	US	Noodle	500

Row Store

Row 1	India
	Chocolate
	1000
Row 2	India
	Ice-cream
	2000
Row 3	Germany
	Chocolate
	4000
Row 4	US
	Noodle
	500

Column Store

Country	India
	India
	Germany
	US
Product	Chocolate
	Ice-cream
	Chocolate
	Noodle
Sales	1000
	2000
	4000
	500

Figure 7. Parquet Columnar Data Storage Format

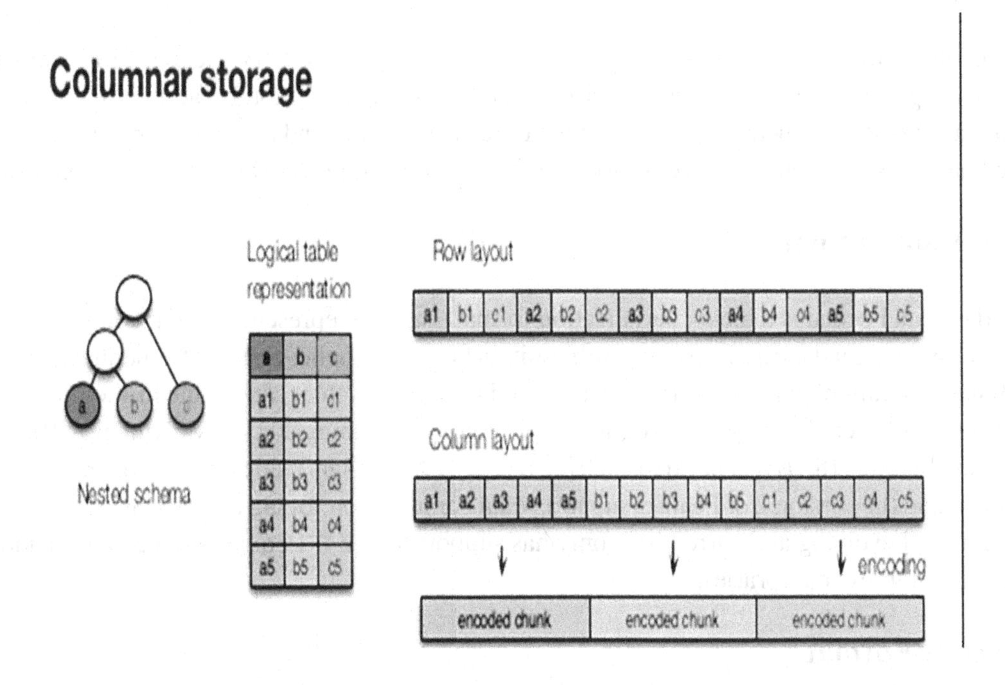

Parquet files also have statistics (such as mean, minimum, maximum etc.) of the data embedded within the file. This feature allows computation engines to be able to skip entire segments of a file based on the data portion the computation engine is looking for (Dremio, n.d.). The feature however allows Parquet to have slower read times during computation. Due to it having embedded metadata, Parquet datasets can be slower to write than row-based data formats such as Avro.

This shortcoming is however offset by it having a very fast read-times due to availability of data statistics and due to being column- optimized. For IIoT and FoF applications at the factory floor with real time needs, Parquet data formats with its fast read times can reduce data processing latency. Parquet however is very case sensitive, and errors can be generated if there are case mismatches in the naming of its columns.

MsgPack Data Format

MsgPack is a binary serialization formats that can be used for inter-process communications and to exchange data between machines and different programming language. It is optimized for speed and size, and it is platform and architecture independent (Sanchez, n.d.). It is schema-less, and it support all data types (float, string, integer, Boolean, array) that JSON supports. However, it requires a fraction of the space that JSON requires, and it is also optimized for its speed of computation (Sanchez, n.d.), (Luis et al., 2021).

Sequence Files Data Format

Sequence file format structure is similar to CSV format, but unlike CSV, Sequence is a binary data format that supports data compression. It also supports schema evolution by adding new fields, but it does not have associated metadata. It is complex to read and parse, and as such, is only used for storing intermediate datasets when sequences of jobs are being executed in Big Data computing environments.

Feather Data Format

The Feather data Format is a binary serialization format that also represents data in columnar format. It is ideal for lightweight data, and as such may be suitable for IIoT applications that generate data intermittently. It has programming bindings from Python and R; and as such, it can enable in-memory exchange of data frames between R and Python programming languages (Ramm, 2021), (McKinney & Richardson, 2020), (Data Flair, 2018), (Apache Arrow, 2019; Dye, 2019). Both its read and write operations are very fast on disk, and the version 2 of Feather supports data compression. Being a relatively new data format, its design is still evolving and currently, it only has support for LZ4 (Lempel-Ziv version 4) and ZSTD (Zstandard) compression libraries.

Pickle Data Format

The Pickle data format is Python specific. It is used in Python programming language to preserve data in binary serialization format for later usage. Pickling have been variously referred to as serialization, flattening and marshalling (Mastromatteo, 2017; Python, Pickle, 2019). The Pickled binary format can later be de-serialized back to Python object. Pickle data however can only be worked with in the Python environment; and cross-serialization using other programming languages is not yet available. Python Pickling exist is two variants viz: ASCII and binary modes. The binary Pickle mode has been known to be more memory efficient when compared to the ASCII Pickle mode. Two serious downsides of using Python Pickling are that it is known to work only with Python, and it can be used by malicious agents to attack IIoT systems. Another negative issue relating to the use of Python Pickling is that it is not backward compatible. As the format of the Pickle file is not defined by the user, if the data is modified to store another variable, the user must create new Pickle file to reflect the change. Else, all existing Pickle files will be broken (Mastromatteo, 2017; Python, Pickle, 2019). Pickle also cannot be accessed in sequential order. Specific chunks located in a specific part of the Pickle file cannot be dynamically accessed in the same manner as Parquet binary format.

MEMORY STORAGE SAVINGS ADVANTAGES RELATING TO THE USE OF DIFFERENT DATA FORMATS – TEST ENVIRONMENT

To test the hypothesis that data format conversion from one format to another in the IIoT factory floor can lead to storage memory space savings, a common PC that is synonymous with widely available IoT devices is used as a test environment. The PC runs on Windows 10 Enterprise 20H2, Intel(R) Xeon(R) E-2176M CPU @ 2.70GHz with 32 GB RAM. The test suite was Jupyter Notebook running on Python 3.6.13 in an Anaconda Navigator data science platform.

Figure 8. Yahoo.com Open-source Stock Market Time Series (1997 - 2022)

	Date	Open	High	Low	Close	Adj Close	Volume
Top Part of Time Series	4/30/1997	17.5	17.5	17	17	15.19202	5800
	5/1/1997	17.5	17.75	17.375	17.75	15.86226	36300
	5/2/1997	17.75	17.875	17.5	17.875	15.97396	17700
	5/5/1997	17.5	18.5	17.5	18.5	16.53249	65100
	5/6/1997	18.625	19.5	18.25	19.5	17.42614	227000
	5/7/1997	20	20.25	19.5	20	17.87297	75600
	5/8/1997	20	20	19.125	19.125	17.09102	5600
	5/9/1997	19	19.625	18.8125	19	16.97931	56400
	5/12/1997	18.75	19.625	18.625	18.875	16.86761	45700
	.						
	.						
	.						
Tail end of Time Series	4/18/2022	51.8	52.65	51.62	52.21	52.21	53100
	4/19/2022	52.05	53.63	52.05	53.14	53.14	68700
	4/20/2022	53.44	53.95	52.71	52.99	52.99	82300
	4/21/2022	53.28	53.68	52.02	52.27	52.27	57600
	4/22/2022	52.02	52.07	49.41	49.59	49.59	262600
	4/25/2022	49.05	49.79	48.35	49.73	49.73	120400
	4/26/2022	49.44	50.42	48.31	48.48	48.48	80400
	4/27/2022	48.53	49.11	48.32	48.62	48.62	114400
	4/28/2022	49.63	49.63	44.36	44.87	44.87	283800
	4/29/2022	44.32	44.6	42.37	42.89	42.89	193100

A stock market time-series dataset containing more than 6000 rows of datapoints (freely available at: https://finance.yahoo.com/) was downloaded and saved into Excel workbook in .csv data format. Each each row of the .csv string have fields containing float, int and string data types. Clippings from top rowa and the last set of rows in the dataset is shown in Figure 8. The time series dataset is used to show the storage memory saving benefits that could be obtained by using each type of data format. It is also used to show the memory saving and compression benefits that could be obtained by converting data formats from one form to another. Python scripts are written to convert the data formats from one form to another. Full details of the Python scripts and how users may be able to extend its applications are given in Part II of this chapter.

RESULT AND DISCUSSION

Results of the memory savings and compression benefits that could be obtained from changing data formats from one form to another are shown in Figure 9 through Figure 17.

In Figure 9, the original .csv time series size on memory was 311 kB. By using Python scripts to convert the .csv to Pickle ASCII, Pickle Binary, JSON, HDF5, YAML, TSV, Feather and XML, there were increases in storage memory requirements. As shown in Table II and Figure 9, HDF5, a binary serialization data format increases the storage memory requirement the most. From the original .csv data size of 311 kB, HDF5 data format increases the memory storage requirement to 1410 kB. This is a 350% increment as shown in Figure 10. As mentioned by researchers in (Stack Overflow, 2019), the large increment witnessed in the HDF5 data format storage may be due to Python Pandas dumping other extraneous overheads into the HDF5 file.

Table 2. Data formats that increase storage space requirements

Original Timeseries (.csv) Size (kB)	Converted Pickle Binary Size (kB)	Converted Pickle ASCII Size (kB)	Converted JSON Size (kB)	Converted HDF5 Size (kB)	Converted YAML Size (kB)	Converted TSV Size (kB)	Converted Feather Size (kB)	Converted XML Size (kB)
311	345	345	1260	1410	375	317	375	1210

Figure 9. Data Formats that Increases Storage Memory Requirements

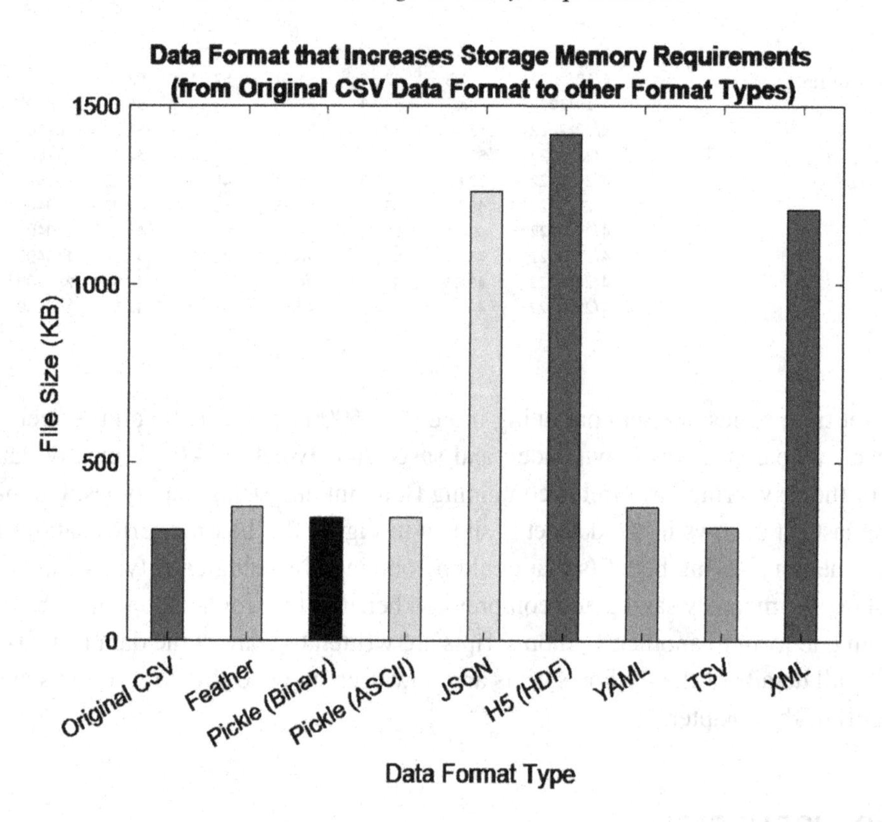

Figure 10. Percentage Increase of Data Format Types that Increases Storage Memory Requirements

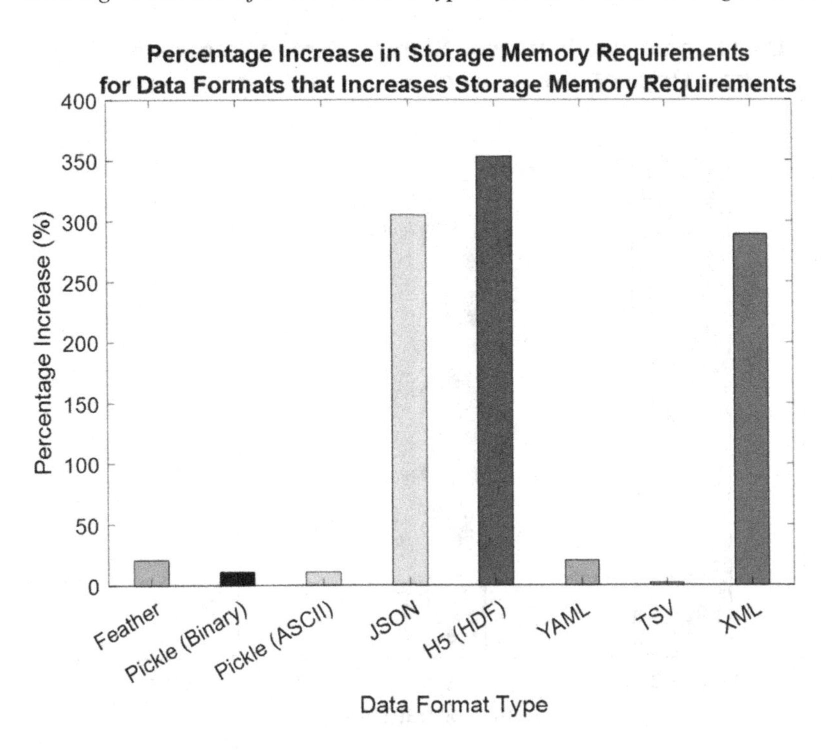

Table 3. Data formats that decrease storage space requirements

Original Timeseries (.csv) Size (kB)	Converted Parquet Size (kB)	Converted Avro Size (kB)
311	164	276

From the foregoing, it seems there might not be a memory saving or compression benefit by converting from CSV to HDF5 format. Other data formats such as Pickle, JSON, YAML, Feather and XML also increases storage memory footprints by varying amounts as shown in Fig. 10; and this indicates that there may not be a direct memory saving benefits by converting to these formats.

Figure 11. Data Formats that Decreases Storage Memory Requirements

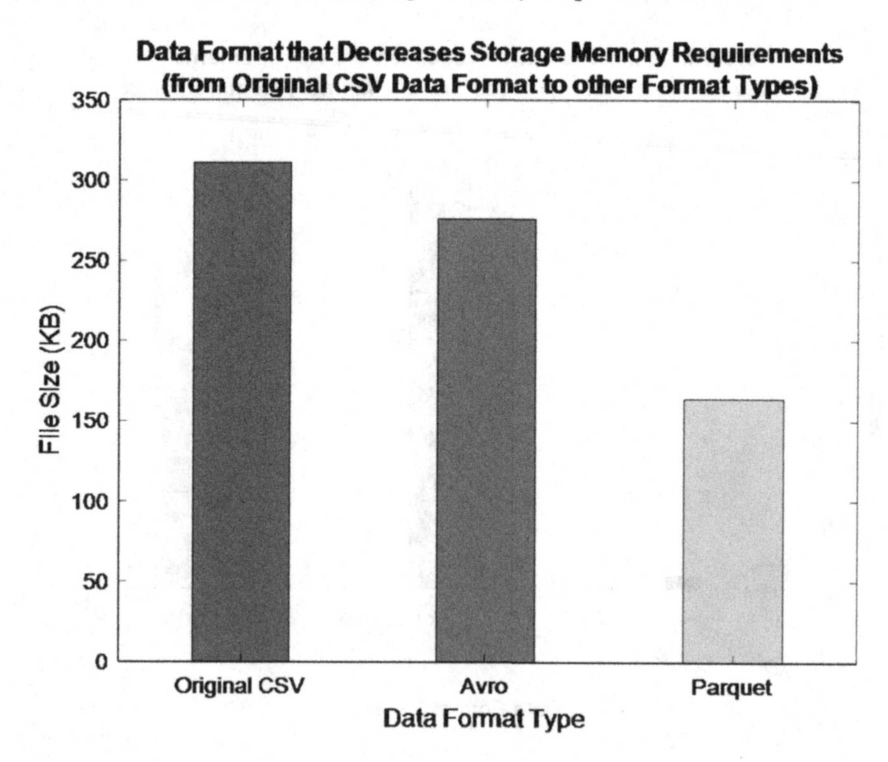

However, as shown Table 3 and in Figure 11 and Figure 12, there are significant decrease in storage memory requirements when the original CSV format was converted to Parquet and Avro formats. From the original CSV format size of 311 kB, the time series dataset was converted to 164 kB and 276 kB for Avro and Parquet formats respectively. As shown in Figure 12, this is approximately 47% and 12% file size decrease for Parquet and Avro formats respectively. This size reduction benefit signifies that there will be a huge savings of about 47% in data storage payment for the cloud Bronze Table when data is converted to Parquet format on the factory floor before sending to the cloud. There will also be additional benefits of the data having schema and metadata information when the data is deposited into the cloud Bronze Table; thus, preventing the incidence of a Data Swamp in the cloud Bronze Table.

Figure 12. Percentage Decrease of Data Format Types that Increases Storage Memory Requirements

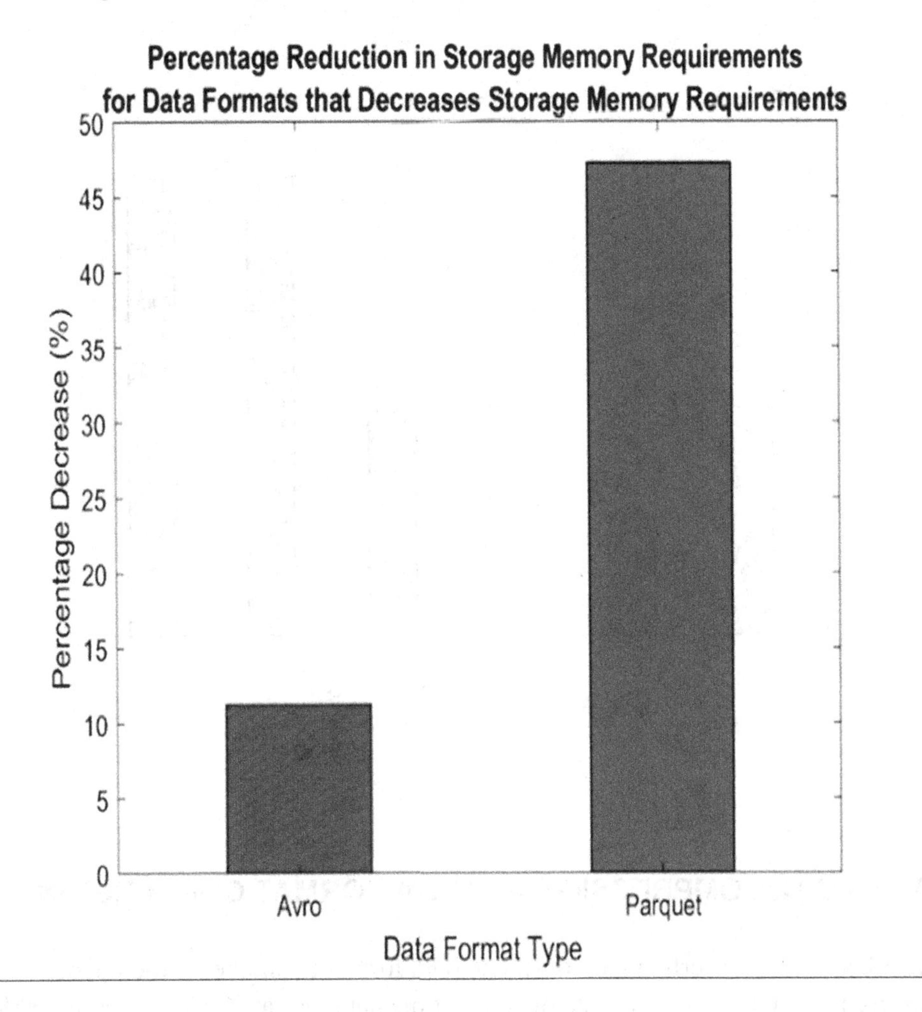

Table 4. Method of JSON data compression by data format conversion

Original Timeseries (.csv) Size (kB)	Parquet to CSV Conversion: Size (kB)	Parquet to TXT Conversion: Size (kB)	CSV to JSON Conversion: Size (kB)	Parquet to JSON Conversion: Size (kB)
311	311	311	1260	625

Figure 13. Effects of Converting Data Formats on Storage Memory Requirements (Parquet & JSON to other types)

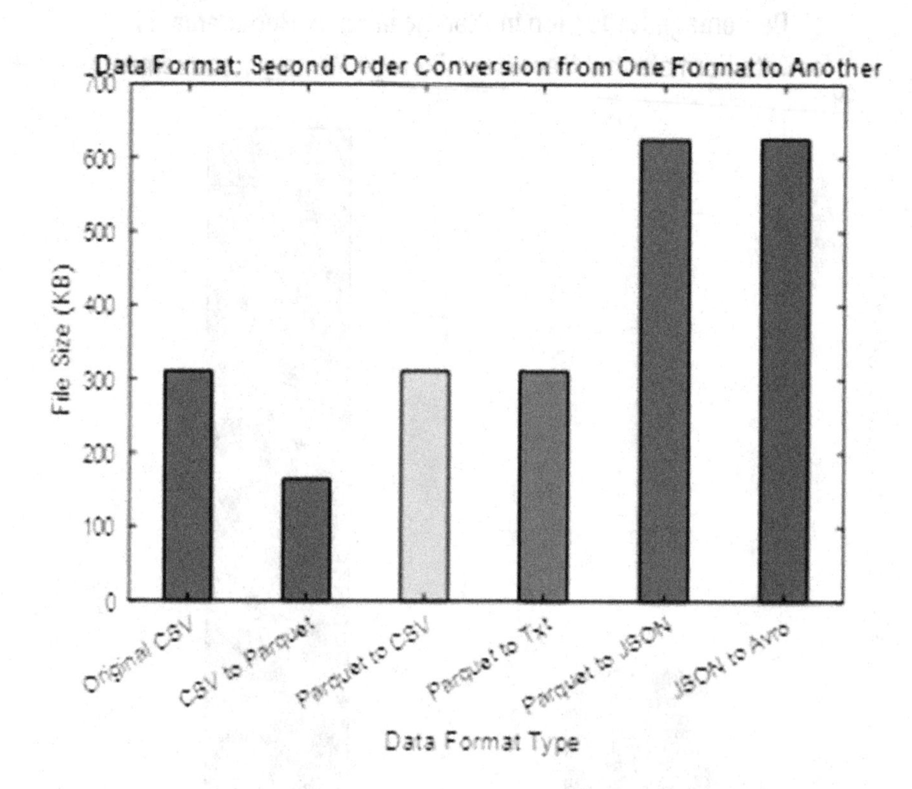

METHOD OF DATA COMPRESSION BY DATA FORMAT CONVERSION

Another direct benefit of converting data from one data format to another is the opportunity of being able to freely use the compression attribute of some of the data format. As discussed in Table 4.

Section 3, the Parquet data format supports data compression. The direct benefit of converting from human-readable format to binary formats that supports compression could be observed in Table 4 and Figure 13. As shown in Figure 13, JSON format can be highly compressed if data is initially converted from CSV to Parquet, and then from Parquet to JSON. As shown in Table 4, the initial size of JSON on disk, when there is conversion from CSV to JSON is 1260 kB. However, there is significant compression on the JSON data file that results from conversion from CSV to Parquet, and then to JSON. The size of JSON file that results from this double conversion process is 625 kB.

Also as shown in Figure 14, it is even possible to reduce the size of JSON data on file further by converting from Parquet to TXT and then to JSON. From Figure 14, it could be observed that the final JSON size on file when there is Parquet to TXT and to JSON conversion is 407 kB. This is an impressive compression compared to the Parquet to JSON process that yields 625 kB in Figure 13. In addition to these compression benefits, conversion process that follows data formats changing steps that are discussed in this chapter does not generally result into a zipped file that other common compression process always yields. As is generally known, using compression methods that results to a zipped file means that further unzipping processes are always needed before the file can be used for other analytics purposes. Also, it is possible to convert from CSV to JSON and then from JSON to Avro and, from

JSON to MsgPack as shown in Figure 14 It is also possible to reduce the size of MsgPack data format on disk by converting from MsgPack to Parquet as shown in Figure 15. In Figure 16, it is shown that converting form one human-readable data format to another human-readable data format such as from CSV to TSV, and from TSV to JSON do not yield a readily observable storage memory benefit.

Figure 14. Compression Benefits of Converting Data Formats on Storage Memory Requirements (Parquet to TXT to JSON)

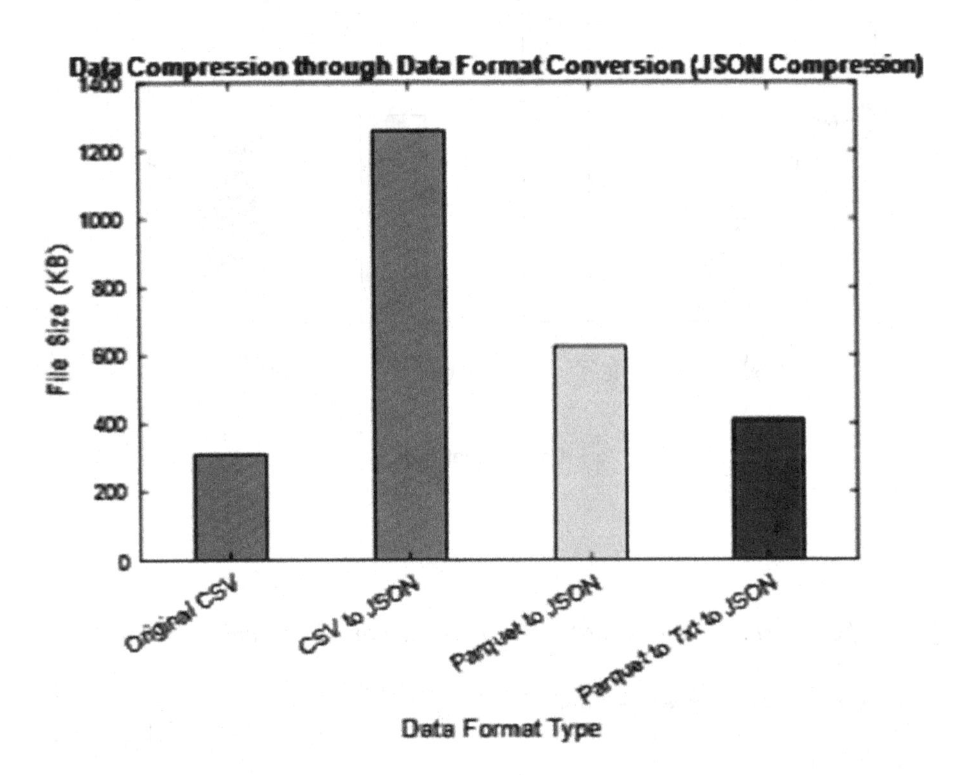

Figure 15. Compression Benefits of Converting Data Formats on Storage Memory Requirements (JSON to other types)

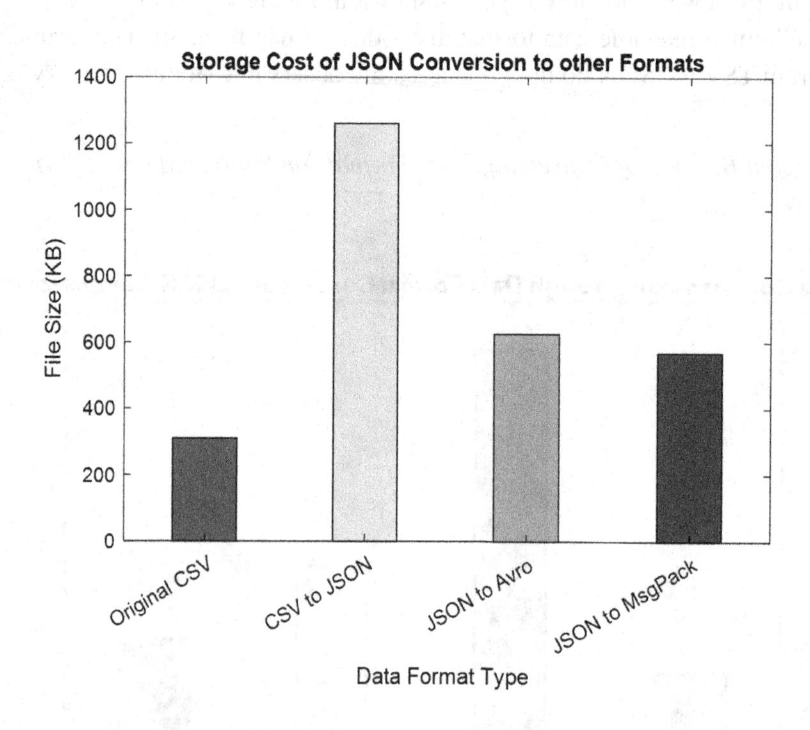

Figure 16. Effects of Converting Data Formats on Storage Memory Requirements (TSV to CSV and JSON)

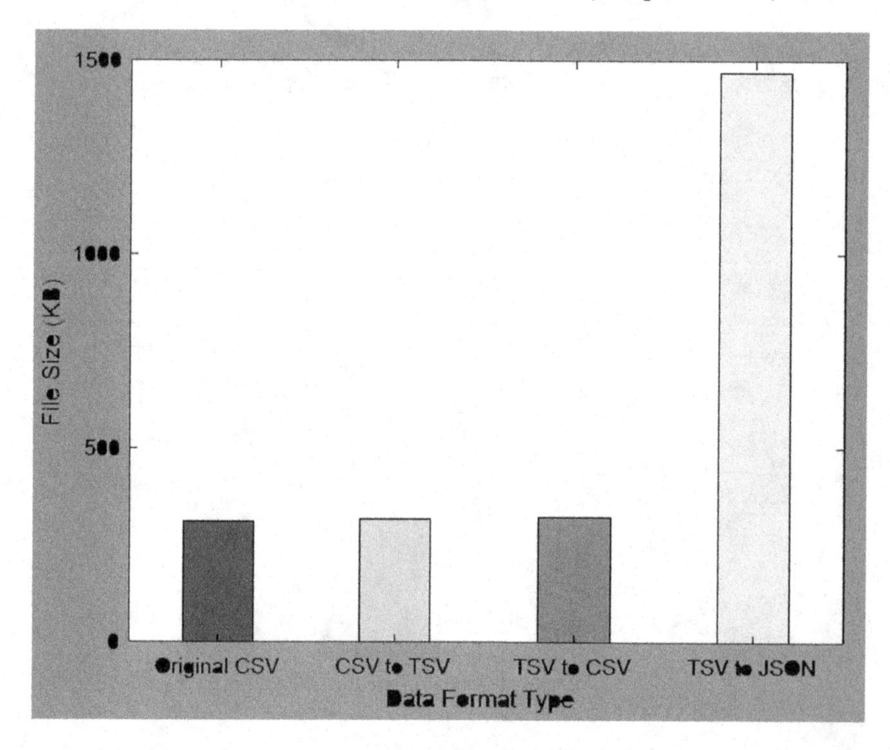

Figure 17. Compression Benefits of Converting Data Formats on Storage Memory Requirements (MsgPack to Parquet)

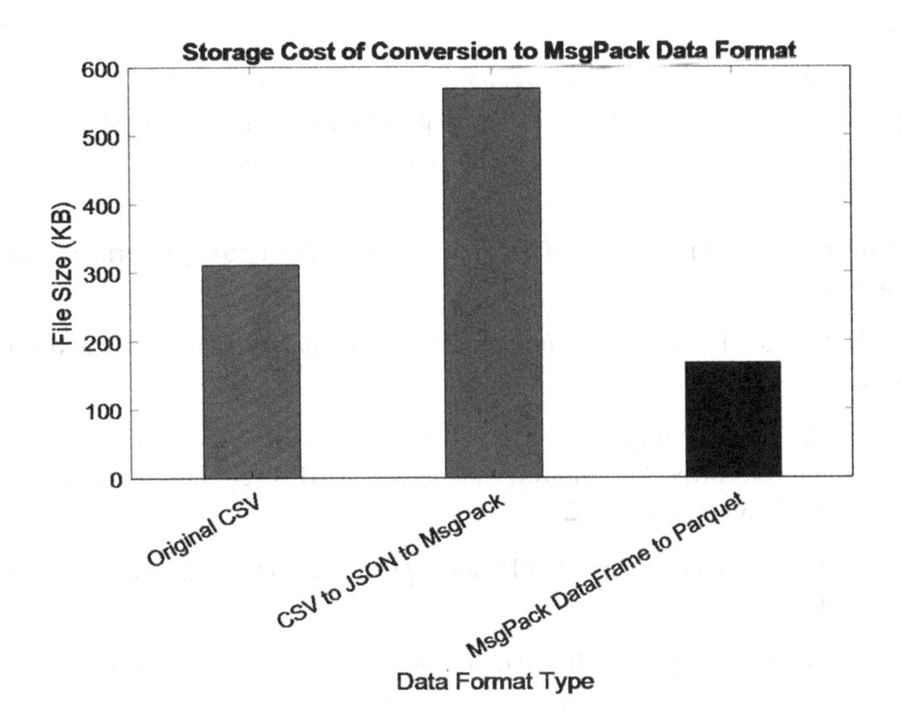

CONCLUSION

In this chapter, the compression and storage memory benefit that may be derived from convert data from one format to another is thoroughly discussed. From factory floor to the cloud, numerous benefits that can result from data format conversion including metadata generation, evolving schema that can aid data governance and democratization. Additional benefits that are discussed includes prevention of Data Swamp issue in the cloud Bronze Table, better coordination and robust data exchange between machines on the factory floors. In Part II of the paper, Python codes and that can be used to achieve data format conversion from one form to another are presented. Additional benefits that may result from data format conversion are also discussed.

REFERENCES

Accenture. (n.d.). Closing the Data-value Gap: How to Become Data Driven and Pivot to the New. *White Paper, Accenture*. https://www.accenture.com/_acnmedia/pdf-108/accenture-closing-data-value-gap-fixed.pdf

Ackerman, H., & King, J. (2019). *Operationalizing the Data Lake – Building and Extracting Value from a Data Lake with a Cloud Native Data Platform*. O'Reilly Media, Incorporated.

Ahmed, S., Ferzund, J., Rehman, A., Usman Ali, A., Sarwar, M., & Mehmood, A. (2017). Modern Data Formats for Big Bioinformatics Data Analytics. *Int'l Journal of Advanced Computer Sc. & Applications (IJACSA), 8*(4).

Apache Arrow. (2019). *Feather File Format*. Apache Arrow. https://arrow.apache.org/docs/python/feather.html#:~:text=There%20are%20two%20file%20format,available%20in%20Apache%20Arrow%200.17

Belov, V., Tatarintsev, A., & Nikulchev, E. (2021). Choosing a Data Storage Format in the Apache Hadoop System. *Symmetry, 13*.

Bhatia, R. (2021). Big Data File Format. *White Paper, Clairvoyant*. https://www.clairvoyant.ai/blog/big-data-file-formats

Chehaibi, M. (2017). Parquet Data Format Used in Thing Worx Analytics. *PTC Community*. https://community.ptc.com/t5/IoT-Tech-Tips/Parquet-Data-Format-used-in-ThingWorx-Analytics/td-p/535228

Data Flair. (2018). What are the File Format in Hadoop. *Data Flair*. https://data-flair.training/forums/topic/what-are-the-file-format-in-hadoop/

Databerg. (2019). Why is High-Quality Data Governance a Key Tool in Industry 4.0? *White Paper, Databerg*.https://blog.datumize.com/why-is-high-quality-data-governance-a-key-tool-in-industry-4.0

Dremio. (n.d.). "What is Apache Parquet, online, Available: https://www.dremio.com/resources/guides/intro-apache-parquet/

Dye, S. (2019). Feather Files: Faster Than the Speed of Light. *Medium*. https://medium.com/@steven.p.dye/feather-files-faster-than-the-speed-of-light-d4666ce24387

Fogg, E. (2020). 5 Steps to Bring Your Legacy System Online with IIoT. *White Paper, Machine Metrics*. https://www.machinemetrics.com/blog/legacy-systems-online-iiot

IBM. (2021). XML Schemas Overview. *White Paper, IBM*. https://www.ibm.com/docs/en/control-desk/7.6.1?topic=schemas-xml-overview

Idreos, S., Alagiannis, I., Johnson, R., & Ailamaki, A. (2011). Here are my Data Files. Here are my Queries: Where are my Results? *5th Biennial Conf. on Innovative Data System Research (CIDR 2011)*; Asilomar, California, USA.

Kamiya, T., Kolesnikov, A., Murphy, B., Watson, K., & Widell, N. (2021). Characteristics of IIoT Information Models. *White Paper, Industrial Internet Consortium.* https://www.iiconsortium.org/pdf/Characteristics-of-IIoT-Information-Models.pdf

Kancharla, C. R., (2021). Augmented Reality Based Machine Monitoring for Legacy Machines: a retrofitting use case. *2021 XXX International Scientific Conference Electronics (ET)*, (pp. 1-5). IEEE. 10.1109/ET52713.2021.9579936

Khandelwal, D. (2020). An Introduction to Big Data Formats. *Big Data and Cloud Practice.* https://bd-practice.medium.com/an-introduction-to-big-data-formats-450c8db3d29a

Lee, D., & Heintz, B. (2019). Productionizing Machine Learning with Delta Lakes. *Databricks.* https://databricks.com/blog/2019/08/14/productionizing-machine-learning-with-delta-lake.html

Levy, E. (2022). What is Parquet File Format and Why You Should Use It. *White Paper, Upsolver.* https://www.upsolver.com/blog/apache-parquet-why-use

Lima, F., Massote, A. A., & Maia, R. F. (2019). IoT Energy Retrofit and the Connection of Legacy Machines Inside the Industry 4.0 Concept. *IECON 2019 - 45th Annual Conference of the IEEE Industrial Electronics Society*, (pp. 5499-5504). IEEE. 10.1109/IECON.2019.8927799

Luis, A., Casarez, P., Cuadrado-Gallego, J., & Patricio, M. (2021). PSON: A Serialization Format for IoT Sensor. *Sensors (Basel)*, *21*(13), 4559. doi:10.339021134559 PMID:34283115

Mastromatteo, D. (2017). The Python Pickle Module: How to Persist Objects in Python. *Real Python.* https://realpython.com/python-pickle-module/

McKinney, W., & Richardson, N. (2020). Feather V2 with Compression Support in Apache Arrow 0.17.0, *White Paper, URSA Labs.* https://ursalabs.org/blog/2020-feather-v2/

Mishra, S. (2021). Demystifying Delta Lake. *Analytics Vidya.* https://medium.com/analytics-vidhya/demystifying-delta-lake-d15869fd3470

MPHY0021. (2022). *Scientific File Formats.* University College London. http://github-pages.ucl.ac.uk/rsd-engineeringcourse/ch01data/070hdf5.html#:~:text=HDF5%20is%20the%20current%20version,hierarchy%2C%20similar%20to%20a%20filesystem

Naidu, V. (2022). Performance of Using Appropriate File Formats in Big Data Hadoop Ecosystem. *Int'l Research Journal of Engnr & Tech, 9*(1).

Oyekanlu, E. (2017). Predictive Edge Computing for Time Series of Industrial IoT and Large Scale Critical Infrastructure based on Open-source software Analytic of Big Data, *2017 IEEE International Conference on Big Data (Big Data)*, (pp. 1663-1669). IEEE. 10.1109/BigData.2017.8258103

Oyekanlu, E. (2018a). Fault-Tolerant Real-Time Collaborative Network Edge Analytics for Industrial IoT and Cyber Physical Systems with Communication Network Diversity. *2018 IEEE 4th International Conference on Collaboration and Internet Computing (CIC)*, (pp. 336-345). IEEE. 10.1109/CIC.2018.00052

Oyekanlu, E. (2018b). Osmotic Collaborative Computing for Machine Learning and Cybersecurity Applications in Industrial IoT Networks and Cyber Physical Systems with Gaussian Mixture Models. *2018 IEEE 4th International Conference on Collaboration and Internet Computing (CIC)*, (pp. 326-335). IEEE. 10.1109/CIC.2018.00051

Oyekanlu, E. (2018c). Distributed Osmotic Computing Approach to Implementation of Explainable Predictive Deep Learning at Industrial IoT Network Edges with Real-Time Adaptive Wavelet Graphs. *2018 IEEE First International Conference on Artificial Intelligence and Knowledge Engineering (AIKE)*, (pp. 179-188). IEEE. 10.1109/AIKE.2018.00042

Oyekanlu, E., & Scoles, K. (2018a). Towards Low-Cost, Real-Time, Distributed Signal and Data Processing for Artificial Intelligence Applications at Edges of Large Industrial and Internet Networks. *2018 IEEE First International Conference on Artificial Intelligence and Knowledge Engineering (AIKE)*, (pp. 166-167). IEEE. 10.1109/AIKE.2018.00037

Oyekanlu, E., & Scoles, K. (2018b). Real-Time Distributed Computing at Network Edges for Large Scale Industrial IoT Networks. *2018 IEEE World Congress on Services (SERVICES)*, (pp. 63-64). IEEE. 10.1109/SERVICES.2018.00045

Plase, D., Niedrite, L., & Taranovs, R. (2017). A Comparison of HDFS Compact Data Formats: Avro Versus Parquet. *Elektronika ir Elektrotechnika*, *9*(3), 267–276.

Python, Pickle. (2019). *Python Object Serialization*. Python. https://docs.python.org/3/library/pickle.html#:~:text=serialization%20and%20deserialization.-,Data%20stream%20format,to%20reconstruct%20pickled%20Python%20objects

Ramm, J. (2021). Feather Documentation. *White Paper, Build Media*. https://buildmedia.readthedocs.org/media/pdf/plume/stable/plume.pdf

Sanchez, A. (n.d.). MessagePack, Racket White Paper, online. Available: https://docs.racket-lang.org/msgpack/index.html#:~:text=MessagePack%20is%20an%20efficient%20binary,addition%20to%20the%20strings%20themselves

Sharma, A. (2022). What is YAML? A Beginner's Guide. *White Paper, Circleci*. https://circleci.com/blog/what-is-yaml-a-beginner-s-guide/#:~:text=YAML%20is%20a%20digestible%20data,that%20JSON%20can%20and%20more;

Silva, B., Sousa, J., & Alenya, G. (2021). Data Acquisition and Monitoring System for Legacy Injection Machines. *2021 IEEE International Conference on Computational Intelligence and Virtual Environments for Measurement Systems and Applications (CIVEMSA)*, (pp. 1-6). IEEE. 10.1109/CIVEMSA52099.2021.9493675

Singh, C. (2018). Advantages and Disadvantages of XML. *Beginners-Book*. https://beginnersbook.com/2018/10/advantages-and-disadvantag es-of-xml/

Snyder, J. (2019). Data Cleansing: An Omission from Data Analytics Coursework. *Information Systems Education Journal (ISEDJ), 17*(6).https://files.eric.ed.gov/fulltext/EJ1224578.pdf

Stack Overflow. (2019). Why do my hdf5 files seem so unnecessarily large? *Stack Overflow*. https://stackoverflow.com/questions/65119241/why-do-my-hdf5-files-seem-so-unnecessarily-large

Staubli, G. (2017). *Real Time Big Data Analytics: Parquet (and Spark) + Bonus*. Linkedin. https://www.linkedin.com/pulse/real-time-big-data-analytics-parquet-spark-bonus-garren-staubli/

Timescale. (2020). Time-Series Compression Algorithms, Explained. *Timescale*. https://www.timescale.com/blog/time-series-compression-algor ithms-explained/

Wasser, L. (2020). Hierarchical Data Formats – What is HDF5? *Neon Science*. https://www.neonscience.org/resources/learning-hub/tutorials /about-hdf5#:~:text=About%20Hierarchical%20Data%20Formats%20 %2D%20HDF5,-The%20Hierarchical%20Data&text=HDF5%20uses%20a%2 0%22file%20directory,metadata%20making%20it%20self%2Ddescrib ing

Xavier, L. (2021). Evaluation and Performance of Reading from Big Data Formats, [Bachelor's Thesis, Federal Univ. of Rio Grande do Sul, Brazil]. https://www.lume.ufrgs.br/bitstream/handle/10183/223552/0011 27314.pdf?sequence=1.

Yavus, B., Armbrust, M., Das, T., & Condie, T. (2017). Working with Complex Data Format with Structured Streaming in Apache Spark 2.1. *Databricks*. https://databricks.com/blog/2017/02/23/working-complex-data-formats-structured-streaming-apache-spark-2-1.html

Chapter 2

Data Engineering for the Factory of the Future, Multimedia Applications and Cyber-Physical Systems:

Part 2 - Algorithms and Python-Based Software Development for Time-Series Data Format Conversion

Emmanuel Oyekanlu
Corning Incorporated, New York, USA

David Kuhn
Corning Incorporated, New York, USA

Grethel Mulroy
Corning Incorporated, New York, USA

ABSTRACT

This chapter is the companion chapter to "Part 1: State-of-the-Art Time-Series Data Formats Performance Evaluation." In this chapter, algorithms for converting data from one format to other formats are presented. To implement the algorithms, existing open-source Python libraries are used extensively, and where needed, new Python routines for converting data formats are developed. It is envisaged that the algorithms and Python libraries and routines that are freely provided in this chapter will be useful for data engineers, data scientists, and for industrial IoT, cyber-physical systems (CPS), multimedia, and big data practitioners who are on the quest to use different types of data formats that are compatible with memory-constrained factory floor IoT devices. It will also be useful for Delta Lake and big data engineers, who are on the quest for delivering robust bronze, silver, and gold data lakes in the cloud.

DOI: 10.4018/978-1-7998-7852-0.ch002

Copyright © 2023, IGI Global. Copying or distributing in print or electronic forms without written permission of IGI Global is prohibited.

INTRODUCTION

In Part I (Oyekanlu et al., 2022), an extensive study regarding the benefits of using different types of existing legacy and state-of-the-art data formats was conducted. In this chapter, algorithms that can facilitate data formats conversion, metadata generation, storage footprint reduction, schema design, schema integration and schema evolution for time series data are provided. Based on extensive literature search, this chapter provides the most extensive but detailed algorithms regarding how different types of data formats can be changed from one format to another. Python implementations for the algorithms are also provided. Our contribution in this chapter will enable Data Engineers, Software Engineers, Cyber-Physical Systems (CPS) Engineers, industrial Multimedia Content Developers, Data Scientists, Cloud Engineers, DevOp Engineers, IoT and Big Data practitioners, etc., to easily and affordable be able to enjoy the benefits of using the different types of available data formats for time series data analytics and for time series-based AI applications.

Without loss of generality, the usefulness of our algorithms, software implementation, and data format conversion approaches in this chapter can be extended to other types of data that are different from time-series data sets. It is instructive to mention that, due to our usage of Python, different type of available IoT devices may be able to easily use our algorithms and software implementation codes to work with different type of data formats for smart manufacturing, medical, multimedia and CPS applications.

Researchers have always been on the quest for providing easy and affordable means by which different data sets can be made more interoperable. Methods of robustly converting data from one form to another is also in high demand. In (Pivarski et al., 2020), authors present Awkward Array, a Python-based, Numpy-like interface that can be used to handle JSON-like data in similar ways by which Numpy handles rectilinear arrays of numbers. Awkward Array is a generalization of Numpy's core function to cater to the needs of nested record, variable-length lists and most other data sets that have JSON-like constructs. Awkward Arrays can also be effectively used to handle JSON-like data with columnar structures. Authors in (Belov et al., 2021) presented a comparative analysis of Avro, comma separated values (CSV), JavaScript Object Notation (JSON), Optimized Row Columnar (ORC) and Parquet data formats on Apache Spark framework. Results of an evaluation of these data formats in terms of volume, and processing speeds for different analytics operations such as sorting, grouping, reading unique values, and filtering operations are also presented.

In (Ahmed et al., 2017), researchers explore different data formats that are appropriate for use on Big Data platforms. Results of the analysis presented in (Ahmed et al., 2017) can be used by different practitioners to select data formats that are most suitable for their project needs. In (Ye et al., 2022), authors discuss the Asset Administration Shell (AAS), a new approach for implementing data interoperability across different CPS automation pyramids. The AASX data format, which represents AAS information; and supports data communication via CPS Open Platform Communications United Architecture (OPC UA) was also presented. Also, in (Ye et al., 2022), an AASX-based solution for bidirectional data exchange between enterprise and control applications in CPS was discussed. An approach for interoperable data format exchange between AASX and Excel spreadsheets applications was also presented.

In (Oyekanlu et al., 2017), authors developed CSV2RDF, a protocol that focuses on how Semantic Web technologies are used to convert CSV data into Resource Definition Language (RDF). Techniques by which CSV data can be parsed into RDF triples are also discussed. As discussed by researchers in (Oyekanlu, 2017; Oyekanlu, 2018a; Oyekanlu, 2018b; Oyekanlu, 2018c; Oyekanlu, 2018d; Oyekanlu, Onidare, & Oladele, 2018; Oyekanlu & Scoles, 2018a; Oyekanlu & Scoles, 2018b; Oyekanlu, Scoles, &

Oladele, 2018; Oyekanlu et al., 2020), in most industrial IoT (IIoT) use cases, there are crucial needs for low-cost and power-use aware approaches for handling big IIoT data for CPS, multimedia and industrial applications. These needs mostly necessitate introducing optimal data formats conversion approaches. Such approaches are vital in helping data practitioners to be able to effectively handle Big Data using data formats that supports power constrained, low-cost IIoT devices and smarter data storage strategies. In particular, a robust and interoperable data format for industrial smart manufacturing applications must be operating system agnostic. It must also support fast and smarter processing in addition to low-cost storage, metadata and data schema generation; from the industrial floor all the way into the cloud layer (Oyekanlu et al., 2022). These qualities and data format attributes also holds true for data needed for reliable and reactive real time multimedia and CPS applications.

Our contribution in this chapter includes the following:

- Algorithms and software implementation by which CSV time series data can be transformed into several other time series data formats such as MessagePack, Feather, JSON, Parquet, Pickle, Avro, Numpy arrays and YAML (YAML Ain't Markup Language). In most cases, the conversion method provided is symmetric since methods are also provided by which the resulting data format can be converted back to the original CSV time series data formats. For example, through our algorithm, a MessagePack data set or dataframe resulting from initial CSV data set can be transformed back into the original CSV data set.
- Extensive and detailed discussion on data format needs and data interoperability for factory of the future, smart manufacturing, CPS and multimedia applications.
- Evaluation of the data frames resulting from implementing our algorithms. The evaluation is in terms of implementing important data analytics activities such as querying (data filtering), sorting, grouping and finding unique values.
- Analyses of the speed of the data analytics activities (querying, sorting etc.) in terms of wall and CPU speed times.
- Introduction of new schema generation algorithms for CSV based time series data.
- Algorithms and methods of converting time series data to tensors. New algorithms and methods of converting CSV time series to multidimensional tensors are discussed. Python implementations are shown.
- Introduction of new metadata generation algorithms for CSV based time series data sets.
- Algorithms and methods of facilitating reliable conversion of Pickle-based time series data to other formats.
- As stated in (Tokcan et al., 2020), only a few data formats such as CSV, JSON, XML, SQL and YAML can be readily mapped by RML (RDF Mapping Language) into RDF for semantic web ontologies. Our contribution in this paper will readily allow the conversion of more data formats into those data formats that are favored by RDF. This will allow easy incorporation of those data formats into formats ingestible by RDF.

Python was chosen as the implementation programming language since it can be readily deployed on most available operating systems and on most low cost IoT devices such as Raspberry Pi and Nvidia Nano. By using Python, low cost IoT devices can be deployed as middle layer data format conversion devices between the factory floor and the cloud layer.

The rest of this chapter is organized as follows. Section II is an in-depth review of the benefits of selecting appropriate data formats for extraction, conversion, and loading (ETL) phases of data engineering, business analytics and business intelligence use cases. Discussion also includes data interoperability challenges for manufacturing CPS, medical CPS, and multimedia systems. Section III presents the algorithms, Python implementations and results of data formats conversion strategies discussed in this chapter. Section IV presents new methods and algorithms for metadata generation for CSV based time series data. In Section V, algorithms, and Python implementation for tensorizing CSV time series are introduced. New methods for working with CSV data sets transformed to tensors are also introduced. In section VI, algorithms, and Python implementation for converting Pickle to other data formats are presented. In section VII, results of important Python based analytics activities such as querying (filtering), grouping, finding unique values and sorting; on the designed algorithms are discussed. Further discussion regarding data formats conversion and conclusion of the chapter is presented in section VIII.

BRIEF OVERVIEW ON THE IMPORTANCE OF DATA FORMAT INTEROPERABILITY AND CONVERSION FOR MULTIMEDIA APPLICATIONS IN CYBER PHYSICAL SYSTEMS

In data engineering, CPS and business intelligence projects, after specifying the customer's needs, selecting the optimal data format for ETL, data analytics, data storage and other crucial multimedia applications processes is a very important project phase (Oliveiros, 2016; Oyekanlu et al., 2022). The choice of data formats can greatly impact the project in terms of execution speed, storage space requirements, compression, and the success of crucial project deliverables. It is worthy to emphasize that, it is not all available ETL tools that readily supports all available data formats. To accommodate some data formats, data engineers may need to write additional converters and parsers, leading to greater project complexity (Pivarski et al., 2020). For example, text (TXT) files, due to its encoding, uses extra computation during data parsing. It also does not have schema and metadata embeddings (Oyekanlu et al., 2022). For data that are in XML and JSON *formats, there can be issues that limits data splitability when processing such data sets in (Oliveiros, 2016) Hadoop Distributed* File Systems (HDFS). Data sets that are not splitable often cannot be processed in parallel in HDFS based systems. The format of data sets also affects querying speed. In general, columnar data formats such as Parquet and ORC offers faster processing advantages when only a subset of the data columns is needed. Columnar data also have compression advantages, leading to a saving in terms of file loading times and storage space savings. However, it is important to be mindful of the specific use case and the needs of the data engineering project at hand. Each of the data formats have its own advantages, and demerits. Hence, the selection of a particular data format should be constrained on the needs of the project at hand. Formats such as TXT, TSV and CSV are good for human readability, but they can add to the project computation query times if the data set is big. Formats such as Petastorm, Avro and Parquet are good for schema generation, schema integration and for faster processing speed. A full treatise on the merits and demerits regarding using some data formats in practice is presented in (Oyekanlu et al., 2022). Many existing systems including smart manufacturing, factory of the future (FoF), Robots, intelligent buildings, implantable medical devices, aircraft systems, autonomous or self-driving vehicles, intelligent rail systems and other System of Systems (SoS) can be categorized as CPS (NIST, n.d.; Sztipanovits, 2013). Majority of these systems uses existing multimedia data as means of establishing semantic coordination, interoperability, and data commonality among their underlying sub-systems. Also, in many CPS, different multimedia applications are used extensively for communication, control, data storage, analytics and for data retrieval. In this section, an overview of the challenges of differences in data formats and data interoperability for multimedia applications in CPS is given.

Data Format and Interoperability Challenges for Multimedia Applications in Cyber-Physical Systems

Interconnected Cyber Physical Manufacturing (CPMS) Systems (Figure 1) and other CPS such as Cyber Physical Medical (CPMeS) Systems (Figure 2) and their underlying multimedia applications generates big data in different data formats. An ecosystem of the different data formats available for use in such complex CPS is tabulated in Table 1. These CPS make extensive use of integrated networking, information processing, wide area system sensorization and actuation capabilities that allows a synergy of at-scale and economic production.

Figure 1. Components and sub-systems of cyber physical manufacturing systems (CPMS) generates big data with different types of multimedia data in different data formats

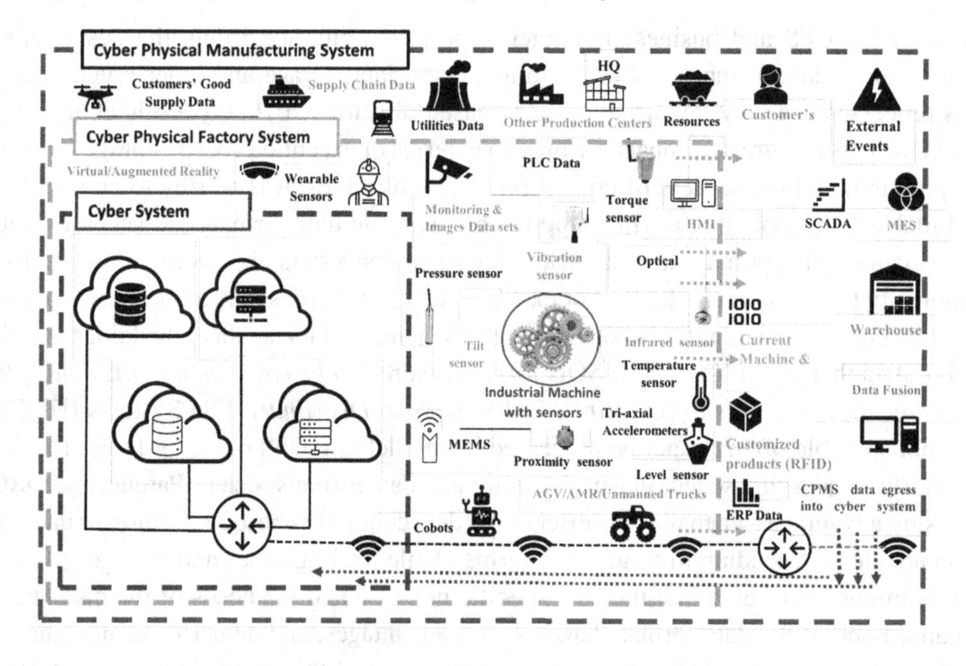

This synergy will include new capabilities that allows physical devices to operate in dynamic environments. These new capabilities will invariably use many sophisticated and interconnected parts that must instantaneously exchange, parse, and act on detailed data in a highly coordinated manner; under unforeseen circumstances, and often, with humans in the loop (Sztipanovits, 2013). As an integral part of CPS ecosystem, FoF will increasingly rely upon numerous cross-functional technologies including robotics, multimedia applications and computer-controlled manufacturing processes. These technologies must be linked to automated design tools, along with integrated, inter-networked, broad-based, dynamic,

Figure 2. Components and sub-systems of Cyber Physical Medical Systems (CPMeS) generates big data with different types of multimedia data in different formats including multidimensional data sets such as tensors

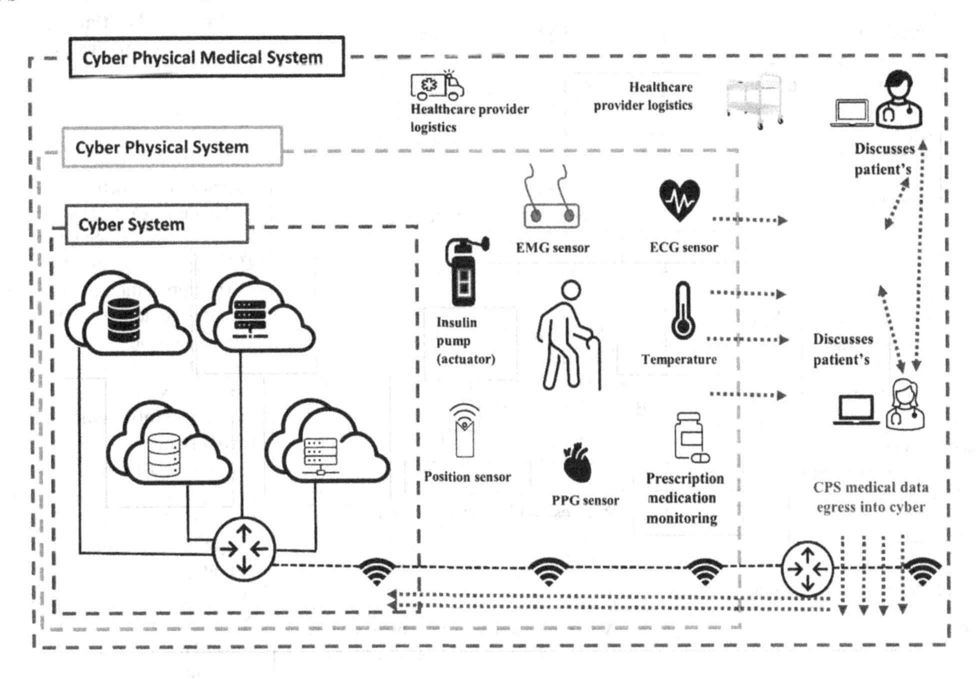

and possibly real-time management of production lines, factories, and supply chains. To adequately measure system performance, equally broad-based and energy efficient, sustainable and futuristics performance metrics that allows environmentally friendly initiatives will be needed. Realizing the full promise of reactive and robust CPS will require seamless interoperability among heterogeneous components, multimedia applications and systems, that are supported by new reference architectures using shared vocabularies and definitions that aids data interoperability and commonality of data semantics. Further, successfully addressing the challenges and harnessing opportunities of stable and robust CPS will invariably require broad consensus in foundational concepts that are centralized on data semantics and interoperable data formats. For most Data Science applications, correct determination of data semantics is crucial (Hulsebos, 2019). This determination is pivotal to the effective accomplishments of other tasks such as data discovery, data cleaning, schema matching (Hulsebos, 2019) and metadata generation. Example of CPS and their enabling multimedia data formats are also listed in Table 1 (Baheti & Gill, 2011; John & Oyekanlu, 2010; Sztipanovits, 2013).

As mentioned earlier, a CPS may have different underlying subsystems (NIST, n.d.). As a result, systems included in a particular CPS may bridge multiple purposes, as well as time and data domains. CPS subsystems may thus require methods of data translation, data accommodation and semantic communication among their different logical domains and among their numerous different data sets. Additionally, CPS architectures should support legacy component integration and migration. Legacy devices may have physical artifacts, software, protocols, syntax, and semantics that exist due to dated design decisions; and they may be inconsistent with the current architectural requirements.

Table 1. Examples of innovative systems, cyber-physical Systems, and their enabling multimedia applications

Innovative Products & Applications	Cyber-Physical System	Enabling Multimedia Devices & Applications
Smart Manufacturing & Production - Supply chain management - Agile connectivity	- Intelligent controls - Human-robot collaboration - Process & assembly automation	Embedded hardware, IoT devices, Augmented & Virtual Reality applications, Computer hardware & Software, Implantable devices, sensors, RFID tags, Computer Aided Designs (CAD) applications, Haptics, tactile and touch-based devices and applications, digital twins' applications, video, audio, graphics, images and text-based applications etc. **Enabling Multimedia Data Formats** **Text/Binary** .csv, .proto, .tfrecords, .parquet, .xslx, .avro, .orc, .petastorm, .txt, .docx, .pdf, .tsv, .json, .hdf5, .pb, .nc, .npy, .npz. etc. **Graphics (images/animated images formats)** .png, .jpg, .xcf,.svg, .bmp, .tiff, .pcx, .dib, .gif, .tga, etc. **Video/Audio** .mov, .mp4, .m4a, .m4v, .mkv, .midi, .mpg/.mpeg, .wav, .wma, .wmv, .avi, .flv, .3gp, .3gpp, h.263, h.264, .3g2, .3gp2, ogg, .wav, .aiff, .rm, .ram,.ra, etc
Energy - Electricity systems - Oil & gas production - Renewable energy systems	- Smart grid - Electric vehicle charging system - Smart oil and gas distribution grid	
Transportation Systems - Vehicle-to-vehicle (V2V) applications - Vehicle-to-Infrastructure (V2I) applications - Vehicle-to-Everything (V2E) applications - Autonomous (self-driving) vehicle (air, surface, space, water) - Intelligent Railway Infrastructure System	- Drive by wire vehicle systems - Plug ins and smart cars - Interactive traffic control systems - Next-generation air transport system (Baheti & Gill, 2011) - Next-generation air transport control - Train-borne monitoring system	
Healthcare - Medical devices - Personal care equipment - Disease diagnosis and prevention	- Wireless body area networks - Assistive healthcare systems - Wearable sensors and implantable devices	
Civil Infrastructure - Bridges and dams - Municipal water and wastewater treatment	- Active monitoring and control system - Smart grids for water and wastewater - Early warning systems	
Building & Structures - High performance residential and commercial buildings - Net-zero energy buildings - Appliances	- Whole building controls - Smart HVAC equipment - Building automation systems - Networked appliance systems	

Methods of sharing data across different domains including data format conversion will thus be needed to cater to the needs of Big Data that are generated in both modern and legacy multimedia data formats. Such methods of data translation that caters to the needs of modern and legacy multimedia data sets that are generated by different CPS subsystems are yet to be fully developed. Due to the need for data interoperability, data conversion and semantic data modeling, a CPS Public Working Group (PWG) was launched by the National Institute of Standards and Technology (NIST) in 2014.

Data Formats Challenges for CPS, Multimedia Systems' AI and Machine Learning Applications

FoF, multimedia technologies and other systems that span CPS uses artificial intelligence (AI) applications extensively.

Example of AI-based CPS are safety-critical CPS such as self-driving cars (Litvin, 2019; Uber, n.d.).

As discussed in (Litvin, 2019) and (Uber, n.d.), state-of-the-art data formats such as Parquet and Petastorm data formats are used to harvest and organize data that are generated from several subsystem including: Lidar, camera, sensors and GPS (Figure 3) that are integral to better understanding and development of self-driving cars. In this section, high dimensionality data issues, semantic interoper-

Figure 3. Safety-critical autonomous CPS like self-driving cars from Uber ATG generates Big Data from several sub-systems. Data formats such as Petastorm and Parquet are used extensively for multimedia AI solutions (Litvin, 2019), (Uber, n.d.) for such safety-critical CPS

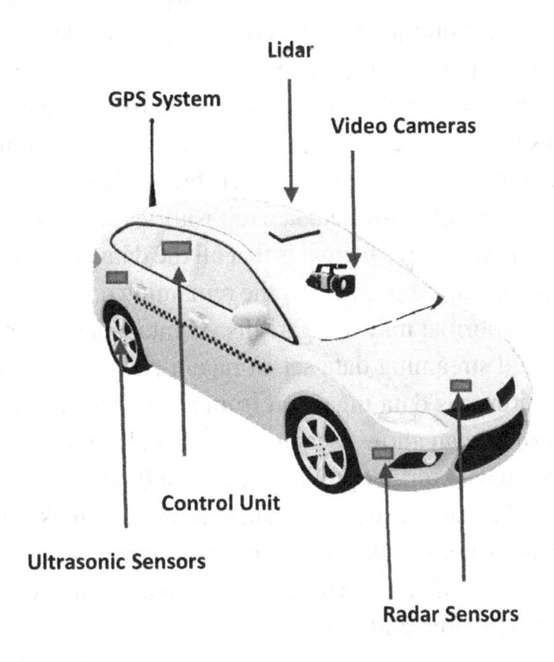

ability issues and data flow impedance mismatch between modern data formats and machine learning frameworks; and how those issues affect the development of multimedia and machine learning solutions for CPS are discussed.

Challenges with CPS High-Dimensionality Data

To interact intelligently with the physical world, modern, autonomous, safety-critical CPS such as self-driving cars depends on machine learning (ML) algorithms. Examples are deep learning algorithms (Dowling, 2019). Sub-systems' data that are included in the data space of such CPS may, in most cases generates high-dimensional images and video data sets. In high dimensional data sets, the number of feature sets may supersede the number of observation data points. This always led to extreme model training difficulties when deterministic systems' results are needed to further develop ML based solutions for safety-critical CPS.

Challenges with CPS Data Storage – Multifile vs Record-Streaming Data Storage

Data from safety-critical CPS may be arranged in the form of a *multifile data sets*, or *record streaming data sets* (Uber, n.d.). In the case of multifile data representation, data points are stored in separate files or directories using different multimedia data formats such as CSV, JPEG, NPZ, PNG, etc. (Uber, n.d.). For CPS such as the Uber ATG, if data sets used for deep learning model training are stored in multifile format, there may exist more than 100 million files for each case of a complete Uber ATG data set. The benefit of the multifile approach is that users and developers may have easy random access to any data set row. An ML training system may however have to make numerous computation-costly roundtrips to relevant file directories to be able to fetch needed ML training data set. The implication is that CPS applications based on multifile storage option may be hard to implement at scale due to cost of computation and cost of data retrieval from the training location up to the data cloud (such as S3 buckets and HDFS) where final system data sets are stored. The alternative approach is to use record streaming data set storage in which sets of rows that represent different features and subsystems are grouped together into one or more files using data formats such as Protocol Buffer (TFRecord), HDF5 and Python Pickle formats. This approach scales well with cloud-based data repositories such as S3 and HDFS. However, a shortcoming of using record streaming storage format is that all the data of a particular column may have to be fetched onto memory and then, unused segment of the remaining data set may have to be discarded.

Using Petastorm or Parquet data format may mitigate disadvantageous computing or memory issues relating to using multifile or record streaming data set storage formats. Petastorm or Parquet data formats can facilitate robust and continuous data ingestion from cloud-based storage repositories such as S3 or HDFS. Also, these data formats can allow users to select chunks of data from columns or rows of interest (Dowling, 2019). Petastorm data format provides a simple function that can be used to augment a standard Parquet data set with a Petastorm specific metadata, thereby making Parquet data sets to be compatible with Petastorm data repositories. Deep learning frameworks such as Pytorch or Tensorflow does not have native support for Parquet data formats. Petastorm data format can thus serve as a conduit for ingesting Parquet data sets into Pytorch and Tensorflow deep learning platforms (Dowling, 2019).

Data representation and storage challenges are pervasive in most safety-critical, factory systems and other CPS and multimedia systems in the whole world. Most of the data sets that are generated by several systems in the world are not always stored in formats that can be used directly to train ML models

Figure 4. Data exists in different types of formats from factory floor to the cloud repositories. Machine learning model training frameworks are also designed to ingest data of specific formats.

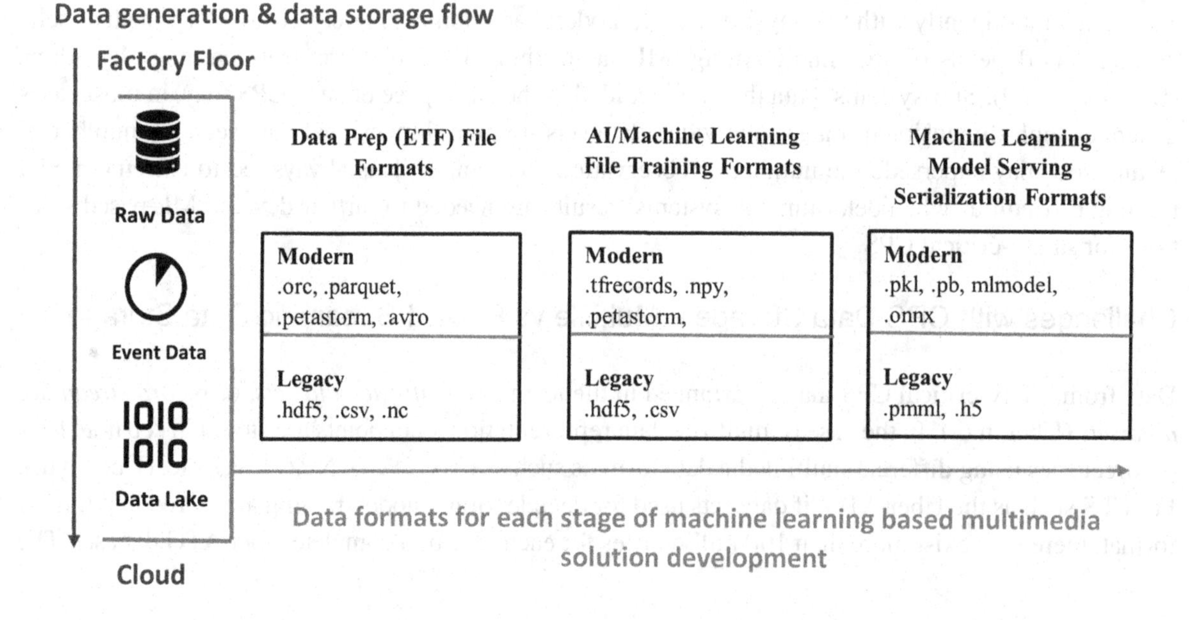

(Dowling, 2019). For factory CPS, modes of data generation and sequence of data storage from most legacy CPS to modern CPS applications is graphed in Figure 4. The graph is also representative of the data flow mode of most other common CPS. repository. To support effective machine learning-based CPS solutions, factory CPS data sets must be represented in the most appropriate data format. Legacy data formats such as CSVHDF5 and NetCDF (Network Common Data Format) may not be compressible, and they may not be splitable. Also, using legacy data formats for ETL applications can be challenging as the data sets may not be convertible into formats that is compatible with training many data processors in parallel. In addition, legacy data formats may not support processes that are geared towards combining different data sets. For ETL applications, formats that are scheme-able, support metadata generation processes; and that readily supports easily scalable machine learning training, such as Avro, Petastorm and Parquet data formats are very suitable. Also, data formats such as tfrecords, .npy, .petastorm, .pkl, .pb, mlmodel, and .onnx (Open Neural Network Exchange) are more suitable for deep learning training and for deep learning models servings as opposed to using more legacy data formats such as .pmml, .csv, .h5 and .hdf5 data formats. Petastorm files are large and splitable. Also, they can compress with data readers for deep learning frameworks such as Pytorch and Tensorflow. They are thus useful for training several GPU data shards in parallel. Unlike legacy data formats, Petastorm files do not bottleneck on the data input/output frontend. Through its Unischema framework, Petastorm have provision for multidimensional data ingestion which is useful for ML model training. Moreover, through Unischema, Petastorm can be used to natively store tensors (multi-dimensional data) in Parquet data format.

When using GPU, tensors are more suitable than arrays for faster processing of multidimensional data when the TensorFlow framework is considered (Pykes, 2021). Unischema provides an API through which Petastorm data can be used with Pytorch and TensorFlow deep learning frameworks (Dowling, 2019). In Table 2 to Table 4 (Dowling, 2019), different data formats and their suitability to different CPS AI initiatives are tabulated.

Table 2. Machine learning feature engineering file formats (Dowling, 2019)

File Formats	File Extension	Binary or Text	Compression Application	Splitable	Schema	Schema Evolution	Column Filtering
Petastorm	.petastorm	Binary	Dict, Snappy, RLE	Yes	Yes	Yes	Yes
Avro	.avro	Binary	Deflate, Snappy	Yes	Yes	Yes	No
Orc	.orc	Binary	Dict, Zlib, RLE	Yes	Yes	Yes	Yes
Parquet	.parquet	Binary	Dict, Snappy, RLE	Yes	Yes	Yes	Yes

Table 3. Machine learning model training data formats (Dowling, 2019)

File Formats	File Extension	Binary or Text	Compression Application	Splitable	Schema	Filtering (Column)
Petastorm	.petastorm	Binary	Dict, Snappy, RLE	Yes	Yes	Yes
Csv	.csv	Text	Gzip	No	No	No
Numpy	.npy	Binary	Gzip	No	No	No
Tfrecords	.tfrecords	Binary	Gzip	Yes	Limited	No
Hdf5	.h5	Binary	Zlib, Szip	No	No	Yes

Table 4. Machine learning model-serving data serialization formats (Dowling, 2019)

File Formats	File Extension	File Type	Compression Application	Quantization	Schema	ML Frameworks
Protobuf	.pb	Binary	Gzip, Varint	Yes	Yes	Tensorflow
Pickle	.pkl	Binary	Gzip	No	No	Scikit-Learn
MLeap	.zip	Binary	Zip	No	No	Pyspark
Apple ML Model	.mlmodel	Binary	Supports compression	Yes	No	iOS Core ML
H5	.h5	Binary	Supports compression	Yes	No	Keras
Onnx	.onnx	Binary	Supports compression	Yes	No	Pytorch
PMML	.pmml	XML	Supports compression	No	Yes	Scikit-Learn
Torch Script	.pt	Binary	Supports compression	Yes	Yes	Pytorch

IMPORTANCE OF TENSOR TYPE DATA SETS AND THEIR CHALLENGES FOR SUPPORTING CPS, BIOMEDICAL AND MULTIMEDIA SYSTEMS AI APPLICATIONS

Due to its structured representation of data formats and its advantages in reducing multidimensional data arrays complexity, tensors are being applied in more fields than ever before. Tensors have been applied to the problem of dictionary learning, image deblurring, spectral data classification, magnetic resonance imaging (Ji et al., 2019), generating inference in latent variable modeling, hidden Markov models, independent component analyses, topic models etc.

Tensors vs Vectors

Tensors are basically a generalization of multidimensional arrays of numerical values. They are mostly produced by generalizing matrices or vectors to multiple dimensions (Ji et al., 2019). However, while arrays or vectors are mutable, tensors are generally not mutable (Pykes, 2021). A pictorial representation of tensors from a 1D tensor to a 6D tensor is shown in Figure 5.

A scalar quantity can be considered a zero-order tensor. Also, as shown in Figure 5, a 1D tensor is a vector while a 2D tensor is a matrix. A 3D tensor is a cube or cuboid based on the type or matrix rank of data being represented. A 4D tensor can be imagined as a vector of cubes. Another way to view a fourth-order tensor is as an extension of the third-order tensor along one dimension. Similarly, a 5D tensor is a matrix of cubes extending along the lateral and longitudinal dimension; and 6d-tensor is a cube of cubes.

Tensor based algorithms have been known to achieve lower computational complexities and better accuracy when compared with traditional vector-based algorithms (Ji et al., 2019). In particular, tensors and their decompositions are very beneficial when deployed to solve unsupervised machine learning problems. They have also been known to produce superior results in temporal and problem relating to multi-relational data analysis (Rabanser et al., 2017). Due to its effectiveness in extracting structural features of data, many applications and algorithms are based on tensor decomposition.

Figure 5. Tensors as a representation of multidimensional arrays of vectors

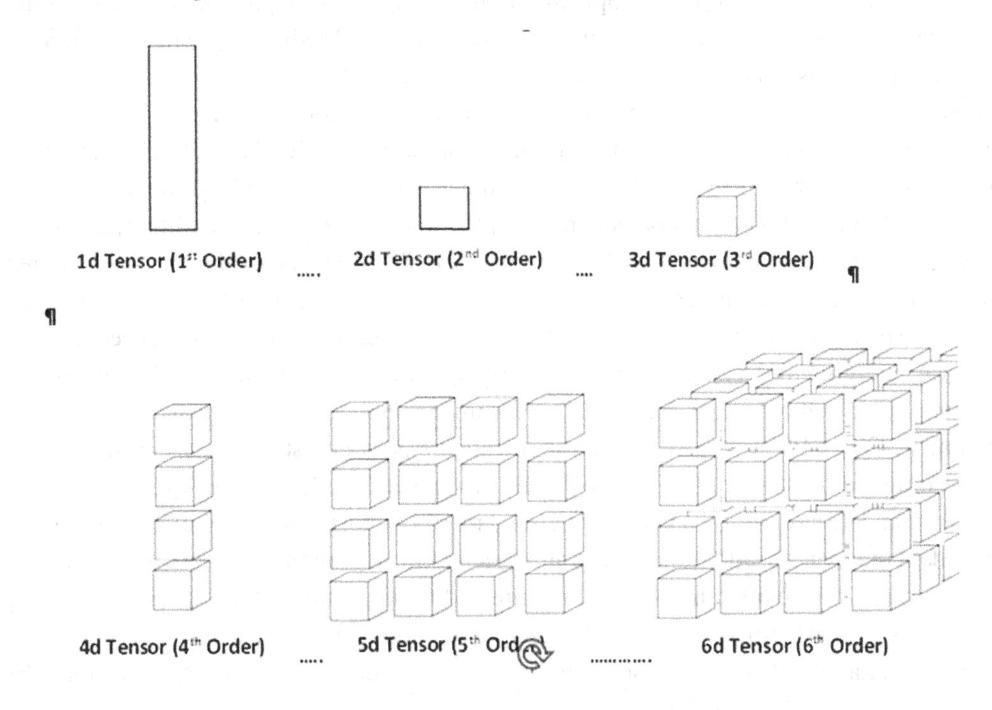

Tensors generally represents higher order statistics. Due to the need for most machine learning (ML) algorithms to use a large amount of structured high-dimensional input data for model training, tensors are being considered as suitable for training many ML algorithm models.

Issues Relating to Using Matrices for Multi-Dimensional Data Sets

High dimensional or multi-dimensional data generally refers to data sets that have more than two dimensions. An example is a 3D image data. Multi-dimensional data always have intrinsic interconnections between the data points in the multidimensional space. A common but naïve approach to modeling and analyzing such complex-structures data is to flatten them into matrices or vectors. After flattening, standard statistical techniques are then used to analyze the resulting data sets. Such common analyses methods however always end up destroying the intrinsic interconnections between the data points in

the multi-dimensional data space. Thus, with the flattening approach, tightly interconnected systems features may end up being treated as independent features (Ji et al., 2019). Another negative outcome of reducing multi-dimensional data sets into vectors or matrices is that the resulting data sets always yield an exponential increase in the number of parameters to be estimated. Using tensors to model multi-dimensional data sets instead of using matrices will help solve some of these data dimensionality issues.

Issues Relating to Using Matrices for Biomedical Applications

In Figure 2 and in Table 1, some important biomedical and healthcare CPS applications are mentioned. Numerous other healthcare and biomedical CPS that relies on accurate and timely data exchange include intelligent operating rooms and hospitals, image-guided surgery and therapy, fluid flow control for medicine and biological assays, and the development of physical and neural prostheses. As a reconfigurable CPS, healthcare and sub-systems that are under Cyber Physical Medical Systems (CPMeS) are often networked, and they must match the needs of patients with special circumstances (Baheti & Gill, 2011). For example, devices such as infusion pumps for sedation, ventilators and oxygen delivery systems for respiration support, and other sensors that are used for monitoring patient's condition are in use in many medical emergency rooms worldwide. Often, these devices must be assembled into a new system configuration to match the needs of specific patients or operational procedures. In addition, specific patient's data or operational procedures which always generates big data must be evaluated almost in real time.

Due to the massive data being generated by biomedical and healthcare CPS, AI and analytics initiatives that are based on matrix data representations are often inadequate in terms of speed, complete data representation and reliability. Also, when we have higher dimension data, it is challenging to generalize methods that works well for matrices, which are 2-dimensional arrays, to tensors of order 3 or higher (Tokcan et al., 2020). CPS systems that cater to healthcare needs must be concise, timely and reliable. Matrices decomposition may not be suitable for the uniqueness needed for such (Rabanser et al., 2017) high-dimensional data system. Besides, matrices and matrix factorizations are often non-unique. To ensure matrix data uniqueness, additional constraints such as orthogonality and positive definiteness must be ensured. Tensors, on the other hand do not require additional positive-definiteness and orthogonality constraints to ensure that factorizations result to unique decompositions. Also, due to tensors being useful for representing data in higher dimensions, they provide better data identifiability conditions (Brandi & Matteo, 2021; Rabanser et al., 2017).

Challenges with Using Tensor Data Formats for CPS Applications

Some challenges of using tensors include the fact that tensors are not readily representable in most existing common data formats such as .txt, .csv, .json etc. Also, tensors in most cases have unique data representation formats such as .tns, .proto, .rb (Rutherford-Boeing format), .mtx (Matrix Market (Coordinate) Format) (Duff & Lewis, n.d.; FROSTT, n.d.; TACCO, n.d.; TensorFlow, n.d.). Thus, there is a need for algorithms that can assist in converting high-dimensional data that exists in common data formats such as .csv, .tsv, Numpy Arrays, Parquet, Avro formats to applicable tensor data formats such as .proto, .tns and .mtx formats.

CHALLENGES OF TENSOR BASED AI APPLICATIONS AND PROSPECTS OF DATA FORMAT CONVERSION AS A SOLUTION

Tracking Objects in Cyber Physical Systems

Tracking algorithms are used extensively in CPS applications (Chen et al., 2010; Rashidibajgan et al., 2020). Tensors are used extensively to represent data for these CPS tracking systems. However, most existing tensor-based tracking algorithms often cannot completely detect intrinsic local geometry and discriminant structure of the image block in tensor form. Due to this issue, data relating to the background are often lost. This often leads to reduction in the accuracy of target tracking algorithm.

Data Format Conversion as a Possible Solution

To solve the problem of background information losses in target tracking, another algorithm that can separately read in background data may be needed. The read background information may exist in another data format. The separate background information may then be converted into a data format that can be merged with the original target tensor data.

Object Classification in Cyber Physical Systems

Solving machine learning classification problems often constrain tensor decomposition to require more training and feature parameters. A large number of data points are always needed for algorithm training. Using a large number of tensor data sets can cause classification algorithms to slowly converge, or not to converge at all.

Data Format Conversion as a Possible Solution

Before tensorizing data, data formats that supports data compression without losing data and without introducing non-useful data artifacts can be used as an intermediate data conversion step.

Shortage of Training Data

For some CPS applications, how to obtain the huge amount of data needed for training AI based solutions is always a problem. Due to the limited available data sample, researchers sometimes, often choose to experiment with simulated data (Ji et al., 2019). As a result, model accuracy is not always fully guaranteed when tensorizing data to a higher dimension.

Multimedia Data Conversion as a Possible Solution

Multimedia solutions, additional CPS sensors and using reliable methods of data conversion can enable more data to be available from a single data source. To improve availability of data sets, different types of sensors can be used to obtain data from the same source in different multimedia formats e.g., video, sound (natural language data sets) and images data sets can be obtained from the same data source using different sensors. If applicable, the different data sets can be translated into a common data format. Resulting data sets can also be converted into various data formats for tensor applications.

Effects of Tensor Decomposition Algorithms

As part of data processing, some tensor algorithms often directly decompose the input data features into multiple dimensions (Ji et al., 2019). The multiple dimensions generated sometimes ends up combining these features with other useless features (Ji et al., 2019). Often, it is challenging to ignore the useless combination of features while accurately extracting useful information in the decomposition.

Data Format Conversion as a Possible Solution

Comparison of the performances of various tensor decomposition methods are needed along with trying out various data translation and conversion methods. However, trying out various data translation as part of examining possible combination of data features can lead to even more data storage problems. As a possible form of solution, converting all data sets into formats that supports data compression will help to limit the data storage issues that can result from trying out various tensor decomposition methods.

Data Modalities for ML Applications

Data modality refers to a certain type of information and/or the representation format in which information is stored (Morency & Baltrusaitis, n.d.). For example, as shown in Figure 6, data can be ingested into

Figure 6. Increasing the modalities of multimodal machine learning (MML) algorithms training data may include transforming the base training data into other different data formats (e.g. .png (image), .CSV (Text), .mp4(MPEG))

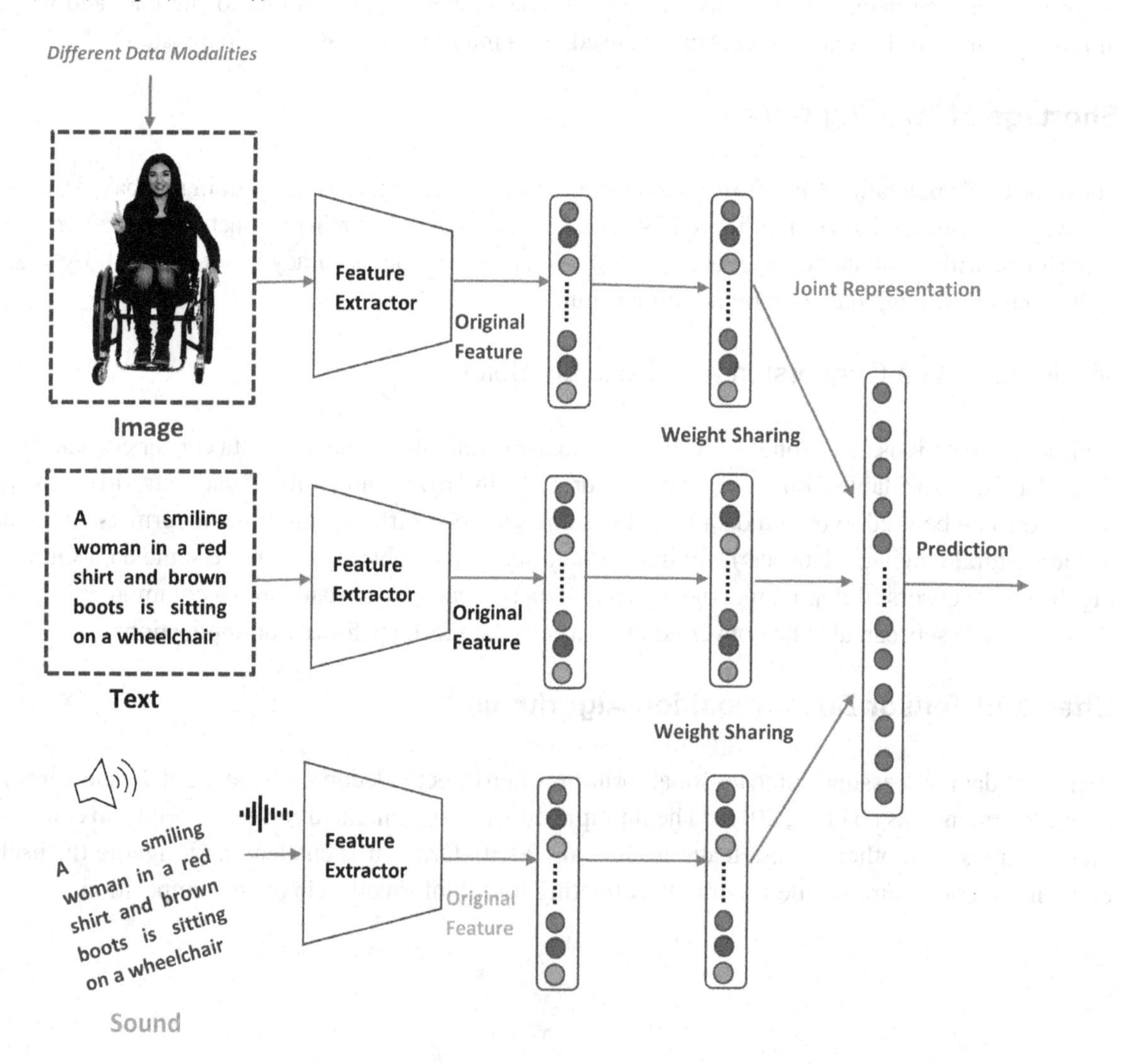

a deep learning training network in the form of an image. The same data can be represented in terms of text data and fed into a separated deep learning LSTM (long short-term memory) network while another LSTM network can ingest the data in from of an audio data set.

Modality have also been associated with the sensory modalities which represent our primary channels of communication and sensation, such as human vision or touch. A research problem or data set is therefore characterized as multimodal when it includes multiple such modalities.

The advantage of using multiple modalities in AI model training is that when the quality of some modalities is low, other modalities can provide effective guidance for the model (Pool Party, n.d.). Current work from literature have shown that the more input modalities that exist, the better the ML model. Core research challenges in deep multimodal ML include Representation, Alignment, Fusion, Translation and Co-Learning (Morency & Baltrusaitis, n.d.).

For AI models to be able to make progress in understanding the world around us, it needs to be able to contextualize, interpret and reason about multimodal data sets. Multimodal ML aims to build models that can process and relate information from multiple sources and modalities. Obtaining data from multimodal sources offers the possibility of capturing correspondences between such modalities. It can also lead to an in-depth understanding of natural phenomena in CPS (Morency & Baltrusaitis, n.d.).

Other challenges in using multi modal data sets to drive sub systems of CPS include how to represent and summarize multimodal data in a way that exploits the complementarity and redundancy of multiple modalities. The heterogeneity of multimodal data sets makes it challenging to construct such representations. For example, language is often symbolic while audio and visual modalities will be represented as signals (Baltrusaitis et al., 2017; Morency & Baltrusaitis, n.d.). Finding a method of translating multi-modal data sets into data basis of similar data formats may be a way to establish multi-modal data correspondence and representation.

Semantics and Ontologies Challenges for AI Integration in CPS and Multimedia Applications

To realize the vision of a highly reactive CPS and IIoT system, sub-systems, and devices in IIoT and CPS must unambiguously understand, interact, and exchange information. For systems to be interoperable, due to different design schemes and mode of operations, formalized ontologies, data semantics and data exchange rules must be integrated into IIoT and CPS systems' design. Semantic interoperability, i.e., the unambiguous access and interpretation of data by different communicating sub-systems, will ensure a common understanding of exchanged data sets. Achieving semantic interoperability in IIoT and sub-systems of CPS will allow billions of devices to communicate and exchange information to perform jobs (Nagowah et al., 2018). However, the increasing diversity and dynamicity of devices in the CPS environment, and the resource-constrained nature of the CPS and IIoT devices have contributed to establishing interoperability as a research field of major importance. Across different countries and across many fields, different manufacturers produce devices without following any specific standards. Thus, the CPS ecosystem makes interoperability of devices to be a very challenging endeavor. However, for a reliable and truly reactive CPS, achieving interoperability is vital so that billions of sensors, actuators, tiny and smart devices embedded in the IIoT environment can communicate and understand each other. Also, with interoperability, many data preparation and information retrieval tasks such as data cleaning, schema matching, data discovery, semantic search, and data visualization will benefit from detecting the semantic types of data columns in relational tables (Zhang, n.d.).

Structured Data, Unstructured Data, and Data Modalities Applications for Semantic AI

Across many CPS ecosystems, devices and sub-systems possess and constantly generate data that are distributed across various database systems. When it comes to the implementation of new AI use cases, usually very specific data is needed. Semantic AI systems can address the need for providing interpretable and meaningful data. If the semantic system is properly designed, it can also provide technologies to create the type of data needed from the very beginning of a data lifecycle.

In addition, using semantic modeling in AI can assist in providing a linkage between structured and unstructured data sets (Pool Party, n.d.). Most machine learning algorithms work well either with unstructured or with structured data, but those two types of data are rarely combined to serve as a model training data set. Semi-structured data types, such as JSON and MessagePack are very rarely used in AI model training. Data format conversion processes and using well-designed semantic data models can help bridge this gap. Using data format conversion, together with Semantic AI, links and relations between business and data objects of all formats such as XML, JSON, MessagePack, relational data, CSV, and unstructured text can be made available for further analysis and for training AI models. This will allow a linkage of data across heterogeneous data sources. Such linkage will serve to provide data objects

Figure 7. Automatic semantic and atomic type data detection will enable the development more reactive and intelligent cyber physical system

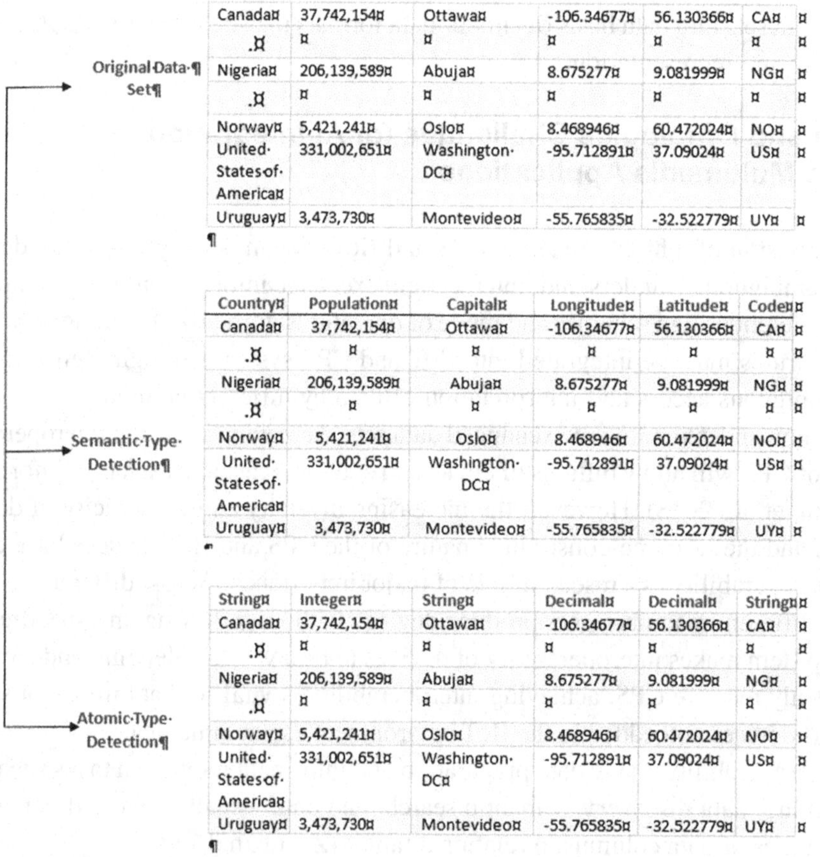

Original Data Set

Canada	37,742,154	Ottawa	-106.34677	56.130366	CA
.					
Nigeria	206,139,589	Abuja	8.675277	9.081999	NG
.					
Norway	5,421,241	Oslo	8.468946	60.472024	NO
United States of America	331,002,651	Washington DC	-95.712891	37.09024	US
Uruguay	3,473,730	Montevideo	-55.765835	-32.522779	UY

Semantic Type Detection

Country	Population	Capital	Longitude	Latitude	Code
Canada	37,742,154	Ottawa	-106.34677	56.130366	CA
.					
Nigeria	206,139,589	Abuja	8.675277	9.081999	NG
.					
Norway	5,421,241	Oslo	8.468946	60.472024	NO
United States of America	331,002,651	Washington DC	-95.712891	37.09024	US
Uruguay	3,473,730	Montevideo	-55.765835	-32.522779	UY

Atomic Type Detection

String	Integer	String	Decimal	Decimal	String
Canada	37,742,154	Ottawa	-106.34677	56.130366	CA
.					
Nigeria	206,139,589	Abuja	8.675277	9.081999	NG
.					
Norway	5,421,241	Oslo	8.468946	60.472024	NO
United States of America	331,002,651	Washington DC	-95.712891	37.09024	US
Uruguay	3,473,730	Montevideo	-55.765835	-32.522779	UY

Figure 8. Interaction between semantic, contextual and data modalities will assist in the development more reactive and intelligent cyber physical system

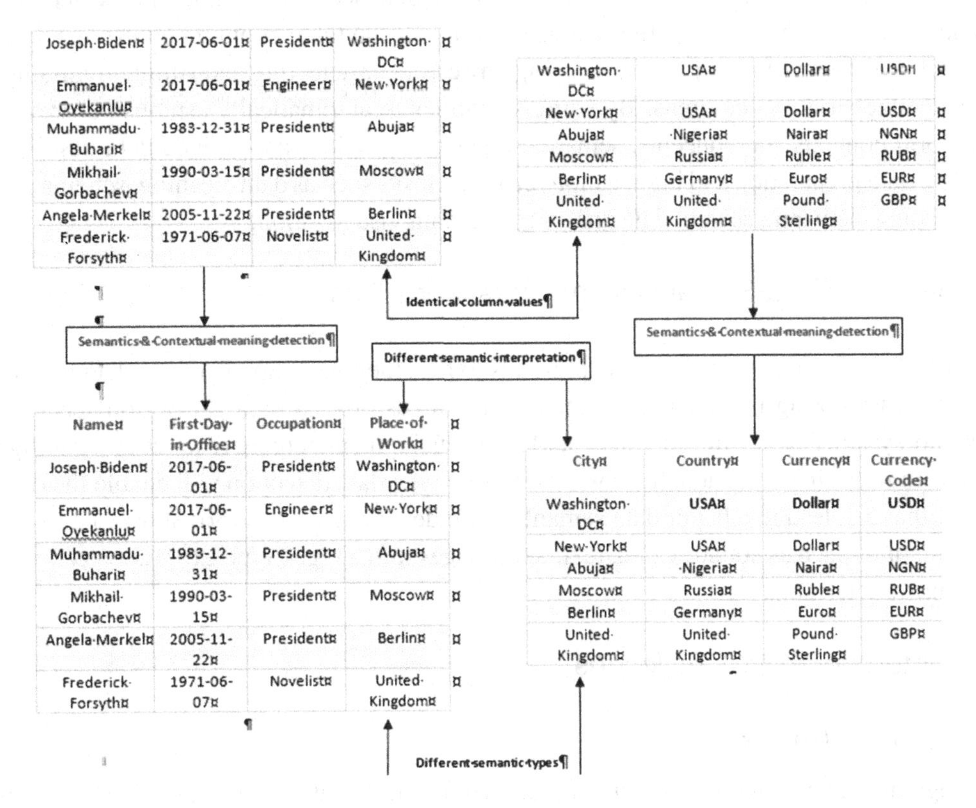

which are composed of information from non-structured data, semi-structured and structured data sets, together with their contexts. At the same time, all the combined data set can be exposed to AI models.

As shown in Figure 7 and Figure 8, from the same data set, well-designed semantic data systems can automatically detect atomic data types needed to reliably operate and improve user experience. Atomic types can include string, integer, and decimal data types. As shown in Figure 7, semantic types such as country, population, and latitude can also be detected. This can help provide finer-grained, operational information that can be useful for further improving and automating many CPS AI and data analytics tasks.

Using semantic analysis, rules can be extended to detect the most meaningful inference regarding a column even if all the data in the columns are unknown. For example, in the first table of Figure 8, well known series of combined characters e.g., 'Joseph Biden' and 'Angela Merkel' will enable the semantic inference system to deduce that the first column in Figure 8 contain names. Also, since many of the data in each row of the first table in Figure 8 contains names of well-known public office holders, inference regarding appropriate column ID to assign to the remaining columns of Figure 8 can be derived. In like manner, semantic inferences regarding column names can also be derived for each column of the second table of Figure 8.

By using semantics AI, automated data cleaning initiatives would also significantly benefit from detecting the semantic data types in each column of the data set. For example, in the second table of Figure 8, knowing that the first and the second columns of the table contain country names and their capitals will hugely simplify the job of correcting any errors or detecting and filling in missing country and capitals values in these columns. By using semantics to automatically establish correspondences with real-world contexts and concepts, semantic types can provide fine-grained data descriptions and thus, improve data preparation and information retrievals tasks such as data cleaning, schema generation, schema matching, semantic search, data analytics and data visualization.

Linkage between Data Semantics and Data Modalities

Another area of research that is yet to be explored is to find the interaction between data modalities and data semantics. Exploring linkage between these vital areas of CPS data generation spaces and their respective sub-systems' data will generally enable the development of more reactive and intelligent CPS. An approach to matching data modalities with semantic type data detection will enable the detection of data context that can lead to a better data semantic type detection. A data format translation from various data modalities such as sound, video, text, and images into a regex type expression can also help to simplify context determination.

Semantic Interoperability Challenges

Standardized Data Models

Problems regarding how to model data in a standardized format that is understandable by different subsystems and stakeholders in the IIoT and CPS environment is a prevailing issue (Megagon Labs, n.d.). Also, assigning relevant meaning and contexts to different data instances remains a prevailing and quite challenging problem.

Contextualized Data Models

Humans can readily understand contexts and they can readily make inference regarding contexts of fuzzy data and thus understand hidden meanings. Similar benefit of intrinsic understanding is not easily attributable to AI models and machines. The problem of how to assign relevant contexts to the same data sets in different instances and domains remains quite challenging. For example, in Figure 8, the column representing 'Place of Work' has similar data to the column representing 'City'. However, the context of use and the semantic interpretations are quite different.

Limitations of Regex Matching-Based Approaches

Most existing AI systems generally use matching-based regular expression (regex) approaches for semantic type detection (Hulsebos, 2019). However, using regex expressions delimits semantic system's versatility. Systems that are designed using only regex expressions in most cases underperform. Such systems are often not robust enough to handle noisy real-world data, and they support only a limited number of semantic data types.

Data Format Challenges for Data Cloud and Lakehouse Applications

Challenges with JSON Objects Data Formatting in the Databricks Lakehouse

In most medical and industrial manufacturing processes, many data sets are generated with custom applications, devices and APIs that constrained more than one JSON objects to be written to the same file (Bynum, 2022). As an example, in Figure 9, a single data file is used to log JSON objects from multiple devices (1A1, 1A2 etc.). A most straightforward approach that many Data Scientist and Data Engineers may adopt to process Figure 9 data set is to use the PySpark json.load() function to read in the data set for further processing (Bynum, 2022). PySpark is a Python API for Apache Spark. It is used to process large datasets in a distributed cluster and in cloud Databricks Lakehouse applications. A problem though is that if the data set shown in Figure 9 were to be parsed using the PySpark json.load() function, the first-row data will be treated as the top-line data definition, while every other record after the first device_id record will get discarded (Bynum, 2022). In other words, every other record after the first device_id record will be prevented from being read (Bynum, 2022) if the PySpark json.load() function is used. Essentially, a JSON file is invalid if it contains more than one JSON object when the json.load() function is used to parse the JSON file.

One way to fix this issue is to redesign the data formatting method at the data generating sources. However, it may be very challenging or nearly impossible to redesign the data generating sources due to the possible usage of custom API or legacy system at each source. One way to solve the problem however may include integrating a data format conversion method or application between the custom API and the cloud Lakehouse data system.

Challenges with Support for Multiple Data Formats in the Cloud Lakehouse Tables

As emphasized in (Armbrust et al., 2020), Data storage format for a Delta table is either a JSON or a Parquet file format. Also, as explained in (Peternel, 2021), cloud Bronze Tables accepts only JSON, XML and TXT data formats. There are various methods by which data can be converted from one format to another in the cloud data ingestion layer. In addition, methods of converting data from one format to another exists between the Bronze and the Silver layers of most cloud data Lakehouses.

In the typical data Lakehouse architecture, the Bronze, Silver and Gold tables serve important purposes in ensuring that cleansed data is available for several CPS use cases. The Bronze layer often ingest operational data from upstream streaming pipeline such as the Amazon Kinesis or the Apache Kafka (Peternel, 2021). Data can also be piped from object storage landing zone such as the Amazon S3 buckets into the Bronze Table. As mentioned earlier, the CSV, XML and JSON formats can readily be ingested into Bronze tables. However, not all data formats that exist at the factory floor can be easily converted to formats that are supported at the cloud data ingestion layer. This problem is even more pronounced in the CPS ecosystem wherein different CPS sub-systems serve as host to several non-standardized devices that produces factory data sets in different formats (Figure 1 and Figure 2). The data format conversion problem can however be solved by converting data from legacy formats to different formats that can be easily ingested into the cloud layer.

In this chapter, several methods and algorithms through which data can be easily converted from one form to another are discussed. Their Python program implementation are also provided.

Figure 9. Several JSON objects corresponding to data from different devices can be logged onto the same data files


```
[
  {
    "device_id" : "1A1"
    "reading" : [{
    "type" : "temp sensor",
     "date" : "20210112",
     "timezone" : "EST",
    "timestamp" : "16:02",
     "location" : "NYEST"
   }],
      "equipment" : [{
      "CAT SCAN Machine" : "1A1",
      "temperature" :"80.7",
       "unit" : "Farenheit",
    }]

   }
{
    "device_id" : "1A2"
    "reading" : [{
    "type" : "vibr sensor",
     "date" : "20210112",
     "timezone" : "EST",
     "timestamp" : "16:07",
     "location" : "NYEST"
   }],
       "equipment" : [{
       "CAT SCAN Machine" : "1A1",
     "freq" : "602",
       "unit" : "MHz",
    }]

   }
{
    "device_id" : "1A9"
    "reading" : [{
    "type" : "vibr sensor",
     "date" : "20210112",
     "timezone" : "EST",
     "timestamp" : "16:14",
     "location" : "NYEST"
   }],
      "equipment" : [{
       "CAT SCAN Machine" : "1A1",
       "freq" : "119",
       "unit" : "GHz",
    }]

   }
]
```

Data Flow Impedance Mismatch between Cloud's Modern Data Formats and Machine Learning Frameworks

In general, ML frameworks are always set up to ingest training data as a sequence of samples. It is thus always desired to have ML training data in easily ingestible layouts with little or no data flow mismatch between the data storage platform and the programming language's input/output application programming interface (API). In general, most ML (especial deep learning) algorithms are data hungry. For effective CPS deep learning solutions, using binary data formats instead of legacy data format such as CVS will reduce the impedance of the training data pipeline. Thus, binary data formats such as. pd, .onnx and

Algorithm 1. Convert CSV Time series into MessagePack Data Format

```
Input:      CSV Timeseries Data without Schema
Output:     MessagePack timeseries and DataFrame
Data:       New York Stock Exchange Time series (1997-2021)
begin

         Input CSV time series data

         Convert each CSV row into Python dictionary

         Append subsequent rows to existing rows with each row appended as a
record

         Output each record as a newline record
         Dump all converted record as a JSON string and save onto local memory
         Load the dumped data using json.load() method
         Convert from JSON into MessagePack format using msgpack.pack()method
and
         save onto local storage

         Read from storage location using msgpack.unpack() method and the
         json.load() method

         Convert from msgpack.unpack() format to dictionary key:value pair
         format

         /* This enables the entire time series to be treated as an element of
         MessagePack format. It is also useful as an intermediate step to con-
vert
         from dictionary to list based format to ensure easier conversion to a
         Python dataframe*/

         Access the dictionary and then convert to a Python list format
         Use Python Pandas to convert to a dataframe

         /* Convert back to CSV if need be. This step can be used to convert
from
         MessagePack to CSV format */
end
```

.pkl in most cases may be suitable for ML model training when the data set is big. If data are stored in multifile format, the throughput of each pipeline will also be improved since binary data formats uses lesser storage space (Dowling, 2019; Oyekanlu et al., 2022) and are generally speedier to ingest than most legacy data formats. Deep learning algorithms generally favors using binary data formats for algorithm training. As the use of deep learning algorithms for multimedia and CPS applications grows, the usage of binary data formats for model training will also grow.

However, most ML frameworks such as ScikitLearn supports legacy data formats such as .csv, .npy and .npz; while in Data Lakes, structured data may be stored in relatively modern data formats such as .parquet or .orc (Dowling, 2019). So, the challenge of converting data formats form one form to another for CPS ML applications persists in several layers in the industry.

METHODS AND ALGORITHMS OF DATA FORMATS CONVERSION

To enjoy the many benefits of the different data formats that are available, in subsequent sections of this chapter, we have provided algorithms by which data formats can be converted from one type to another without any loss of data. To implement the algorithms provided, a stock market time series data set containing more than 6000 rows of datapoints (freely available at: https://finance.yahoo.com/) was downloaded and saved into Excel workbook in CSV data format (Oyekanlu et al., 2022). Each row of the CVS string has fields containing float, int and string data types. Clippings of data from top rows and the last set of rows in the data set is shown in Figure 10. In this chapter, the stock market time series data set is used to show how CSV time-series data can be efficiently converted from the CSV data format to other data formats. Algorithms that can convert CSV time series to both legacy and state-of-the-art data formats are covered in depth. However, algorithms that are provided can be leveraged to convert data formats different from CSV to other desired data formats. Our methods can also be used to convert

Figure 10. Yahoo.com Open-source Stock Market Time Series (1997 - 2022)

	Date	Open	High	Low	Close	Adj Close	Volume
Top Part of Time Series	4/30/1997	17.5	17.5	17	17	15.19202	5800
	5/1/1997	17.5	17.75	17.375	17.75	15.86226	36300
	5/2/1997	17.75	17.875	17.5	17.875	15.97396	17700
	5/5/1997	17.5	18.5	17.5	18.5	16.53249	65100
	5/6/1997	18.625	19.5	18.25	19.5	17.42614	227000
	5/7/1997	20	20.25	19.5	20	17.87297	75600
	5/8/1997	20	20	19.125	19.125	17.09102	5600
	5/9/1997	19	19.625	18.8125	19	16.97931	56400
	5/12/1997	18.75	19.625	18.625	18.875	16.86761	45700
	......						
Tail end of Time Series	4/18/2022	51.8	52.65	51.62	52.21	52.21	53100
	4/19/2022	52.05	53.63	52.05	53.14	53.14	68700
	4/20/2022	53.44	53.95	52.71	52.99	52.99	82300
	4/21/2022	53.28	53.68	52.02	52.27	52.27	57600
	4/22/2022	52.02	52.07	49.41	49.59	49.59	262600
	4/25/2022	49.05	49.79	48.35	49.73	49.73	120400
	4/26/2022	49.44	50.42	48.31	48.48	48.48	80400
	4/27/2022	48.53	49.11	48.32	48.62	48.62	114400
	4/28/2022	49.63	49.63	44.36	44.87	44.87	283800
	4/29/2022	44.32	44.6	42.37	42.89	42.89	193100

non-time-series data sets from one format to another. Based on extensive literature survey, our contribution in this chapter provides the most in-depth data format conversion treatise available in literature. In addition to the algorithms, analyses of the algorithms and their Python implementations for important data analytics activities such as querying, finding unique values, sorting, grouping etc., are also provided in subsequent sections.

Algorithm and Python Implementation for Converting CSV Time-Series to MessagePack Time-Series

As discussed in Part I and in (MessagePack, n.d.), MessagePack is an efficient binary serialization format. Its structure is similar to JSON, but the syntax enables it to be faster and smaller than JSON. It is supported by over fifty programming languages. However, a simple and efficient method by which CSV-based time-series data can be converted into MessagePack is yet to be presented in literature. Here, we present Algorithm 1. It is an algorithm by which CSV data format time series can be converted to MessagePack data format time series. Python implementation of Algorithm 1 is shown in Figure 11.

Figure 11. Python implementation of algorithm 1 – Conversion of CSV timeseries to messagepack data format

```
CSVFileInput = r'C:\Users\...........\Yahoodata.csv'
JSONFileOutput = r'C:\Users\.........\Yahoodata2.json'
def FileInput_to_json(CSVFileInput, JSONFileOutput):
    fileToStoreConvertedArray = []
    with open(CSVFileInput,  newline='') as inputfile:
        dialect = csv.Sniffer().sniff(inputfile.read(1024))
        inputfile.seek(0)
        csvReader = csv.DictReader(inputfile)
        #convert each csv row into python dict
        for row in csvReader:
            fileToStoreConvertedArray.append(row)
      with open(JSONFileOutput, 'w', newline='') as jsonfile:
        jsonString = json.dumps(fileToStoreConvertedArray, indent=4)
        jsonfile.write(jsonString)
start = time.perf_counter()
FileInput_to_json(CSVFileInput ,JSONFileOutput)
finish = time.perf_counter()
with open("Yahoodata2.json", "r") as read_file:
    datajson = json.load(read_file)
MsgPackdata = datajson
# Write msgpack file
with open('MsgPackdata.msgpack', 'wb') as inputMsgPackdata:
    msgpack.pack(MsgPackdata, inputMsgPackdata)
# Read MsgPack in from storage location
with open('MsgPackdata.msgpack', "rb") as MsgPackdata_file:
    # MsgPackdata_loaded = json.load(MsgPackdata_file)
    MsgPackdata_loaded = msgpack.unpack(MsgPackdata_file)
newloadedMsgPack = {"convertMSGPckToDF": MsgPackdata_loaded}
MsgPackDF = pd.DataFrame(newloadedMsgPack['convertMSGPckToDF'])
MsgPackDF
```

Figure 12. Top part of the resulting timeseries data for Python implementation of algorithm 1 – Conversion of CSV timeseries to messagepack. compare to original timeseries in figure 10

```
# Output of Step 9 in Algorithm 1 (i.e., output at the point:
newloadedMsgPack = {"convertMSGPckToDF": MsgPackdata_loaded} in the Python
code)

{'convertMSGPckToDF': [{'Date': '4/30/1997',
   'Open': '17.5',
   'High': '17.5',
   'Low': '17',
   'Close': '17',
   'Adj Close': '15.192017',
   'Volume': '5800'},
  {'Date': '5/1/1997',
   'Open': '17.5',
   'High': '17.75',
   'Low': '17.375',
   'Close': '17.75',
   'Adj Close': '15.862257',
   'Volume': '36300'},
  {'Date': '5/2/1997',
   'Open': '17.75',
   'High': '17.875',
   'Low': '17.5',
   'Close': '17.875',
   'Adj Close': '15.97396',
   'Volume': '17700'},
  {'Date': '5/5/1997',
   'Open': '17.5',
   'High': '18.5',
   'Low': '17.5',
   'Close': '18.5',
   'Adj Close': '16.532488',
   'Volume': '65100'},
  {'Date': '5/6/1997',
   'Open': '18.625',
   'High': '19.5',
   'Low': '18.25',
   'Close': '19.5',
   'Adj Close': '17.426138',
   'Volume': '227000'},
```

Any Python version from Python 3.6 and above can be used to implement the algorithm. Python 3.6.8 in Anaconda environment have been used for the current implementation. Output of the algorithm showing a small segment of the original CSV data (of Figure 10) that has been converted to MessagePack format using the Python implementation shown in Figure 11 is shown in Figure 12. Also, a small segment of the CSV time-series data after conversion to MessagePack-based time series *data frame* is shown in Figure 13. Python libraries that are needed for implementing Algorithm 1 are shown in Figure 14. For any intending practitioner willing to deploy the algorithm, these Python libraries should be included as part of the code in Figure 12.

Figure 13. Top part of dataframe output for python implementation of algorithm 1 – Conversion of CSV time-series to messagepack

```
# Output of the 2 final steps in Algorithm 1 (i.e., output of the code part:
MsgPackDF = pd.DataFrame(newloadedMsgPack['convertMSGPckToDF'])
MsgPackDF in the Python code (Fig. 11))
```

	Date	Open	High	Low	Close	Adj Close	Volume
0	4/30/1997	17.5	17.5	17	17	15.192017	5800
1	5/1/1997	17.5	17.75	17.375	17.75	15.862257	36300
3	5/5/1997	17.5	18.5	17.5	18.5	16.532488	65100
105	9/29/1997	17.5	18.25	17.5	17.5	15.638843	4300
106	9/30/1997	17.5	17.75	17.5	17.5	15.638843	21000
108	10/2/1997	17.5	17.5	17.25	17.4375	15.582988	6800
115	10/13/1997	17.5	17.625	17.25	17.4375	15.582988	9200
135	11/10/1997	17.5	17.5	17	17	15.192017	3100
168	12/29/1997	17.5	17.625	17	17	15.192017	3300
498	4/22/1999	17.5	19	17.5	19	16.979313	160800
529	6/7/1999	17.5	17.625	17.375	17.625	15.750552	53200
565	7/28/1999	17.5	17.5	17.25	17.375	15.527136	27400
4039	5/20/2013	17.5	17.74	17.4	17.469999	16.068762	78200
4051	6/6/2013	17.5	18.24	17.5	18.23	16.767805	101500

Figure 14. Python libraries needed for the implementation of algorithm 1

```
import msgpack
import matplotlib.pyplot as plt
import pandas as pd
import numpy as np
import json
import csv
```

Based on extensive literature searches, our implementation is the first reported full algorithm that shows method by which cleansed CSV based time-series can be efficiently converted to MessagePack data format using Python. Procedures used for implementing the algorithm also provides avenues by which the resulting MessagePack format can be easily converted to other data format and their respective data frames by allowing ample opportunities for method reuse.

Algorithm and Python Implementation for Converting CSV Time-Series to JSON Format Time-Series

The two broad approaches to structuring data for analytics, data transfer and exchange are the binary and text-based formats. The JSON format is text based and is thus human readable. It is language independent, and it can be used with most modern programming languages. JSON was created as a means of solving the coding agility issues that always disadvantages the XML data format (Oyekanlu et al., 2022).

JSON, which is derived from a subset of JavaScript programming language, have thus, to a large extent, replaced the XML as a format for data sharing between clients and servers and as a data analytics format. XML data format was formerly the de-facto method of exchanging data between Internet clients and servers. However, JSON is more compact than XML, it allows faster execution than XML, and its syntax is simple and self-describing. By being self-describing, JSON data structure can in some cases be interpreted by applications that does not formerly have knowledge of the data to expect. An added advantage of JSON-based server-side parsing is that it increases server responsiveness. Clients can thus get faster responses to their queries. It is for this reason that JSON is widely adopted as a standard data exchange format for most client-server Internet applications. JSON is easy to parse, and no additional code is needed for JSON data parsing. Data in JSON exist in key-value pairs. With the key-value pair arrangement, a colon separates the field name and value. XML syntax however is stricter, and it has support for schemas and namespaces (Ezeelive Technologies, n.d.; Progress, 2017).

In JSON, a complete object begins and ends with curly braces, and object members consists of strings and values that are comma separated. Arrays in JSON begins and ends with braces. JSON values can be a literal, an array, a string, or an object. In JSON, unlike CSV, unstructured, complex or hierarchical data types can be easily parsed due to JSON format versatility. Also, many simple editors can readily display JSON due to its simple text nature (Ezeelive Technologies, n.d.; Progress, 2017). Large size video, audio and images data sets can be shared easily using JSON arrays, thus making JSON a versatile data format suitable for web API's and for web development.

JSON Data Format Disadvantages

While JSON is compact when compared to XML, it is still quite verbose when compared to other data formats. In terms of Internet security, JSON can be quite dangerous if used with an untrusted browser or service. A JSON call returns a response wrapped in a function call which if executed can expose the receiver to a hacking attack. Writing comments on JSON data can also be very challenging if not outrightly impossible for some JSON data sets. JSON also does not support error handling for JSON calls. dynamic insertion scripts are present in a JSON call, responses will be adequately returned. If these are not inserted however, a 404 error from a server may not be detected and timeouts that does not alert the user will occur after some time. This can lead to a waste of time for time-critical processes. It can also lead to confusion and lot of debugging time wasted on tracing failure points (Ezeelive Technologies, n.d.; Progress, 2017; Python, n.d.; Tyson, n.d.).

In this chapter, two algorithms are provided by which CSV time series can be converted into JSON time series. If the delimiters parameter is optional. However, if it is given, it is interpreted as a string containing possible valid delimiter characters (Python, n.d.).

Algorithm 2: CSV Timeseries Conversion to JSON Format Timeseries – DictReader() & Sniffer() Methods

```
Input:    CSV Timeseries Data
Output:   JSON Timeseries and Dataframe
Data:     New York Stock Exchange Timeseries (1997-2021)
begin
          Input CSV timeseries data

          Use csv.Sniffer() method to detect the format of the CSV file.

          Use csv.Sniffer() method to format each row into a record.

          /* Sniffer method can be used to detect the parameters of a CSV file.
It
          will detect the presence of delimiters, the number of preamble rows,
the
          presence of CSV file headers, the presence or otherwise of byte
transfer
          (quote signs), whether the records in a CSV file are of the same
length and
          the data type of each field in the CSV file */

          Read in each row as a record into a dictionary. Use the csv.DictRead-
er()
          class for this activity. /* csv.DictReader() will  infer field names
from
          the first record of the CSV file */

          Store each directory record into an array. /* this will put each row
of record
          into proper JSON format */

          Dump the array records with some indents using the json.dump() meth-
od.

          Cast the array into a single JSON file using the to_json() method.

          Save onto local disk.

          Read from local disk using json.load() method.

          Render (or display) the JSON file

          Convert to a dataframe if there is need for analysis.

end
```

Algorithm 2: CSV-JSON Conversion Method I: DictReader() and csv.Sniffer() Method

The first CSV to JSON algorithm (Algorithm 3) provided in this chapter works through using Python csv.DictReader() class and the csv.Sniffer() method. csv.DictReader() creates an object that operates like a regular reader but maps the information in each row to a dictionary whose keys are given by the optional *fieldnames* parameter. It also uses csv.Sniffer() method. The Sniffer class is used to detect the format of a CSV file. On one hand, the sniff (*sample, delimiters=None*) method is used to analyze the given *sample* and it returns a dialect subclass that reflects out the parameters found. On the other hand,

Algorithm 3: CSV Timeseries to JSON Timeseries Format Conversion – to_json() Method

```
Input:      CSV Timeseries Data
Output:     JSON Timeseries and Dataframe
Data:       New York Stock Exchange Timeseries (1997-2021)
begin

         Input CSV timeseries data
         Convert CSV timeseries data to a Pandas dataframe
         Convert from Pandas dataframe to JSON using the 'to_json' method and
save
         onto local memory

         /*Important -  At this stage, use JSON dataframe 'index' format only.
The
         JSON dataframe index format is the only format that can yield a Pan-
das
         dataframe that can be queried */

         Load JSON dataframe from local memory using the 'json.load' method
         Dump data using the 'json.dump' method.
         Use Python 'eval' method to convert data from JSON string to proper
JSON
         format
         Convert to a Pandas dataframe
         Transform the dataframe from row to column orientation using the
'transpose'
         method
         /*If need be, convert back to CSV format. Can serve as a means of
converting
         JSON Pandas dataframe to CSV data format */
         Analyze dataframe  (sort, group, obtain unique values, filter)

end
```

Figure 15. Python implementation of algorithm 2 – conversion of CSV timeseries to JSON data format using dictreader() and sniffer() methods

```
CSVFileInput = r'C:\Users\...........\Yahoodata.csv'
JSONFileOutput = r'C:\Users\.........\Yahoodata2.json'
def FileInput_to_json(CSVFileInput, JSONFileOutput):
    fileToStoreConvertedArray = []
    with open(CSVFileInput,  newline='') as inputfile:
        dialect = csv.Sniffer().sniff(inputfile.read(1024))
        inputfile.seek(0)
        csvReader = csv.DictReader(inputfile)
        #convert each csv row into python dict
        for row in csvReader:
            fileToStoreConvertedArray.append(row)
      with open(JSONFileOutput, 'w', newline='') as jsonfile:
        jsonString = json.dumps(fileToStoreConvertedArray, indent=4)
        jsonfile.write(jsonString)
start = time.perf_counter()
FileInput_to_json(CSVFileInput ,JSONFileOutput)
finish = time.perf_counter()
print(f"Conversion of all TimeSeries file rows completed successfully in {fin-
ish - start:0.4f} seconds")
# Read in JSON data
with open("Yahoodata2.json", "r") as read_file:
    datajson = json.load(read_file)
datajsonPandasDataFrame = pd.DataFrame(datajson)
```

Its syntax and development look verbose and less concise than the second CSV-JSON conversion method discussed in this chapter. The second method (method II) uses the to_json() method immediately after converting the CSV time series to a Pandas data frame. The keen reader will however note that Method I also uses the to_json() method, and it also convert CSV time series to Pandas data frame; but only as part of later stages of the time series data conversion process. The advantage of the syntax given in method I is that it gives the developer better control and leverage over the direct use of the to_json() method. When using method I, developers are able to interact with the inner details of the algorithm such as csv.Reader() and the csv.Sniffer() methods. They are thus able to exert more control over different CSV file types. In addition, they are able to modify the CSV-JSON conversion process to suit the needs of each file. Full implementation of Algorithm 2 is shown in Figure 15. The resulting JSON format features proper JSON record format as shown in Figure 16. The Pandas dataframe that results from Algorithm 2 is shown in Figure 17.

Algorithm 3: CSV-JSON Conversion Method II: pd.DataFrame.to_json Method

In Algorithm 3, conversion of CSV timeseries to JSON data format is also accomplished. However, in the case of Algorithm 3, the 'to_json()' method that is provided as part of the Python JSON library (import json) is used. Full implementation of Algorithm 3 is shown in Figure 18, and details of converting the CSV timeseries to JSON using the to_json() method and further into a dataframe is shown in Figure 19. JSON data format that is generated from Python implementation of Algorithm 3 is shown in

Figure 16. Top part of the resulting timeseries data for python implementation of algorithm 2 – Conversion of CSV timeseries to JSON data format using dictreader() and sniffer() methods

```
(Algorithm 2 output at the point: datajson = json_load(read_file); in the Python code)

[{'Date': '4/30/1997',
  'Open': '17.5',
  'High': '17.5',
  'Low': '17',
  'Close': '17',
  'Adj Close': '15.192017',
  'Volume': '5800'},
 {'Date': '5/1/1997',
  'Open': '17.5',
  'High': '17.75',
  'Low': '17.375',
  'Close': '17.75',
  'Adj Close': '15.862257',
  'Volume': '36300'},
 {'Date': '5/2/1997',
  'Open': '17.75',
  'High': '17.875',
  'Low': '17.5',
  'Close': '17.875',
  'Adj Close': '15.97396',
  'Volume': '17700'},
 {'Date': '5/5/1997',
  'Open': '17.5',
  'High': '18.5',
  'Low': '17.5',
  'Close': '18.5',
  'Adj Close': '16.532488',
  'Volume': '65100'},
 {'Date': '5/6/1997',
  'Open': '18.625',
  'High': '19.5',
  'Low': '18.25',
  'Close': '19.5',
  'Adj Close': '17.426138',
  'Volume': '227000'},
```

using Algorithm 3. However, it could be observed that in some columns such as the 'Open' and 'High' columns, the dataframe (Figure 21) resulting from using Algorithm 3 have more floating-point precision than the dataframe generated by using Algorithm 2 (Figure 17). The 'Date' column format for both Algorithms are also different when their respective dataframes are considered. The choice of algorithm selected for converting the CSV time series to JSON will however depend on the needs of the developer and the constraints imposed by the project at hand.

It is worthy to note the differences in the renderings of the JSON data between outputs generated by Algorithm 2 (Figure 16) and Algorithm 3 (Figure 20) respectively. Also noteworthy are the type of

Figure 17. Top part of dataframe output for python implementation of algorithm 2 – Conversion of CSV timeseries to JSON data format using dictreader() and sniffer() methods

	Date	Open	High	Low	Close	Adj Close	Volume
0	4/30/1997	17.5	17.5	17	17	15.192017	5800
1	5/1/1997	17.5	17.75	17.375	17.75	15.862257	36300
2	5/2/1997	17.75	17.875	17.5	17.875	15.97396	17700
3	5/5/1997	17.5	18.5	17.5	18.5	16.532488	65100
4	5/6/1997	18.625	19.5	18.25	19.5	17.426138	227000
...	...	...	...	...	...	...	...
6288	4/25/2022	49.049999	49.790001	48.349998	49.73	49.73	120400
6289	4/26/2022	49.439999	50.419998	48.310001	48.48	48.48	80400
6290	4/27/2022	48.529999	49.110001	48.32	48.619999	48.619999	114400
6291	4/28/2022	49.630001	49.630001	44.360001	44.869999	44.869999	283800
6292	4/29/2022	44.32	44.599998	42.369999	42.889999	42.889999	193100

6293 rows × 7 columns

Pandas dataframes resulting from implementing both Algorithm 2 and Algorithm 3. The overriding benefit of using Algorithm 2 is that Algorithm 2 yields outputs with proper JSON format. It also provides the developer different choices, means and methods of determining the format and arrangement of the JSON timeseries that are generated. The developer has more control over the time series format and outputs as compared to data frame resulting from the implementation of the algorithm for the time series of Figure 10 is shown in Figure 24.

Figure 18. Python implementation of algorithm 3 – Conversion of CSV timeseries to JSON data format using to_json() method

```
YahooPricedata = pd.read_csv('/Users/……./YahooData.csv')
#Convert the CSV explicitly to Pandas Dataframe
YahooPriceDataFrame = pd.DataFrame(YahooPricedata)
#Convert the data Pandas Dataframe to JSON format
YahooPriceDataFrame.to_json('YahooPriceDataFrameToJSON.json')
Read in the JSON file formulated with Pandas
with open("YahooPriceDataFrameToJSON.json", "r") as read_file:
    datajsonPandasMethod = json.load(read_file)
datajsonPandasMethod
```

Figure 19. Python implementation of algorithm 3 – Conversion of CSV timeseries to JSON data format using to_json() method and then pandas dataframe

```
YahooPricedata = pd.read_csv('/Users/......./YahooData.csv')
#Convert the CSV explicitly to Pandas Dataframe
YahooPriceDataFrame = pd.DataFrame(YahooPricedata)
#Convert the data Pandas Dataframe to JSON format
YahooPriceDataFrame.to_json('YahooPriceDataFrameToJSON.json')
Read in the JSON file formulated with Pandas
with open("YahooPriceDataFrameToJSON.json", "r") as read_file:
    datajsonPandasMethod = json.load(read_file)
datajsonPandasMethod
```

Algorithm and Python Implementation for Converting CSV Time-Series to Feather Format Time-Series

Feather is a portable file format for fast, language-agnostic binary data frame storage for Python Pandas and R programming languages. Two Feather versions are currently available. The first version, V1 does not support compression; however, version 2 (V2) supports compression. V2 can be represented on computer disks as an Arrow IPC file format (Apache Arrow, n.d.; Ramm, 2021). Feather is powered by Apache Arrow; and it is essentially a cross language development for in-memory design.

Feather has high read-write performance. It also has minimal application programming interface (API) and it is lightweight. Hence it can push out data frames out of memory faster. Feather is also language agnostic; meaning that its data frames are the same regardless of the language that is used to read the files. Due to these advantages in addition to the Feather advantages listed in (Oyekanlu et al., 2022), it will be greatly beneficial if there exist an algorithm by which practitioners can convert CSV and other file formats to Feather format. Algorithm 4 is designed to accomplish just that. Python libraries that are used to implement Algorithm 4 are displayed in Figure 22. A concise implementation of the algorithm is as shown in Figure 23.

Algorithm and Python Implementation for Converting CSV Time-Series to YAML Format Time-Series

Due to JSON and XML being quite verbose when used to parse data (Stanford University, n.d.), (Tutorialspoint, n.d.), other less verbose data format that have similar syntax and data structure to JSON were developed. YAML is an example of such compact data formats. As opposed to markup languages such as XML that focuses on text documents, YAML is primarily a data serialization format that integrates very well with most common data types that are native to many programming languages (Zaczyński, n.d.). The YAML file is supported in all programming languages. As such, developers can write codes that generates a YAML file in one language, and the file can be used in other languages without any modifications.

YAML is different from markup languages in that, there exists no inherent text in YAML since it is used to present only data in human-readable form. Due to its readability, YAML has extensive application as a data exchange format. It is also widely used for creating network configuration files (Zaczyński, n.d.). For example, in managed cloud applications such as IBM Cloud Paks AI applications, YAML data

Figure 20. Top part of the resulting timeseries data for python implementation of Algorithm 3 – Conversion of CSV timeseries to JSON data format using to JSON Method


```
(Algorithm 3 output at the point: datajsonPandasMethod = json_load(read_file);
datajsonPandasMethod; in the Python code)
{'Date': {'0': '4/30/1997',
  '1': '5/1/1997',
  '2': '5/2/1997',
  '3': '5/5/1997',
  '4': '5/6/1997',
  '5': '5/7/1997',
  '6': '5/8/1997',
  '7': '5/9/1997',
  '8': '5/12/1997',
  '9': '5/13/1997',
  '10': '5/14/1997',
  '11': '5/15/1997',
  '12': '5/16/1997',
  '13': '5/19/1997',
  'Volume': '227000'},
………..
'995': '4/10/2001',
  '996': '4/11/2001',
  '997': '4/12/2001',
  '998': '4/16/2001',
  '999': '4/17/2001',
  ...},

'Open': {'0': 17.5,
  '1': 17.5,
  '2': 17.75,
  '3': 17.5,
  '4': 18.625,
  '5': 20.0,
  '6': 20.0,
  '7': 19.0,
  '8': 18.75,
  '9': 18.875,
  '10': 19.0,
…………
'997': 3.0,
  '998': 3.0,
  '999': 3.12,
  ...},
'High': {'0': 17.5,
  '1': 17.75,
  '2': 17.875,
……….
```

Figure 21. Top part of dataframe output for python implementation of algorithm 3 – Conversion of CSV timeseries to JSON data format using to JSON () method

	Date	Open	High	Low	Close	Adj Close	Volume
0	1997-04-30	17.500000	17.500000	17.000000	17.000000	15.192017	5800
1	1997-05-01	17.500000	17.750000	17.375000	17.750000	15.862257	36300
2	1997-05-02	17.750000	17.875000	17.500000	17.875000	15.973960	17700
3	1997-05-05	17.500000	18.500000	17.500000	18.500000	16.532488	65100
4	1997-05-06	18.625000	19.500000	18.250000	19.500000	17.426138	227000
...	...	...	...	...	...	...	...
6288	2022-04-25	49.049999	49.790001	48.349998	49.730000	49.730000	120400
6289	2022-04-26	49.439999	50.419998	48.310001	48.480000	48.480000	80400
6290	2022-04-27	48.529999	49.110001	48.320000	48.619999	48.619999	114400
6291	2022-04-28	49.630001	49.630001	44.360001	44.869999	44.869999	283800
6292	2022-04-29	44.320000	44.599998	42.369999	42.889999	42.889999	193100

6293 rows × 7 columns

Algorithm 4: Conversion of CSV Time Series into Feather DataFrame

```
Input:   CSV Timeseries Data
Output:  Feather Data Frame
Data:    New York Stock Exchange
         Timeseries (1997-2021)
begin
     Input CSV timeseries data

     Convert CSV timeseries data
     to a pandas dataframe.

     Convert the Pandas data frame to
     a Feather data frame and save
     onto local memory. Use the
     to_feather() method.

     Load Feather data frame from
     local memory. Use pd_read()
     method

     /*If need be, convert back to
     CSV format. Can serve as a means
     Of converting Feather to CSV */

     /* Analyze Feather data (sort,
     group, query, obtain unique
     values, filter)*/
end
```

Figure 22. Python libraries needed for the implementation of algorithm 4

```
import msgpack
import pandas as pd
import csv
import pyarrow.feather as feather
```

Figure 23. Python implementation of algorithm 4 – Conversion of CSV timeseries to feather data format

```
Yahoodata = pd.read_csv('/Users/.................../YahooData.csv')
Yahoodatadf = pd.DataFrame(Yahoodata)
Yahoodatadf.to_feather('YahoodataFeather.feather')
YahoodataFeather = pd.read_feather('YahoodataFeather.feather')
YahoodataFeather
```

files are used for storing customers' network settings to ensure automation and consistent results when network devices are changed (IBM, n.d.a; IBM, n.d.b). It is also used extensively in DevOps applications such as Ansible, Docker and Kubernetes. YAML is used in Ansible to specify the desired state of a remote infrastructure, to orchestrate IT processes, and to manage entire network configuration. Ansible Playbooks are commonly written in YAML (Sharma, 2021).

Figure 24. Top part of dataframe output for python implementation of algorithm 4 – Conversion of CSV time-series to feather data format

```
# Output of the final step in Fig. 23
```

	Date	Open	High	Low	Close	Adj Close	Volume
0	4/30/1997	17.500000	17.500000	17.000000	17.000000	15.192017	5800
1	5/1/1997	17.500000	17.750000	17.375000	17.750000	15.862257	36300
2	5/2/1997	17.750000	17.875000	17.500000	17.875000	15.973960	17700
3	5/5/1997	17.500000	18.500000	17.500000	18.500000	16.532488	65100
4	5/6/1997	18.625000	19.500000	18.250000	19.500000	17.426138	227000
...	...	...	...	...	...	...	...
6288	4/25/2022	49.049999	49.790001	48.349998	49.730000	49.730000	120400
6289	4/26/2022	49.439999	50.419998	48.310001	48.480000	48.480000	80400
6290	4/27/2022	48.529999	49.110001	48.320000	48.619999	48.619999	114400
6291	4/28/2022	49.630001	49.630001	44.360001	44.869999	44.869999	283800
6292	4/29/2022	44.320000	44.599998	42.369999	42.889999	42.889999	193100

For Docker, YAML is used to describe the microservices that populates Dockerized applications. In Kubernetes, YAML is used to define objects in computer clusters and for resource deployment. It is also used to orchestrate and manage applications (Zaczyński, n.d.). YAML files can also be added to source code control platforms such Github so that changes can be tracked and audited. It is also suitable for building continuous integration and continuous delivery (CI/CD) pipelines. RESTful APIs are also described on OpenAPI platforms using YAML (Zaczyński, n.d.). YAML is a superset of JSON since it will support any data formation and application that is supported by JSON. However, YAML stripped-down syntax enables it to be more concise than JSON. To enable data to be more concise, YAML uses indentations for hierarchical data arrangement as opposed to braces that are used in JSON. Also, while JSON mapping keys can be unique, in YAML, they *must* be defined as unique key mappings.

Python does not have a native in-built support for YAML data format (Zaczyński, n.d.). As such, due to the versatility and usefulness of YAML, it conformance to YAML syntax are slightly different. Also, the data frames that they yield, even though the data frames were all constructed with the aid of

Algorithm 5a: CSV Timeseries to YAML Data Format Conversion (Dictionary Orientation = Records)

```
Input:     CSV Timeseries Data without Schema
Output:    YAML timeseries and DataFrame
Data:      New York Stock Exchange Timeseries (1997-2021)
begin

        Input CSV timeseries data. Use the Pandas pd.read_csv() method.

        Serialize the data into YAML format. Use the yaml.dump() method from PyYAML
        Library

        Specify the first argument of the  yaml.dump() method as a dataframe to Python
        dictionary type.
        /* Use  df.to_dict() method here */

        Specify 'record' as the orientation argument of the df.to_dict() method.

        Specify your named YAML file as an 'outfile'   /* If this stage is omitted,
        the named output file will not be generated */

        Prettify the output YAML data by specifying other arguments of the yaml.dump()
        method
        /* At this stage, specify the width, indent, sort_keys, default_flow_style
        and the canonical arguments */
        /* 'default_flow_style' is used to display the contents of the nested blocks
        with proper indentation. The default value is True. When 'True', values inside
        the nested lists are shown in
        the flow style but setting this tag to False will display the block style's
        contents with proper indentation [65] */
        /* 'sort_keys' is used to sort the keys in alphabetical order. Default value
        is 'True'. When 'False', the insertion order is maintained. 'Indent' is used
        to specify the preferred indentation. 'Width' is used to specify the preferred
        width. 'Canonical' is used to force the preferred style for scalars
        and collections [65] */

        Dump the YAML data onto local memory
        Load from local memory. /*Use YAML safe loading to prevent dangerous data
        from being serialized */

        Render serialized data
        Convert to a dataframe if need be. /*Use Pandas dataframe method*/

end
```

Python Pandas library are different. Some of the data frames in some of the algorithms, for example Algorithm 5a and Algorithm 5b produces data columns with data that have more decimal precisions and number scales than other Algorithms; for example, Algorithm 5c. will be quite useful if there exist algorithms by which various type of data sets and data formats can be converted to YAML data format. This is accomplished in this chapter.

Five CSV to YAML data conversion algorithms (Algorithm 5a, Algorithm 5b, Algorithm 5c, Algorithm 5d and Algorithm 5e) are provided in this chapter. The syntaxes are essentially similar, but their outputs in terms of YAML structure, outline, and Python libraries that are used to implement the algorithms are shown in Figure 25. The 'import yaml' library could be added by doing a pip install of PyYAML (PyYAML 6.0, n.d.). The differences in the algorithms are essentially accomplished by exploiting the slight variations in the 'df.to_dict()' method of the Pandas library. The orientation of the method is fully exploited to show the different strategies that could be used to parse CSV data sets to YAML data formats. From YAML, algorithms are further provided in this chapter to transform the data sets to other highly useful data formats. The 'df.to_dict()' orientation formats that were considered for converting CSV time series to YAML data format in this chapter include: 'records', 'list', 'index', 'dict' and 'split' orientations respectively.

Figure 25. Python libraries needed for the implementation of algorithm 5a

```
import pandas as pd
import csv
import yaml
```

Figure 26. Python implementation of algorithm 5a – Conversion of CSV timeseries to YAML data format

```
df = pd.read_csv('/Users/................./YahooData.csv')
with open('YahoodataYAMLformat.yml', 'w') as outfile:
    yaml.dump(
        df.to_dict(orient='records'),
        outfile,
        canonical=True
        sort_keys=True,
        default_flow_style=False,
        indent=6,
        width=22,
    )
# We can read the YAML file using the PyYAML module's yaml.load() function.
This function parse and converts a YAML object to a Python dictionary (dict
object). This process is known as Deserializing YAML into a Python

with open('/Users/oyekanlea2/Desktop/TimeSeriesCodes/YahoodataYAMLformat.yml')
as f:
    yamldata = yaml.safe_load(f)
#Render serialized data
pprint(yamldata)
YahoodataYAMLDF = pd.DataFrame(yamldata)
YahoodataYAMLDF
```

Figure 27. Top part of the resulting timeseries data for python implementation of Algorithm 5a – Conversion of CSV timeseries to YAML data format (df.to_dict(orient='records') option)

```
(Algorithm 5a output at the point: pprint(yamldata); in the Python code)

[{'Adj Close': 15.192017000000002,
  'Close': 17.0,
  'Date': '4/30/1997',
  'High': 17.5,
  'Low': 17.0,
  'Open': 17.5,
  'Volume': 5800},
 {'Adj Close': 15.862257000000001,
  'Close': 17.75,
  'Date': '5/1/1997',
  'High': 17.75,
  'Low': 17.375,
  'Open': 17.5,
  'Volume': 36300},
 {'Adj Close': 15.97396,
  'Close': 17.875,
  'Date': '5/2/1997',
  'High': 17.875,
  'Low': 17.5,
  'Open': 17.75,
  'Volume': 17700},
```

Figure 28. Top part of dataframe output for python implementation of algorithm 5a – Conversion of CSV timeseries to YAML data format using 'df.to_dict(orient='records')' option

	Adj Close	Close	Date	High	Low	Open	Volume
0	15.192017	17.000000	4/30/1997	17.500000	17.000000	17.500000	5800
1	15.862257	17.750000	5/1/1997	17.750000	17.375000	17.500000	36300
2	15.973960	17.875000	5/2/1997	17.875000	17.500000	17.750000	17700
3	16.532488	18.500000	5/5/1997	18.500000	17.500000	17.500000	65100
4	17.426138	19.500000	5/6/1997	19.500000	18.250000	18.625000	227000
...	...	...	...	...	...	...	...
6288	49.730000	49.730000	4/25/2022	49.790001	48.349998	49.049999	120400
6289	48.480000	48.480000	4/26/2022	50.419998	48.310001	49.439999	80400
6290	48.619999	48.619999	4/27/2022	49.110001	48.320000	48.529999	114400
6291	44.869999	44.869999	4/28/2022	49.630001	44.360001	49.630001	283800
6292	42.889999	42.889999	4/29/2022	44.599998	42.369999	44.320000	193100

6293 rows × 7 columns

¶

Figure 29. Python implementation of algorithm 5b– List dictionary orientation option for YAML time-series data format

```
df = pd.read_csv('/Users/................../YahooData.csv')
with open('YahoodataYAMLformat.yml', 'w') as outfile:
    yaml.dump(
        df.to_dict(orient='list'),
        outfile,
        canonical=True
        sort_keys=True,
        default_flow_style=False,
        indent=6,
        width=22,
    )
# We can read the YAML file using the PyYAML module's yaml.load() function.
This function parse and converts a YAML object to a Python dictionary (dict
object). This process is known as Deserializing YAML into a Python

with open('/Users/....................../YahoodataYAMLformat.yml') as f:
    yamldata = yaml.safe_load(f)
#Render serialized data
pprint(yamldata)
YahoodataYAMLDF = pd.DataFrame(yamldata)
YahoodataYAMLDF
```

Figure 30. Top part of the resulting timeseries data for Python implementation of algorithm 5b – Conversion of CSV timeseries to YAML data format (df.to_dict(orient='list') option)

```
(Algorithm 5b output at the point: pprint(yamldata); in the Python code)

{'Adj Close': [15.192017000000002,
               15.862257000000001,
               ..........
               44.869999,
               42.889998999999996],
 'Close': [17.0,
           17.75,
           17.875,
           ..............
           44.869999,
           42.889998999999996],

 'Date': ['4/30/1997',
          '5/1/1997',
          ..............
          '4/28/2022',
          '4/29/2022'],
          ..............
```

Figure 31. Top part of dataframe output for python implementation of algorithm 5b – Conversion of CSV timeseries to YAML data format using 'df.to_dict(orient='list')' option

	Adj Close	Close	Date	High	Low	Open	Volume
0	15.192017	17.000000	4/30/1997	17.500000	17.000000	17.500000	5800
1	15.862257	17.750000	5/1/1997	17.750000	17.375000	17.500000	36300
2	15.973960	17.875000	5/2/1997	17.875000	17.500000	17.750000	17700
3	16.532488	18.500000	5/5/1997	18.500000	17.500000	17.500000	65100
4	17.426138	19.500000	5/6/1997	19.500000	18.250000	18.625000	227000
...	...	...	...	...	...	...	...
6288	49.730000	49.730000	4/25/2022	49.790001	48.349998	49.049999	120400
6289	48.480000	48.480000	4/26/2022	50.419998	48.310001	49.439999	80400
6290	48.619999	48.619999	4/27/2022	49.110001	48.320000	48.529999	114400
6291	44.869999	44.869999	4/28/2022	49.630001	44.360001	49.630001	283800
6292	42.889999	42.889999	4/29/2022	44.599998	42.369999	44.320000	193100

6293 rows × 7 columns

Figure 32. Python implementation of algorithm 5c– Index dictionary orientation option for YAML timeseries data format

```
df = pd.read_csv('/Users/...................../YahooData.csv')
with open('YahoodataYAMLformat.yml', 'w') as outfile:
    yaml.dump(
        df.to_dict(orient='index'),
        outfile,
        canonical=True
        sort_keys=True,
        default_flow_style=False,
        indent=6,
        width=22,
    )
# We can read the YAML file using the PyYAML module's yaml.load() function.
This function parse and converts a YAML object to a Python dictionary (dict
object). This process is known as Deserializing YAML into a Python

with open('/Users/......................../YahoodataYAMLformat.yml') as f:
    yamldata = yaml.safe_load(f)
#Render serialized data
pprint(yamldata)
YahoodataYAMLDF = pd.DataFrame(yamldata).transpose()
YahoodataYAMLDF
```

Figure 33. Top part of the resulting timeseries data for python implementation of algorithm 5c – Conversion of CSV timeseries to YAML data format (df.to_dict(orient='index') option)

```
(Algorithm 5c output at the point: pprint(yamldata); in the Python code)
{0: {'Adj Close': 15.192017000000002,
     'Close': 17.0,
     'Date': '4/30/1997',
     'High': 17.5,
     'Low': 17.0,
     'Open': 17.5,
     'Volume': 5800},
 1: {'Adj Close': 15.862257000000001,
     'Close': 17.75,
     'Date': '5/1/1997',
     'High': 17.75,
     'Low': 17.375,
     'Open': 17.5,
     'Volume': 36300},
 2: {'Adj Close': 15.97396,
     'Close': 17.875,
     'Date': '5/2/1997',
     'High': 17.875,
     'Low': 17.5,
     'Open': 17.75,
     'Volume': 17700},
```

Figure 34. Top part of dataframe output for python implementation of algorithm 5c – Conversion of CSV timeseries to YAML data format using 'df.to_dict(orient='index')' option

	Adj Close	Close	Date	High	Low	Open	Volume
0	15.192	17	4/30/1997	17.5	17	17.5	5800
1	15.8623	17.75	5/1/1997	17.75	17.375	17.5	36300
2	15.974	17.875	5/2/1997	17.875	17.5	17.75	17700
3	16.5325	18.5	5/5/1997	18.5	17.5	17.5	65100
4	17.4261	19.5	5/6/1997	19.5	18.25	18.625	227000
...	...	...	...	...	...	...	...
6288	49.73	49.73	4/25/2022	49.79	48.35	49.05	120400
6289	48.48	48.48	4/26/2022	50.42	48.31	49.44	80400
6290	48.62	48.62	4/27/2022	49.11	48.32	48.53	114400
6291	44.87	44.87	4/28/2022	49.63	44.36	49.63	283800
6292	42.89	42.89	4/29/2022	44.6	42.37	44.32	193100

6293 rows × 7 columns

Figure 35. Python implementation of algorithm 5d – Dict dictionary orientation option for YAML time-series data format

```
df = pd.read_csv('/Users/......................../YahooData.csv')
with open('YahoodataYAMLformat.yml', 'w') as outfile:
    yaml.dump(
        df.to_dict(orient='dict'),
        outfile,
        canonical=True
        sort_keys=True,
        default_flow_style=False,
        indent=6,
        width=22,
    )

with open('/Users/........................../YahoodataYAMLformat.yml') as f:
    yamldata = yaml.safe_load(f)
#Render serialized data
pprint(yamldata)
YahoodataYAMLDF = pd.DataFrame(yamldata)
YahoodataYAMLDF
```

Figure 36. Top part of the resulting timeseries data for python implementation of algorithm 5d – Conversion of CSV timeseries to YAML data format (df.to_dict(orient='dict') option)

```
(Algorithm 5d output at the point: pprint(yamldata); in the Python code)

{'Adj Close': {0: 15.192017000000002,
               1: 15.862257000000001,
               2: 15.97396,
               ..........
               6291: 44.869999,
               6292: 42.889998999999996},
'Close': {0: 17.0,
          1: 17.75,
          ..............
          6291: 44.869999,
          6292: 42.889998999999996},

'Date': {0: '4/30/1997',
         1: '5/1/1997',
         ..............
         6291: '4/28/2022',
         6292: '4/29/2022'},
```

Python implementations of each algorithm and their respective results are given in Figure 26 to Figure 39. It is noteworthy that due to the 'split' orientation of Algorithm 5e in its 'df.to_dict()' method implementation, it is not feasible to generate a data frame for Algorithm 5e using Python Pandas. It is possible however, to generate Pandas's data frames for all other YAML conversion algorithms (Algorithms 5a – Algorithm 5d).

Figure 37. Top part of dataframe output for python implementation of algorithm 5d – Conversion of CSV Timeseries to YAML data format using 'df.to_dict(orient='dict')' option

	Adj Close	Close	Date	High	Low	Open	Volume
0	15.192017	17.000000	4/30/1997	17.500000	17.000000	17.500000	5800
1	15.862257	17.750000	5/1/1997	17.750000	17.375000	17.500000	36300
2	15.973960	17.875000	5/2/1997	17.875000	17.500000	17.750000	17700
3	16.532488	18.500000	5/5/1997	18.500000	17.500000	17.500000	65100
4	17.426138	19.500000	5/6/1997	19.500000	18.250000	18.625000	227000
...	...	...	...	...	...	...	...
6288	49.730000	49.730000	4/25/2022	49.790001	48.349998	49.049999	120400
6289	48.480000	48.480000	4/26/2022	50.419998	48.310001	49.439999	80400
6290	48.619999	48.619999	4/27/2022	49.110001	48.320000	48.529999	114400
6291	44.869999	44.869999	4/28/2022	49.630001	44.360001	49.630001	283800
6292	42.889999	42.889999	4/29/2022	44.599998	42.369999	44.320000	193100

6293 rows × 7 columns

Figure 38. Python implementation of algorithm 5e – Split dictionary orientation option for YAML time-series data format. The Split orientation method does not support dataframe generation

```
df = pd.read_csv('/Users/oyekanlea2/Desktop/TimeSeriesCodes/YahooData.csv')
with open('YahoodataYAMLformat.yml', 'w') as outfile:
    yaml.dump(
        df.to_dict(orient='split'),
        outfile,
        canonical=True
        sort_keys=True,
        default_flow_style=False,
        indent=6,
        width=22,
    )

with open('/Users/oyekanlea2/Desktop/TimeSeriesCodes/YahoodataYAMLformat.yml')
as f:
    yamldata = yaml.safe_load(f)
#Render serialized data
pprint(yamldata)
YahoodataYAMLDF = pd.DataFrame(yamldata)
YahoodataYAMLDF
```

Figure 39. Top part of the resulting timeseries data for Python implementation of algorithm 5e – Conversion of CSV timeseries to YAML data format (df.to_dict(orient='split') option). Split orientation method does not support dataframe production


```
{'columns': ['Date', 'Open', 'High', 'Low', 'Close', 'Adj Close', 'Volume'],
 'data': [['4/30/1997', 17.5, 17.5, 17.0, 17.0, 15.192017000000002, 5800],
          ['5/1/1997', 17.5, 17.75, 17.375, 17.75, 15.862257000000001, 36300],
          ['5/2/1997', 17.75, 17.875, 17.5, 17.875, 15.97396, 17700],
          ['5/5/1997', 17.5, 18.5, 17.5, 18.5, 16.532488, 65100],
          ['5/6/1997', 18.625, 19.5, 18.25, 19.5, 17.426138, 227000],
          ['5/7/1997', 20.0, 20.25, 19.5, 20.0, 17.872971, 75600],
          ['5/8/1997', 20.0, 20.0, 19.125, 19.125, 17.091022, 5600],
          ['5/9/1997', 19.0, 19.625, 18.8125, 19.0, 16.979313, 56400],
          ['5/12/1997', 18.75, 19.625, 18.625, 18.875, 16.867607, 45700],
          ['5/13/1997', 18.875, 19.25, 18.875, 19.125, 17.091022, 1400],
          ['5/14/1997', 19.0, 19.25, 19.0, 19.0, 16.979313, 2000],
```

………………………………..

Algorithm 5b: CSV Timeseries to YAML Data Format Conversion (Dictionary Orientation = List)

```
Input:    CSV Timeseries Data without Schema
Output:   YAML timeseries and DataFrame
Data:     New York Stock Exchange Timeseries (1997-2021)
begin
        Input CSV timeseries data. Use the Pandas pd.read_csv() method.

        Serialize the data into YAML format. Use the yaml.dump() method from PyYAML
        Library

        Specify the first argument of the  yaml.dump() method as a dataframe to Python
        dictionary type.
        /* Use  df.to_dict() method here */

        Specify 'list' as the orientation argument of the df.to_dict() method.

        Specify your named YAML file as an 'outfile'  /* If this stage is omitted,
        the named output file will not be generated */

        Prettify the output YAML data by specifying other arguments of the yaml.dump()
        method

        Dump the YAML data onto local memory
        Load from local memory. /*Use YAML safe loading to prevent dangerous data
        from being serialized */

        Render serialized data
        Convert to a dataframe if need be. /*Use Pandas dataframe method*/

end
```

Algorithm 5c: CSV Timeseries to YAML Data Format Conversion (Dictionary Orientation = Index)

```
Input:      CSV Timeseries Data without Schema
Output:     YAML time series and DataFrame
Data:       New York Stock Exchange Timeseries (1997-2021)
begin
         Input CSV timeseries data. Use the Pandas pd.read_csv() method.

         Serialize the data into YAML format. Use the yaml.dump() method from PyYAML
         Library

         Specify the first argument of the  yaml.dump() method as a dataframe to Python
         dictionary type.
         /* Use  df.to_dict() method here */

         Specify 'list' as the orientation argument of the df.to_dict() method.

         Specify your named YAML file as an 'outfile'  /* If this stage is omitted,
         the named output file will not be generated */

         Prettify the output YAML data by specifying other arguments of the yaml.dump()
         method

         Dump the YAML data onto local memory
         Load from local memory. /*Use YAML safe loading to prevent dangerous data
         from being serialized */

         Render serialized data
         Convert to a dataframe if need be. /*Use Pandas dataframe method*/
         Use transpose() method to convert generated dataframe into portrait form.

end
```

Algorithm 5d: CSV Timeseries to YAML Data Format Conversion (Dictionary Orientation = Dict)

```
Input:      CSV Timeseries Data without Schema
Output:     YAML timeseries and DataFrame
Data:       New York Stock Exchange Timeseries (1997-2021)
begin
         Input CSV timeseries data. Use the Pandas pd.read_csv() method.

         Serialize the data into YAML format. Use the yaml.dump() method from PyYAML
         Library

         Specify the first argument of the  yaml.dump() method as a dataframe to Python
         dictionary type.
         /* Use  df.to_dict() method here */

         Specify 'dict' as the orientation argument of the df.to_dict() method.

         Specify your named YAML file as an 'outfile'  /* If this stage is omitted,
         the named output file will not be generated */

         Prettify the output YAML data by specifying other arguments of the yaml.dump()
         method

         Dump the YAML data onto local memory
         Load from local memory. /*Use YAML safe loading to prevent dangerous data
         from being serialized */

         Render serialized data
         Convert to a dataframe if need be. /*Use Pandas dataframe method*/

end
```

Algorithm 5e: CSV Timeseries to YAML Data Format Conversion (Dictionary Orientation = Split)

```
Input:      CSV Timeseries Data without Schema
Output:     YAML timeseries and DataFrame
Data:       New York Stock Exchange Timeseries (1997-2021)
begin
        Input CSV timeseries data. Use the Pandas pd.read_csv() method.

        Serialize the data into YAML format. Use the yaml.dump() method from PyYAML
        Library

        Specify the first argument of the  yaml.dump() method as a dataframe to Python
        dictionary type.
        /* Use  df.to_dict() method here */

        Specify 'split' as the orientation argument of the df.to_dict() method.

        Specify your named YAML file as an 'outfile'  /* If this stage is omitted,
        the named output file will not be generated */

        Prettify the output YAML data by specifying other arguments of the yaml.dump()
        method

        Dump the YAML data onto local memory
        Load from local memory. /*Use YAML safe loading to prevent dangerous data
        from being serialized */

        Render serialized data
        Convert to a dataframe if need be. /*Use Pandas dataframe method*/

end
```

Algorithm and Python Implementation for Converting and Generating Metadata for CSV Time-series Using Parquet Data Format

The Parquet data format has seen widespread adoption as the industry standard, open-source storage format for fast analytical querying of big data (Levy, 2022). Using Parquet format results in reduced file sizes and increased speeds. Extracting data from a Parquet dataframe have significant performance benefits including the benefits of working with column-oriented data format, equipped with self-describing metadata and schema structure. Parquet is unlike other data formats such as Avro and CSV that are row-based. Parquet is essentially optimized to accomplish high speed access and analytics (Levy, 2022), (Duniam et al., 2021), (Oyekanlu et al., 2022).

In terms of analytical performance, when compared to row-based data formats such as CSV, Parquet has significant performance advantages. When running queries, Parquet can be used to focus only on the relevant data part. Thus, the amount of data scanned will be significantly smaller; and it will result in less I/O data fetch. In-depth analysis of the advantages of the Parquet data format is given in (Oyekanlu et al., 2022).

In this chapter, Algorithm 6a is provided as a method of converting CSV time series to Parquet data format time series; and further to Parquet data frame. Also shown in this chapter is a multi-level data format conversion strategy by which data can be converted from CSV to Parquet and vice versa. Data can also be converted from Parquet to TXT and JSON data formats. The strategy is detailed in Algorithm 6a, and the Python implementation is shown in Figure 40. Data frame that is generated when data is converted from CSV data format to Parquet data format is shown in Figure 41.

Algorithm 6a: CSV Timeseries to Parquet Data Format Conversion & Parquet to Multi-data Format Conversion Methods

```
Input:      CSV Timeseries Data without Schema
Output:     Parquet DataFrame
Data:       New York Stock Exchange Timeseries (1997-2021)
begin
            Input CSV timeseries data. Use the Pandas pd.read_csv() method.
            Convert CSV timeseries data to a pandas dataframe
            Convert to a Parquet dataframe and save onto local memory. Use the
            'to_parquet()' method.
            Use 'pyarrow' as the conversion engine
            Save the newly generated Parquet file onto local memory as a '.parquet'
            file.
            Import the '.parquet' file from local memory
            Render the Parquet file using the 'read_parquet()' method.
            /*Convert back to CSV data format using the to_csv() method if need be */
            /*If there is a need, convert from Parquet to JSON data format using the
            to_json() method */
            /*If there is a need, convert from Parquet to text data format using the
            to_csv () method. Save with .txt format */

end
```

Figure 40. Python implementation of algorithm 6a – Conversion of CSV time series to parquet time series and parquet multi-data format (JSON, TXT, CSV) conversion

```
Import pandas as pd
import pyarrow
import csv

df = pd.read_csv('/Users/...................../YahooData.csv')
Yahoodata = pd.read_csv('/Users/........./YahooData.csv')
YahoodataDF = pd.DataFrame(Yahoodata)
YahoodataDF.to_parquet('YahoodataDFParquet.parquet', engine="pyarrow")
YahoodataDFParquet = pd.read_parquet('YahoodataDFParquet.parquet')

# Convert from Parquet back to CSV if there is need
parquetFromPandasDatafame = pd.read_parquet('YahoodataDFParquet.parquet')
parquetFromPandasDatafame.to_csv('ConvertFromParquetBackToCSV.csv', index=False
# Convert From CSV to Parquet to Text File
parquetFromPandasDatafame.to_csv('ConvertParquettoTxtFile.txt', index=False,
sep=' ')
# Convert from CSV to Parquet and to JSON
parquetFromPandasDatafame.to_json('ConvertFromParquetToJSON.json')
```

It is well known that CSV data formats does not naturally have embedded file metadata. It is for this reason that researchers are looking for several methods of creating metadata for CSV data sets. For example, in (Firth, 2022), researchers described a method by which CSV files on the Web (CSVW) can have their metadata generated using a JSON file.

Figure 41. Top part of the resulting timeseries data for python implementation of algorithm 6a – Conversion of CSV timeseries to parquet data format time series

```
# Output of Algorithm 6a (i.e., output of the code part:
YahoodataDFParquet = pd.read_parquet('YahoodataDFParquet.parquet')
```

```
#Read the newly converted Parquet File
YahoodataDFParquet = pd.read_parquet('YahoodataDFParquet.parquet')
YahoodataDFParquet
```

	Date	Open	High	Low	Close	Adj Close	Volume
0	4/30/1997	17.500000	17.500000	17.000000	17.000000	15.192017	5800
1	5/1/1997	17.500000	17.750000	17.375000	17.750000	15.862257	36300
2	5/2/1997	17.750000	17.875000	17.500000	17.875000	15.973960	17700
3	5/5/1997	17.500000	18.500000	17.500000	18.500000	16.532488	65100
4	5/6/1997	18.625000	19.500000	18.250000	19.500000	17.426138	227000
...	...	...	...	...	...	...	...
6288	4/25/2022	49.049999	49.790001	48.349998	49.730000	49.730000	120400
6289	4/26/2022	49.439999	50.419998	48.310001	48.480000	48.480000	80400
6290	4/27/2022	48.529999	49.110001	48.320000	48.619999	48.619999	114400
6291	4/28/2022	49.630001	49.630001	44.360001	44.869999	44.869999	283800
6292	4/29/2022	44.320000	44.599998	42.369999	42.889999	42.889999	193100

6293 rows × 7 columns

Algorithm 6b: A Parquet Based Method of Metadata Generation for CSV Time Series

```
Input:     CSV Timeseries Data without Schema
Output:    Time series statistics and metadata
Data:      New York Stock Exchange Timeseries (1997-2021)
begin
        Input CSV timeseries data. Use the Pandas pd.read_csv() method.
        Convert CSV timeseries data to a pandas dataframe
        Convert to a Parquet dataframe and save onto local memory. Use the
        'to_parquet()' method.
        Save the newly generated Parquet file onto local memory as a '.parquet' file.
        Import the '.parquet' file from local memory
        Use the pyarrow.parquet libaray to read in the Parquet file from local
        memory
        Assign the loaded '.parquet' file as an object to a named file.
        Use the Pyarrow library '.metadata()' method to render the file metadata
        If there is need, render the file statistics as a metadata using the Python
        'print' method.
        /* Use the Parquet metadata row_group() and column()methods at this stage*/
end
```

Figure 42. Python implementation and results of algorithm 6b – Method of metadata generation for CSV time series using parquet data format

```
# Read Parquet File metada
Import pandas as pd
import pyarrow
import csv
import pyarrow.parquet as pq

df = pd.read_csv('/Users/oyekanlea2/Desktop/TimeSeriesCodes/YahooData.csv')
Yahoodata = pd.read_csv('/Users/........./YahooData.csv')
YahoodataDF = pd.DataFrame(Yahoodata)
YahoodataDF.to_parquet('YahoodataDFParquet.parquet', engine="pyarrow")
YahoodataDFParquet = pd.read_parquet('YahoodataDFParquet.parquet')
parquet_file = pq.ParquetFile('YahoodataDFParquet.parquet')
parquetmetadata = parquet_file.metadata
# Render the metadata
parquetmetadata
<pyarrow._parquet.FileMetaData object at 0x00000255FCBE2188>
  created_by: parquet-cpp version 1.5.1-SNAPSHOT
  num_columns: 7
  num_rows: 6293
  num_row_groups: 1
  format_version: 1.0
  serialized_size: 1875

print(parquet_file.metadata.row_group(0).column(1).statistics)

<pyarrow._parquet.RowGroupStatistics object at 0x00000255FCAD7788>
  has_min_max: True
  min: 1.1875
  max: 65.91
  null_count: 0
  distinct_count: 0
  num_values: 6293
  physical_type: DOUBLE
```

In this chapter, we have contributed Algorithm 6b as a method of producing metadata for CSV data sets. In addition to producing metadata, Algorithm 6b can also be used to generate data statistics for CSV data

the CSV time series into Parquet data format and then using the Parquet data formats metadata generating advantages to produce the file's metadata from inherent CSV data characteristics. Python implementation of Algorithm 6b and the output of the algorithm is shown in Figure 42.

Algorithm and Python Implementation for Multi-data format Convertion: CSV to HDF5 Data Format; HDF5 to YAML and HDF5 to MessagePack

Since its development in 1988 at the University of Illinois, the Hierarchical Data Format (HDF) have gone through a series of changes (Gabriel, n.d.). The current version is the HDF-5, and it supports very large file applications. It also has support for parallel I/O interface with bindings for Python, C, Fortran and Java (Gabriel, n.d.). HDF-5 can be used to store very large files. It can also be used to access streaming data patterns using clusters of commodity hardware. To enjoy the many advantages of using HDF-5 data

Algorithm 7a: Multiformat Data Conversion Algorithm: CSV to HDF5 HDF5 to YAML

```
Input:     CSV Timeseries Data without Schema
Output:    HDF5 and YAML DataFrames
Data:      New York Stock Exchange Timeseries (1997-2021)
begin

       /* CSV to HDF5 */
       Input CSV timeseries data
       Convert CSV timeseries data to a pandas dataframe
       Convert to a HDF5 data frame and save onto local memory. Use the 'to_hdf()'
       method.
       Select appropriate key and mode for the 'to_hdf()' method. Default are 'df'
       and 'w' respectively
       Safe the resulting timeseries as a '.h5' data format on local memory.
       Load from local memory and convert to Pandas Dataframe

       /* HDF5 to YAML */
       Load the '.h5' file from local memory and serialize the data into YAML format.
       Use the yaml.dump() method from PyYAML library

       Specify the first argument of the  yaml.dump() method as a dataframe to Python
       dictionary type. /* Use  df.to_dict() method here */

       Specify 'records' as the orientation argument of the df.to_dict() method.

       Assign a name to the YAML file. Specify your named YAML file as an 'outfile'
       /* If this stage is omitted, the named output file will not be generated */
       / * Repeat the remaining steps same as Algorithm 5 */

       Render serialized data
       Convert to a dataframe if need be. /*Use Pandas dataframe method*/
end
```

format, in this chapter, an algorithm (Algorithm 7a) that shows method by which CSV time series data formats can be converted to HDF-5 data format have been provided. Algorithm 7a have been provided as a multi-data format algorithm since the algorithm is also used to show how to convert from HDF5 to YAML. Results of Algorithm 7a are shown in Figure 43 to Figure 46. In Algorithm 7b, further methods and means of multi-data format conversion processes are shown.

In Algorithm 7b, processes of converting HDF5 to MessagePack data format is shown. To accomplish Algorithm 7b, it is necessary to first convert from HDF5 to YAML, after which it is convenient to convert to MessagePack data format. Algorithms 7a and 7b are provided to show how convenient it is to navigate from one data format to another using the various algorithms and methods that are provided in this chapter. Python implementation and results of Algorithms 7b are shown in Figure 47 to Figure 49.

Figure 43. Python implementation and results of algorithm 7a – Multiformat data conversion algorithm (CSV to HDF5; HDF5 to YAML;)

```
import pandas as pd
import csv
import yaml
from pprint import pprint
# Convert from CSV to HDF5 Format
YahooPricedata = pd.read_csv('/Users/..........YahooData.csv')
#Convert the CSV explicitly to Pandas Dataframe
YahooPriceDataFrame = pd.DataFrame(YahooPricedata)
#Convert the data Pandas Dataframe to HDF5 format
YahooPriceDataFrame.to_hdf('YahooPriceDataFrameToHDF.h5', key='df', mode='w')
YahoodataHDF5DataFrame = pd.read_hdf('/Users/........./YahooPriceDataFrameToHDF.h5')
YahoodataHDF5DataFrame
# Convert from HDF5 to YAML format
df = pd.read_hdf('/Users/........................ /YahooPriceDataFrameToHDF.h5')
with open('YahoodataHDF2YAML.yml', 'w') as outfile:
    yaml.dump(
        df.to_dict(orient='records'),
        outfile,
        sort_keys=False,
        width=72,
        indent=4
    )
# read in the HDF2YAML file
with open('/Users/....................YahoodataHDF2YAML.yml') as f:
    HDF2yamldata = yaml.safe_load(f)
pprint(HDF2yamldata)
YahoodataYAMLDF = pd.DataFrame(YahoodataYAML)
YahoodataYAMLDF
```

Figure 44. Top part of the resulting timeseries data for python implementation of algorithm 7a – Conversion of CSV timeseries to HDF5 data format

```
# Output of Algorithm 7a (i.e., output of the point:
YahoodataHDF5DataFrame = pd.read_hdf('/Users/........./YahooPriceDataFrameToHDF.h5')

YahoodataHDF5DataFrame

in the Python code)
```

	Date	Open	High	Low	Close	Adj Close	Volume
0	4/30/1997	17.500000	17.500000	17.000000	17.000000	15.192017	5800
1	5/1/1997	17.500000	17.750000	17.375000	17.750000	15.862257	36300
2	5/2/1997	17.750000	17.875000	17.500000	17.875000	15.973960	17700
3	5/5/1997	17.500000	18.500000	17.500000	18.500000	16.532488	65100
4	5/6/1997	18.625000	19.500000	18.250000	19.500000	17.426138	227000
...	...	...	...	...	...	...	...
6288	4/25/2022	49.049999	49.790001	48.349998	49.730000	49.730000	120400
6289	4/26/2022	49.439999	50.419998	48.310001	48.480000	48.480000	80400
6290	4/27/2022	48.529999	49.110001	48.320000	48.619999	48.619999	114400
6291	4/28/2022	49.630001	49.630001	44.360001	44.869999	44.869999	283800
6292	4/29/2022	44.320000	44.599998	42.369999	42.889999	42.889999	193100

6293 rows × 7 columns

Figure 45. Top part of the resulting timeseries data for python implementation of algorithm 7a – Conversion of HDF5 timeseries to YAML data format timeseries

```
# Output of Algorithm 7a (i.e., output at the point:
pprint(HDF2yamldata)      } in the Python code)
[{'Adj Close': 15.192017000000002,
  'Close': 17.0,
  'Date': '4/30/1997',
  'High': 17.5,
  'Low': 17.0,
  'Open': 17.5,
  'Volume': 5800},
 {'Adj Close': 15.862257000000001,
  'Close': 17.75,
  'Date': '5/1/1997',
  'High': 17.75,
  'Low': 17.375,
  'Open': 17.5,
  'Volume': 36300},
```

Figure 46. Top part of the resulting dataframe for python implementation of algorithm 7a – Conversion of HDF5 to YAML data format.

```
# Output of Algorithm 7a (i.e., output of the point:
YahoodataYAMLDF = pd.DataFrame(YahoodataYAML)

YahoodataYAMLD; in the Python code
```

	Date	Open	High	Low	Close	Adj Close	Volume
0	4/30/1997	17.500000	17.500000	17.000000	17.000000	15.192017	5800
1	5/1/1997	17.500000	17.750000	17.375000	17.750000	15.862257	36300
2	5/2/1997	17.750000	17.875000	17.500000	17.875000	15.973960	17700
3	5/5/1997	17.500000	18.500000	17.500000	18.500000	16.532488	65100
4	5/6/1997	18.625000	19.500000	18.250000	19.500000	17.426138	227000
...	...	...	...	...	...	...	...
6288	4/25/2022	49.049999	49.790001	48.349998	49.730000	49.730000	120400
6289	4/26/2022	49.439999	50.419998	48.310001	48.480000	48.480000	80400
6290	4/27/2022	48.529999	49.110001	48.320000	48.619999	48.619999	114400
6291	4/28/2022	49.630001	49.630001	44.360001	44.869999	44.869999	283800
6292	4/29/2022	44.320000	44.599998	42.369999	42.889999	42.889999	193100

6293 rows × 7 columns

Algorithm 7b: Multiformat Data Conversion Algorithm: HDF5 to MessagePack Data Format

```
Input:     HDF5 Timeseries Data without Schema
Output:    MessagePack DataFrame
Data:      New York Stock Exchange Timeseries (1997-2021)
begin

        /* HDF5 to YAML */
        Load the '.h5' file from local memory and serialize the data into YAML format.
        Use the yaml.dump() method from PyYAML library /* Note that it may be
        necessary to first perform conversion from CSV to HDF5 (Algorithm 7a)

        Specify the first argument of the  yaml.dump() method as a dataframe to Python
        dictionary type. /* Use  df.to_dict() method here */

        Specify 'records' as the orientation argument of the df.to_dict() method.

        Assign a name to the YAML file. Specify your named YAML file as an 'outfile'
        /* If this stage is omitted, the named output file will not be generated */
        / * Repeat the remaining steps same as Algorithm 5 */

        Save the YAML data onto local memory.
        Load the data from local memory. /* Use YAML safe load method */

        Assign as an object to the msppack.pack() method

        Save onto local memory

        Read from storage location using msgpack.unpack() an the json.load()
        method.

        Convert from msgpack.unpack() format to dictionary key:value pair format


        Access the dictionary and then convert to a Python list format
        Use Pandas to convert to a dataframe

        /* Convert back to CSV if need be. This step canbe used to convert from
        MessagePack to CSV format */


end
```

Algorithm and Python Implementation for Converting CSV Time-Series to TSV Data Format

Tab Separated Values (TSV) data formats have been around for a long time, and it has similar features and characteristics as CSV data format. TSV is very efficient for Javascript, Perl and Python to process, without losing any part of the original data. Data in TSV file formats are essentially stored in tabular structures.

In TSV data formats, each record in the table is rendered as one line of the entire file. The field's values in the record are separated by tab characters. TSV files function well as a data exchange format between programs that use structured tables or spreadsheets. The TSV file format is widely supported and is very similar to the CSV data format (W3C, 2012), (Library of Congress, n.d.).

Figure 47. Python implementation of algorithm 7b – Multiformat data conversion algorithm (HDF5 to YAML; YAML to messagepack;)

```
import pandas as pd
import json
import csv
import yaml
import msgpack
from pprint import pprint
# Convert from to HDF5 YAML
df = pd.read_hdf('/Users/......../YahooPriceDataFrameToHDF.h5')
with open('YahoodataHDF2YAML.yml', 'w') as outfile:
    yaml.dump(
        df.to_dict(orient='records'),
        canonical,
        outfile,
        sort_keys=False,
        width=22,
        indent=6
    )
# read in the HDF2YAML file
with open('/Users/........../YahoodataHDF2YAML.yml') as f:
    HDF2yamldata = yaml.safe_load(f)
# Convert to MessagePack
MsgPackdata = HDF2yamldata
# Write msgpack file
with open('MsgPackdata.msgpack', 'wb') as inputMsgPackdata:
    msgpack.pack(MsgPackdata, inputMsgPackdata)
# Read MsgPack in from storage location
with open('MsgPackdata.msgpack', "rb") as MsgPackdata_file:
    # MsgPackdata_loaded = json.load(MsgPackdata_file)
    MsgPackdatafromHDF5_loaded = msgpack.unpack(MsgPackdata_file)
# Convert MsgPack to dictionary format
newloadedMsgPackFromHDF5 = {"convertHDF5ToMessagePack": MsgPackdatafromHDF5_loaded}
# Convert MsgPack to pandas dataframe
MsgPackDFFromHDF5 = pd.DataFrame(newloadedMsgPackFromHDF5['convertHDF5ToMessagePack'])
MsgPackDFFromHDF5
```

However, TSV have not been able to enjoy the widespread acceptance that CSV data format has. Both CSV and TSV data formats are delimiter-separated value formats. As shown in a hypothetical data set in Figure 50, in CSV, each record field values are separated by commas rather than tabular spaces. A disadvantage of having commas in CSV data format is that escaping the comma within the fields can require more computing power.

The amount of processing power required will invariably scale up with increasing file sizes. Also, in CSV data formats, all the extra escaping characters and quotes add more weight to the final file size. This naturally increases the storage space requirements. In equivalent TSV data set shown in Figure 51 and Figure 52, all the disadvantages of using the CSV data formats are not encountered. In this chapter, Algorithm 8a have been provided as a means of converting CSV time series to TSV time series.

Python implementation of the CSV to TSV conversion algorithm is detailed in Figure 53. Output of the algorithm is as shown in Figure 54. In like manner, Algorithm 8b shows means and methods by which TSV data format can be converted to HDF5 data format. Python implementation of the algorithm and the resulting data frame are shown in Figure 54 and Figure 55 respectively.

In Algorithm 8c, methods and means by which the TSV data format can be converted into YAML data format is shown. Resulting YAML timeseries and data frame are shown in Figure 56, Figure 57, and

Figure 48. Timeseries result of algorithm 7b – Multiformat data conversion algorithm (HDF5 to YAML; YAML to messagepack;)

```
# Output of Algorithm 7b (i.e., output at the point:

newloadedMsgPackFromHDF5 = {"convertHDF5ToMessagePack": MsgPackdatafromHDF5_loaded}
newloadedMsgPackFromHDF5} in the Python code)

{'convertHDF5ToMessagePack': [{'Date': '4/30/1997',
   'Open': 17.5,
   'High': 17.5,
   'Low': 17.0,
   'Close': 17.0,
   'Adj Close': 15.192017000000002,
   'Volume': 5800},
  {'Date': '5/1/1997',
   'Open': 17.5,
   'High': 17.75,
   'Low': 17.375,
   'Close': 17.75,
   'Adj Close': 15.862257000000001,
   'Volume': 36300},
  {'Date': '5/2/1997',
   'Open': 17.75,
   'High': 17.875,
   'Low': 17.5,
   'Close': 17.875,
   'Adj Close': 15.97396,
   'Volume': 17700},
```

Figure 49. Top part of the resulting dataframe for python implementation of algorithm 7b – Multiformat data conversion algorithm (HDF5 to YAML; YAML to messagepack;)

```
# Output of Algorithm 7b (i.e., output of the point:
MsgPackDFFromHDF5 =
pd.DataFrame(newloadedMsgPackFromHDF5['convertHDF5ToMessagePack'])

MsgPackDFFromHDF5;     in the Python code
```

	Date	Open	High	Low	Close	Adj Close	Volume
0	4/30/1997	17.500000	17.500000	17.000000	17.000000	15.192017	5800
1	5/1/1997	17.500000	17.750000	17.375000	17.750000	15.862257	36300
2	5/2/1997	17.750000	17.875000	17.500000	17.875000	15.973960	17700
3	5/5/1997	17.500000	18.500000	17.500000	18.500000	16.532488	65100
4	5/6/1997	18.625000	19.500000	18.250000	19.500000	17.426138	227000
...	...	...	...	...	...	...	...
6288	4/25/2022	49.049999	49.790001	48.349998	49.730000	49.730000	120400
6289	4/26/2022	49.439999	50.419998	48.310001	48.480000	48.480000	80400
6290	4/27/2022	48.529999	49.110001	48.320000	48.619999	48.619999	114400
6291	4/28/2022	49.630001	49.630001	44.360001	44.869999	44.869999	283800
6292	4/29/2022	44.320000	44.599998	42.369999	42.889999	42.889999	193100

6293 rows × 7 columns

Figure 50. Structure of CSV data format for a hypothetical data set

```
Name, Country, Occupation
Emmanuel Oyekanlu, USA, AI Engineer
Christiano Ronaldo, Portugal, Soccer Player
David Kuhn, USA, IT Manager
Lionel Messi, Argentina, Soccer Player
Sam Zoubi, USA, IT Director
Grethel Mulroy, USA, IT Manager
Mohammadu Buhari, Nigeria, President
Kylian Mbappe, France, Soccer Player
```

Figure 51. Structure of TSV data format for the CSV data shown in Figure 50

```
Name<TAB>Country<TAB> Occupation
Emmanuel Oyekanlu <TAB>USA<TAB>AI Engineer
Christiano Ronaldo<TAB>Portugal <TAB> Soccer Player
David Kuhn <TAB> USA <TAB> IT Manager
Lionel Messi <TAB> Argentina <TAB> Soccer Player
Sam Zoubi <TAB> USA <TAB> IT Director
Grethel Mulroy <TAB> USA <TAB> IT Manager
Mohammadu Buhari <TAB> Nigeria <TAB> President
Kylian Mbappe<TAB>France<TAB> Soccer Player
```

Figure 52. Structure of TSV data format without the <TAB> for the CSV data shown in Figure 50

```
Name                Country    Occupation
Emmanuel Oyekanlu   USA        AI Engineer
Christiano Ronaldo  Portugal   Soccer Player
David Kuhn          USA        IT Manager
Lionel Messi        Argentina  Soccer Player
Sam Zoubi           USA        IT Director
Grethel Mulroy      USA        IT Manager
Mohammadu Buhari    Nigeria    President
Kylian Mbappe       France     Soccer Player
```

Figure 53. Python implementation and results of algorithm 8a – CSV timeseries to TSV data conversion algorithm

```
import pandas as pd
import csv
with open('/Users/......................../YahooData.csv','r') as inputdata, open('/Users/............................../CSV2TSV.tsv', 'w') as TSVGenerator:
    inputdata = csv.reader(inputdata)
    TSVGenerator = csv.writer(TSVGenerator, delimiter='\t')
    for row in inputdata:
        TSVGenerator.writerow(row)
#Read in converted TSV File as input. Use a CSV Reader. Use a tab (\t) separator in the pd.read_csv Pandas reader
YahoodataTSV = pd.read_csv('/Users/............................../CSV2TSV.tsv' ,sep='\t')
print(YahoodataTSV);
# TSV dataframe
YahoodataTSV
# Convert dataframe back to CSV if the need arises
YahoodataTSV.to_csv('/Users/................../TSVDataFrameBackToCSV.csv', index=False)
```

Figure 54. Top part of the resulting dataframe for python implementation of algorithm 8a – Conversion of CSV to TSV data format.

```
import pandas as pd
import csv

# TSV to HDF5
YahooPricedata = pd.read_csv('/Users/......... /CSV2TSV.tsv' ,sep='\t')
YahooPricedata.to_hdf('YahooPriceTSVToHDF.h5', key='df', mode='w')
YahoodataTSV2HDF5DataFrame = pd.read_hdf('/Users/......... /YahooPriceTSVToHDF.h5')
YahoodataTSV2HDF5DataFrame
```

Figure 58. Also discussed in this chapter are algorithms (Algorithm 8d and Algorithm 8e) by which TSV data sets can be converted to MessagePack and JSON data formats respectively. Python implementation and results of these algorithms are shown in Figure 59 to Figure 62.

Figure 55. Python implementation and results of algorithm 8b – TSV timeseries to HDF5 data conversion algorithm

	Date	Open	High	Low	Close	Adj Close	Volume
0	4/30/1997	17.500000	17.500000	17.000000	17.000000	15.192017	5800
1	5/1/1997	17.500000	17.750000	17.375000	17.750000	15.862257	36300
2	5/2/1997	17.750000	17.875000	17.500000	17.875000	15.973960	17700
3	5/5/1997	17.500000	18.500000	17.500000	18.500000	16.532488	65100
4	5/6/1997	18.625000	19.500000	18.250000	19.500000	17.426138	227000
...	...	...	...	...	...	...	...
6288	4/25/2022	49.049999	49.790001	48.349998	49.730000	49.730000	120400
6289	4/26/2022	49.439999	50.419998	48.310001	48.480000	48.480000	80400
6290	4/27/2022	48.529999	49.110001	48.320000	48.619999	48.619999	114400
6291	4/28/2022	49.630001	49.630001	44.360001	44.869999	44.869999	283800
6292	4/29/2022	44.320000	44.599998	42.369999	42.889999	42.889999	193100

6293 rows × 7 columns

Algorithm 8b: TSV Timeseries Conversion to HDF5 Data Format

```
Input:     TSV Timeseries Data without Schema
Output:    HDF5 Data frame
Data:      New York Stock Exchange Timeseries (1997-2021)
begin

      Input TSV timeseries data
      Apply Pandas reader with separator or delimiter to read-in the TSV timeseries
      Data.
      Cast the TSV data to HDF format using the 'to_hdf()' method. Use appropriate
      key and mode. Default key and mode are 'df' and 'w' respectively.
      Save the '.h5' file onto local memory.
      Read-in the '.h5' file as dataframe from local memory using the 'read_hdf()'
      method.

end
```

Algorithm 8a: CSV Timeseries to TSV Data Format Conversion

```
Input:      CSV Timeseries Data without Schema
Output:     TSV Data frame
Data:       New York Stock Exchange Timeseries (1997-2021)
begin
         Input CSV timeseries data
         Open a file in the format 'file.tsv'
         Apply CSV reader to read-in the CSV data into the file.tsv data file
         Apply CSV writer with delimiter to write the input CSV timeseries
data into
         the TSV file
         Write-in data as records in row after row.
         Save the TSV file onto local memory
         Load the TSV file data from local memory using Pandas
         Render data as a dataframe
         /* Convert TSV back to CSV data format if the need arises */
end
```

Figure 56. Top part of the resulting dataframe for python implementation of algorithm 8b – Conversion of TSV to HDF5 data frame.

```
# TSV to YAML Python Implementation
Import csv
Import yaml
df = pd.read_csv('/Users/................./YahooData.csv')
with open(' YahoodataTSV2YAML.yml', 'w') as outfile:
    yaml.dump(
        df.to_dict(orient='records'),
        outfile,
        canonical=True
        sort_keys=True,
        default_flow_style=False,
        indent=6,
        width=22,
    )
# We can # read in the HDF2YAML file
with open('/Users/........./YahoodataTSV2YAML.yml') as f:
    TSV2YAMLdata = yaml.safe_load(f)
pprint(TSV2YAMLdata)

# TSV to YAML Dataframe
YahoodataTSVtoYAMLDataFrame = pd.DataFrame(TSV2YAMLdata)
YahoodataTSVtoYAMLDataFrame
```

Figure 57. Python implementation of algorithm 8c – : TSV timeseries conversion to YAML data format+


```
[{'Adj Close': 15.192017000000002,
  'Close': 17.0,
  'Date': '4/30/1997',
  'High': 17.5,
  'Low': 17.0,
  'Open': 17.5,
  'Volume': 5800},
 {'Adj Close': 15.862257000000001,
  'Close': 17.75,
  'Date': '5/1/1997',
  'High': 17.75,
  'Low': 17.375,
  'Open': 17.5,
  'Volume': 36300},
 {'Adj Close': 15.97396,
  'Close': 17.875,
  'Date': '5/2/1997',
  'High': 17.875,
  'Low': 17.5,
  'Open': 17.75,
  'Volume': 17700},
```


Figure 58. Timeseries result of algorithm 8c – : TSV timeseries conversion to YAML data format

	Date	Open	High	Low	Close	Adj Close	Volume
0	4/30/1997	17.500000	17.500000	17.000000	17.000000	15.192017	5800
1	5/1/1997	17.500000	17.750000	17.375000	17.750000	15.862257	36300
2	5/2/1997	17.750000	17.875000	17.500000	17.875000	15.973960	17700
3	5/5/1997	17.500000	18.500000	17.500000	18.500000	16.532488	65100
4	5/6/1997	18.625000	19.500000	18.250000	19.500000	17.426138	227000
...	...	...	...	...	...	...	...
6288	4/25/2022	49.049999	49.790001	48.349998	49.730000	49.730000	120400
6289	4/26/2022	49.439999	50.419998	48.310001	48.480000	48.480000	80400
6290	4/27/2022	48.529999	49.110001	48.320000	48.619999	48.619999	114400
6291	4/28/2022	49.630001	49.630001	44.360001	44.869999	44.869999	283800
6292	4/29/2022	44.320000	44.599998	42.369999	42.889999	42.889999	193100

6293 rows × 7 columns

Algorithm 8c: TSV Timeseries Conversion to YAML Data Format

```
Input:     TSV Timeseries Data without Schema
Output:    YAML time series and Data frame
Data:      New York Stock Exchange Timeseries (1997-2021)
begin

      Input TSV timeseries data
      Apply Pandas reader with separator or delimiter to read-in the TSV timeseries
      Data.
      Cast the TSV data as a YAML outfile using the 'yaml.dump()' method.
      Prettify as appropriate by specifying the 'df.to_dict()' parameters including:
      'canonical', 'sort_keys', 'default_flow_style', 'indent' and 'width'
      Save the '.yml' file onto local memory.
      Read-in the '.yml' file as YAML times series from local memory using the
     'yaml.safe_load()' method
      Render the '.yml' time series using appropriate display method.
      /* Render as a data frame using Pandas if needed */

end
```

Figure 59. Dataframe result of algorithm 8c – : TSV Timeseries conversion to YAML data format

```
# Convert from TSV to YAML to MessagePack
with open('/Users/oyekanlea2/Desktop/TimeSeriesCodes/YahoodataTSV2YAML.yml') as f:
    TSV2YAMLdata = yaml.safe_load(f)
MsgPackdata = TSV2YAMLdata
# Write msgpack file
with open('MsgPackdata.msgpack', 'wb') as inputMsgPackdata:
    msgpack.pack(MsgPackdata, inputMsgPackdata)
# Read MsgPack in from storage location
with open('MsgPackdata.msgpack', "rb") as MsgPackdata_file:
    # MsgPackdata_loaded = json.load(MsgPackdata_file)
    MsgPackdatafromTSV_loaded = msgpack.unpack(MsgPackdata_file)
# Convert MsgPack to dictionary format
newloadedMsgPackFromTSV = {"convertTSVtoTAMLtoMessagePack": MsgPackdatafromTSV_loaded}
newloadedMsgPackFromTSV
```

Figure 60. Python implementation of algorithm 8d – : TSV Timeseries conversion to messagepack data format+

```
# Convert from TSV to YAML to MessagePack
{'convertTSVtoTAMLtoMessagePack': [{'Date': '4/30/1997',
    'Open': 17.5,
    'High': 17.5,
    'Low': 17.0,
    'Close': 17.0,
    'Adj Close': 15.192017000000002,
    'Volume': 5800},
  {'Date': '5/1/1997',
    'Open': 17.5,
    'High': 17.75,
    'Low': 17.375,
    'Close': 17.75,
    'Adj Close': 15.862257000000001,
    'Volume': 36300},
  {'Date': '5/2/1997',
    'Open': 17.75,
    'High': 17.875,
    'Low': 17.5,
    'Close': 17.875,
```

Figure 61. Timeseries result of algorithm 8d – : TSV timeseries conversion to messagepack data format

```
# Convert from TSV to YAML to MessagePack
#From TSV to JSON
YahooPricedata= pd.read_csv('/Users/..............CSV2TSV.tsv'  ,sep='\t')
YahooPricefromTSVtoJSON = pd.DataFrame(YahooPricedata)
#Convert the data Pandas Dataframe to JSON format
YahooPricefromTSVtoJSON.to_json('YahooPriceDatafromTSVToJSON.json')
# Read in the JSON file formulated with Pandas
with open("YahooPriceDatafromTSVToJSON.json", "r") as read_file:
    FromTSVtoJSON = json.load(read_file)
    FromTSVtoJSON
```

Figure 62. Python implementation of algorithm 8e – : TSV timeseries conversion to JSON data format

```
{'Date': {'0': '4/30/1997',
  '1': '5/1/1997',
  '2': '5/2/1997',
  '3': '5/5/1997',
  '4': '5/6/1997',
  '5': '5/7/1997',
......................
  '998': '4/16/2001',
  '999': '4/17/2001',
  ...},
 'Open': {'0': 17.5,
  '1': 17.5,
  '2': 17.75,
  '3': 17.5,
......................
```

Figure 63. Timeseries result of algorithm 8e – : TSV timeseries conversion to JSON data format

```
# Algorithm to read from csv to JSON and to generate data schema
Import pandas as pd
Import csv
Import json
Yahoodata = pd.read_csv('/Users/........../YahooData.csv')
Yahoodatadf = pd.DataFrame(Yahoodata)
YahoodataToJSON = Yahoodatadf.to_json(orient="table")
parsed = json.loads(YahoodataToJSON)
YahoodataToJSONRecordFormat = json.dumps(parsed)
# Use eval to remove the single quotes at the back of the [](i.e '[]') and then turn it re-
sulting
YahoodataToJSONRecordFormatEval = eval(YahoodataToJSONRecordFormat)
```

SCHEMA GENERATION ALGORITHMS AND PYTHON IMPLEMENTATIONS

In this section, a list of algorithms that can be used to integrate data schema with CSV and timeseries data sets are provided. CSV data format does not have a way of embedding and integrating data schemas along with CSV data sets. Researchers in (Retter et al., n.d.) described CSV Schema (CSVS), a textual language that can be used to define the data structure, types and rules for CSV data formats. In this chapter, we have introduced three additional Python based algorithms by which CSV timeseries can be enriched with a schema definition.

Algorithm 8d: TSV Timeseries Conversion to MessagePack Data Format

```
Input:     TSV Timeseries Data without Schema
Output:    MessagePack time series and Data frame
Data:      New York Stock Exchange Timeseries (1997-2021)
begin

       Input TSV timeseries data
       Convert the TSV time series to '.yml' data format using Algorithm 8c or any
       other appropriate algorithm.
       Read-in the '.yml' file as YAML times series from local memory using the
      'yaml.safe_load()' method

       Convert from YAML format into MessagePack format using msgpack.pack()method and
          save onto local storage

          Read from storage location using msgpack.unpack() method and the
          json.load() method

          Convert from msgpack.unpack() format to dictionary key:value pair
          format

          /* This enables the entire time series to be treated as an element of
          MessagePack format. It is also useful as an intermediate step to convert
          from dictionary to list based format to ensure easier conversion to a
          Python dataframe*/

          Render as a MessagePack time series
          /* If needed, use Python Pandas to convert to a dataframe */

end
```

Algorithm 8e: TSV Timeseries Conversion to JSON Data Format

```
Input:     TSV Timeseries Data without Schema
Output:    JSON time series and Data frame
Data:      New York Stock Exchange Timeseries (1997-2021)
begin

       Apply CSV reader with a separator or delimiter to read-in the TSV data
       Convert to a dataframe using Python Pandas.
       Cast as a JSON time series using the 'to_json()' method.
       Save onto local memory
       Read the time series from local memory as a JSON file. Use 'json_load()' method
       Render the JSON time series
       /* Convert to a data frame if there is a need. Use Pandas data frame method */

end
```

Algorithm and Python Implementation for JSON Based Schema Generation Algorithm for CSV Time Series Data

The first algorithm for schema generation that is provided in this chapter is Algorithm 9a. Algorithm 9a works through understanding the inherent characteristics of the time series by working with the JSON version of the CSV time series data.

Algorithm 9a: CSV-JSON Based Data Schema Generation

```
Input:     CSV Timeseries Data without Schema
Output:    JSON Timeseries Data with Schema
Data:      New York Stock Exchange Timeseries (1997-2021)
begin

        Input CSV timeseries data
        Convert CSV timeseries data to a Pandas dataframe
        Convert from Pandas dataframe to JSON format. Use the 'to_json()' method.
        Use the 'table' orientation format only as argument for the 'to_json()' method

        /* other JSON based dataframe formats such as 'split', 'column', 'index',
        'records', 'values' does not work for schema generation */

        Use Python 'eval' method to convert data from JSON string to proper JSON
        format
        Render data to reveal data schema

end
```

Python implementation of Algorithm 9a is shown in Figure 64, and its result for the time series of Figure 10 is shown in Figure 65. To implement the algorithm, the 'to_json()' method of the Python 'json' library is used along with the 'table' orientation of the 'to_json()' method. The resulting JSON data was then stripped of the extra hyphen sign so as to convert the data to proper JSON format. This stripping process is accomplished by using the Python 'eval()' method.

Figure 64. Python implementation of JSON based schema generation algorithm (Algorithm 9a) for CSV time series data

```
# JSON based Schema generation for time series data

{'schema': {'fields': [{'name': 'index', 'type': 'integer'},
   {'name': 'Date', 'type': 'string'},
   {'name': 'Open', 'type': 'number'},
   {'name': 'High', 'type': 'number'},
   {'name': 'Low', 'type': 'number'},
   {'name': 'Close', 'type': 'number'},
   {'name': 'Adj Close', 'type': 'number'},
   {'name': 'Volume', 'type': 'integer'}],
  'primaryKey': ['index'],
  'pandas_version': '0.20.0'},
 'data': [{'index': 0,
   'Date': '4/30/1997',
   'Open': 17.5,
   'High': 17.5,
   'Low': 17.0,
   'Close': 17.0,
   'Adj Close': 15.192017,
   'Volume': 5800},
  {'index': 1,
   'Date': '5/1/1997',
   'Open': 17.5,
   'High': 17.75,
   'Low': 17.375,
   'Close': 17.75,
   'Adj Close': 15.862257,
   'Volume': 36300},
.........................
```

Figure 65. Schema generation result for algorithm 9a – JSON based schema generation for CSV time series data

```
# This data schema provision method can also be used to access the timeseries data as a list

import pandas as pd
import json
import csv
CSVFileInput = r'C:\Users\...... \Yahoodata.csv'
JSONFileOutput = r'C:\Users\......\Yahoodata.json'

def FileInput_to_json(CSVFileInput, JSONFileOutput):
    fileToStoreConvertedArray = []
    with open(CSVFileInput,  newline='') as inputfile:
        dialect = csv.Sniffer().sniff(inputfile.read(1024))
        inputfile.seek(0)
        csvReader = csv.DictReader(inputfile)

        #convert each csv row into python dict
        for row in csvReader:
            fileToStoreConvertedArray.append(row)

    with open(JSONFileOutput, 'w', newline='') as jsonfile:
        jsonString = json.dumps(fileToStoreConvertedArray, indent=4)
        jsonfile.write(jsonString)

start = time.perf_counter()
FileInput_to_json(CSVFileInput ,JSONFileOutput)

with open("Yahoodata3.json", "r") as read_file:
    datajson = json.load(read_file)

datajson
```

Figure 66. Convert CSV time series to JSON time series using algorithm 2

```
# Output of datajson (last step) in Fig. 65

[{'Date': '4/30/1997',
  'Open': '17.5',
  'High': '17.5',
  'Low': '17',
  'Close': '17',
  'Adj Close': '15.192017',
  'Volume': '5800'},
 {'Date': '5/1/1997',
  'Open': '17.5',
  'High': '17.75',
  'Low': '17.375',
  'Close': '17.75',
  'Adj Close': '15.862257',
  'Volume': '36300'},

 {'Date': '4/17/2001',
  'Open': '3.12',
  'High': '3.12',
  'Low': '2.95',
  'Close': '2.95',
  'Adj Close': '2.636262',
  'Volume': '11600'},
 ...]
```

Figure 67. JSON time series version of the CSV time series, i.e., 'datajson' from figure 66

```
#Define the Avro Schema separately and then concatenate with the JSON-based version of the
CSV time series (datajson)

# Define the Avro schema for the data set here.

avroschema = {"namespace": "New_York_Stock_Exchange(2007-2021)_Schema.avro",
 "type": "record",
 "name": "User",
 "fields": [
     {"Date": "Date", "type": "string"},
     {"Daily Opening Stock Price": "Open",  "type": ["float", "null"]},
     {"Daily High Stock Price": "High", "type": ["float", "null"]},
     {"Daily Low Stock Price": "Low", "type": ["float", "null"]},
     {"Daily Stock Closing Price": "Close", "type": ["float", "null"]},
     {"Daily Stock Adj Closing Price": "Adj Close", "type": ["float", "null"]},
     {"Daily Stock Traded Volume": "Volume", "type": ["int", "null"]}
 ]}
```

It is worthy to note that based on extensive literature searches, Algorithm 9a is the first reported attempt by which the Python 'eval()' method is used exclusively as a means of revealing inherent data schema of CSV time series data. Our implementation is also unique, and it is different from the CSVS method introduced in (Retter et al., n.d.). In (Retter et al., n.d.), users of the CSVS method have to manually specify the schema of the CSV data. In our own case however, the schema is mined from the existing CSV data. Our method is thus very promising and quite useful for different variety of CSV data sets; including for those that does not have well-defined columnar structures.

The second method for defining schemas for CSV time series provided in this chapter is Algorithm 9b, the: ***Avro-JSON List Decoupling Algorithm***. The algorithm basically works by first implementing Algorithm 2 to convert the CSV time series to JSON time series using the DictReader() and Sniffer() methods. Python implementation of this initial implementation stage is shown in Figure 66, and the result is shown in Figure 67 as 'datajson'

Algorithm 9b: Avro-JSON List Decoupling Algorithm for CSV Time series Schema Generation

```
Input:     CSV Timeseries Data without Schema
Output:    JSON Timeseries Data with Schema
Data:      New York Stock Exchange Timeseries (1997-2021)
begin

         Input CSV timeseries data
         Convert each CSV row into Python dictionary
         Append subsequent rows to existing rows with each row appended as a record
         Output each record as a newline record
         Dump all converted record as a JSON string and save onto local memory
         Load the dumped data using json.load() method
         Specify data schema using Avro schema format
         Concatenate Schema and loaded JSON data as a Python list
         Render data to show generated schema

         /* Benefit of this algorithm when compared to Algorithm 9a is that schema and actual
timeseries data can be separately accessed from the list and analyzed */

end
```

Figure 68. Define avro format-based data schema for the CSV time series

```
# Concatenate  the Avro schema with timeseries (datajson).
{"namespace": "New_York_Stock_Exchange(2007-2021)_Schema.avro",
 "type": "record",
 "name": "User",
 "fields": [
     {"Date": "Date", "type": "string"},
     {"Daily Opening Stock Price": "Open",  "type": ["float", "null"]},
     {"Daily High Stock Price": "High", "type": ["float", "null"]},
     {"Daily Low Stock Price": "Low", "type": ["float", "null"]},
     {"Daily Stock Closing Price": "Close", "type": ["float", "null"]},
     {"Daily Stock Adj Closing Price": "Adj Close", "type": ["float", "null"]},
     {"Daily Stock Traded Volume": "Volume", "type": ["int", "null"]},datajson
 ]
}
```

Figure 69. Concatenate the time series avro format schema with the JSON rendered time series

```
{'namespace': 'New_York_Stock_Exchange(2007-2021)_Schema.avro',
 'type': 'record',
 'name': 'User',
 'fields': [{'Date': 'Date', 'type': 'string'},
  {'Daily Opening Stock Price': 'Open', 'type': ['float', 'null']},
  {'Daily High Stock Price': 'High', 'type': ['float', 'null']},
  {'Daily Low Stock Price': 'Low', 'type': ['float', 'null']},
  {'Daily Stock Closing Price': 'Close', 'type': ['float', 'null']},
  {'Daily Stock Adj Closing Price': 'Adj Close', 'type': ['float', 'null']},
  {'Daily Stock Traded Volume': 'Volume', 'type': ['int', 'null']},
  [{'Date': '4/30/1997',
    'Open': '17.5',
    'High': '17.5',
    'Low': '17',
    'Close': '17',
    'Adj Close': '15.192017',
    'Volume': '5800'},
   {'Date': '5/1/1997',
    'Open': '17.5',
    'High': '17.75',
    'Low': '17.375',
    'Close': '17.75',
    'Adj Close': '15.862257',
    'Volume': '36300'},
```

Figure 70. Avro format-based result for the python implementation of Figure 67

```
# Avro-JSON List Decoupling Algorithm
#Alternatively, define Timeseries and AvroSchema separately and now combine them together in
a list. Let the schema be the 1st item in the list and the timeseries the 2nd item on the
list.
# This approach enables the schema to be easily modifiable
AvroSchemaAndTimeseries = [avroschema, datajson]
AvroSchemaAndTimeseries
```

Figure 71. Avro-JSON list decoupling algorithm

```
[{'namespace': 'New_York_Stock_Exchange(2007-2021)_Schema.avro',
  'type': 'record',
  'name': 'User',
  'fields': [{'Date': 'Date', 'type': 'string'},
   {'Daily Opening Stock Price': 'Open', 'type': ['float', 'null']},
   {'Daily High Stock Price': 'High', 'type': ['float', 'null']},
   {'Daily Low Stock Price': 'Low', 'type': ['float', 'null']},
   {'Daily Stock Closing Price': 'Close', 'type': ['float', 'null']},
   {'Daily Stock Adj Closing Price': 'Adj Close', 'type': ['float', 'null']},
   {'Daily Stock Traded Volume': 'Volume', 'type': ['int', 'null']}]},
 [{'Date': '4/30/1997',
  'Open': '17.5',
  'High': '17.5',
  'Low': '17',
  'Close': '17',
  'Adj Close': '15.192017',
  'Volume': '5800'},
 {'Date': '5/1/1997',
  'Open': '17.5',
  'High': '17.75',
  'Low': '17.375',
  'Close': '17.75',
  'Adj Close': '15.862257',
  'Volume': '36300'},
 {'Date': '5/2/1997',
  'Open': '17.75',
  'High': '17.875',
  'Low': '17.5',
  'Close': '17.875',
  'Adj Close': '15.97396',
  'Volume': '17700'},
```

Figure 72. Schema generation result for the avro-JSON list decoupling algorithm

```
# To convert to pandas dataframe, use the 2nd item on the list. Analysis
from this point onward will default to JSON timeseries analysis

AvrodatajsonPandasFrameData = pd.DataFrame(AvroSchemaAndTimeseries[1])

AvrodatajsonPandasFrameData
```

	Date	Open	High	Low	Close	Adj Close	Volume
0	4/30/1997	17.5	17.5	17	17	15.192017	5800
1	5/1/1997	17.5	17.75	17.375	17.75	15.862257	36300
2	5/2/1997	17.75	17.875	17.5	17.875	15.97396	17700
3	5/5/1997	17.5	18.5	17.5	18.5	16.532488	65100
4	5/6/1997	18.625	19.5	18.25	19.5	17.426138	227000

Figure 73. Dataframe from algorithm 9b (Avro-JSON list decoupling algorithm for CSV time series schema generation)

```
# Convert CSV Time Series to Python Numpy Array
import pandas as pd
import numpy as np
mydata = pd.read_csv('/Users/......... /YahooData.csv')
mydata_array = np.array(mydata)
mydata_array
array([['4/30/1997', 17.5, 17.5, ..., 17.0, 15.192017000000002, 5800],
       ['5/1/1997', 17.5, 17.75, ..., 17.75, 15.862257000000001, 36300],
       ['5/2/1997', 17.75, 17.875, ..., 17.875, 15.97396, 17700],
       ...,
       ['4/27/2022', 48.529999, 49.110001000000004, ..., 48.619999,
        48.619999, 114400],
       ['4/28/2022', 49.630001, 49.630001, ..., 44.869999, 44.869999,
        283800],
       ['4/29/2022', 44.32, 44.599998, ..., 42.889998999999996,
        42.889998999999996, 193100]], dtype=object)
```

The next step of Algorithm 9b is to specify the Avro data format schema for the CSV time series. This step is shown in Figure 68. After the step shown in Figure 68, the Avro schema can then be concatenated with the JSON-based time series (datajson) as shown in Figure 69. Python result of this implementation is shown in Figure 70.

Alternatively, the Avro schema, for the CSV time series (named as 'avroschema' in Figure 68) can be concatenated with 'datajson' in the form of a list. The list can be easily decoupled, and the schema modified if and whenever the CSV time series is updated with a new record or a new column of data. Python implementation of this version of Algorithm 9b is as shown in Figure 71. Result of the schema-enriched time series is as shown in Figure 72. Another advantage of the List-Decoupling Algorithm is that data frame of the time series can also be easily obtained as shown in Figure 73.

ALGORITHM AND PYTHON IMPLEMENTATION FOR TIME SERIES LIST, NUMPY ARRAYS AND TENSORS FOR CSV TIME SERIES.

In this section, in addition to the in-depth discussion regarding tensors applications in section II, the importance, and applications of Python List, Numpy Arrays and Tensor applications for CSV time series are discussed. As explained in (GeeksforGeeks, n.d.) and (Koidan, 2019), Python Lists can be used in the same way as Python arrays. However, the performance of List can be slow due to computer processors having to do more data processing when data sets are represented as a List. Similar to Python List, Numpy arrays are ordered, they are mutable, can be enclosed in squared brackets, and they can be used to store non-unique items. Python methods of converting CSV time series to Numpy array and Python List; and their respective data frames are shown in Figure 74 to Figure 77.

When compared in term of performance, Numpy arrays are great for numerical operations. Also, they can be used to store data more compactly and thus, are efficient for storing large data amount. This could be noticed when the Numpy array of Figure 74 is compared with the Python List of Figure 76. The Python List of Figure 76 have more data hyphens and quotations for all data sets in all columns as opposed to the representation of the same data set as a Numpy array in Figure 74. In Figure 74, only

Figure 74. Conversion of CSV time series data set to Numpy Array

```
# Conversion of Python Numpy Array to data frame
mydata_arrayDF = pd.DataFrame(mydata_array)
```

	0	1	2	3	4	5	6
0	4/30/1997	17.5	17.5	17	17	15.192	5800
1	5/1/1997	17.5	17.75	17.375	17.75	15.8623	36300
2	5/2/1997	17.75	17.875	17.5	17.875	15.974	17700
3	5/5/1997	17.5	18.5	17.5	18.5	16.5325	65100
4	5/6/1997	18.625	19.5	18.25	19.5	17.4261	227000
...	...	...	...	...	...	...	...
6288	4/25/2022	49.05	49.79	48.35	49.73	49.73	120400
6289	4/26/2022	49.44	50.42	48.31	48.48	48.48	80400
6290	4/27/2022	48.53	49.11	48.32	48.62	48.62	114400
6291	4/28/2022	49.63	49.63	44.36	44.87	44.87	283800
6292	4/29/2022	44.32	44.6	42.37	42.89	42.89	193100

6293 rows × 7 columns

Figure 75. Conversion of python array to a data frame

```
# Conversion of CSV time series to Python List
import pandas as pd
import numpy as np
csv_filename = '/Users/........./YahooData.csv'
with open(csv_filename) as f:
    reader = csv.reader(f)
    csvlist = list(reader)
    print(csvlist)
[['Date', 'Open', 'High', 'Low', 'Close', 'Adj Close', 'Volume'], ['4/30/1997', '17.5',
'17.5', '17', '17', '15.192017', '5800'], ['5/1/1997', '17.5', '17.75', '17.375', '17.75',
'15.862257', '36300'], ['5/2/1997', '17.75', '17.875', '17.5', '17.875', '15.97396',
'17700'], ['5/5/1997', '17.5', '18.5', '17.5', '18.5', '16.532488', '65100'], ['5/6/1997',
'18.625', '19.5', '18.25', '19.5', '17.426138', '227000'],......................
```

the date column of the time series has data hyphens and quotations. Thus, it is intuitive to conclude that Numpy List will be more challenging to process than Python Numpy arrays.

A tensor is similar to an n-dimensional array (numpy.ndarray). The difference however is that tensors can easily employ GPUs to help accelerate numeric computation. The Python's 'torch' library can be used to a great effect in converting n-dimensional time series to tensors. As discussed in section II, tensors are basically mathematical objects that generalizes vectors to higher dimensions (Kumar, 2022). They can be used to great effect in representing multidimensional data such as images and videos. They are thus very useful in handling big data sets with dimensions greater than $10^{\wedge 6}$ or more (Kumar, 2022). Algorithm 10a shows a general method by which CSV time series with numerous columns can be converted into tensors with the aid of Python 'torch' while Algorithm 10b shows means and methods by which CSV time series 1D tensors can be easily generalized into greater dimensional tensors using Python 'torch' library.

Figure 76. Conversion of CSV time series to a python list

```
# Conversion of Python Numpy Array to data frame
mydata_arrayDF = pd.DataFrame(mydata_array)
```

	0	1	2	3	4	5	6
0	4/30/1997	17.5	17.5	17	17	15.192	5800
1	5/1/1997	17.5	17.75	17.375	17.75	15.8623	36300
2	5/2/1997	17.75	17.875	17.5	17.875	15.974	17700
3	5/5/1997	17.5	18.5	17.5	18.5	16.5325	65100
4	5/6/1997	18.625	19.5	18.25	19.5	17.4261	227000
...	...	...	...	...	...	...	...
6288	4/25/2022	49.05	49.79	48.35	49.73	49.73	120400
6289	4/26/2022	49.44	50.42	48.31	48.48	48.48	80400
6290	4/27/2022	48.53	49.11	48.32	48.62	48.62	114400
6291	4/28/2022	49.63	49.63	44.36	44.87	44.87	283800
6292	4/29/2022	44.32	44.6	42.37	42.89	42.89	193100

6293 rows × 7 columns

Figure 77. Conversion of CSV list to a data frame

```
import csv
import torch
import pandas as pd
import numpy as np
Yahoodata3 = pd.read_csv('/Users/......./YahooData.csv')
Yahoodata3df = pd.DataFrame(Yahoodata3)
# Select Col 1 to Col 7 only and leave out the date column. Remove other string columns too
or cast the string column as an integer or float column if there will be no data loss
YahoodataDFCol2To7Only = Yahoodata3df.iloc[:, np.r_[1:7]]
YahoodataDFCol2To7Only
# Tensorize all the cols of the dataframe and assign the resulting tensored cols to a vari-
able
YahoodataTensorizedCols = [torch.tensor(YahoodataDFCol2To7Only[column]) for column in Yahoo-
dataDFCol2To7Only]
YahoodataTensorizedCols
# Append the tensorized cols with the pruned date cols if need be
YahoodataTensorizedColsWithDates = [YahoodataTensorizedCols, Yahoodata3dfCol0]
YahoodataTensorizedColsWithDates
# Convert to data frame if needed
ConvertTensorToNumpyArrayDF = pd.DataFrame(YahoodataTensorizedCols)
ConvertTensorToNumpyArrayDF
```

Python implementation of Algorithm 10a is shown in Figure 77 and the CSV time series of Figure 10 without the date column is shown in Figure 79. 1D tensor results of Algorithm 10a using time series of Figure 10a is shown in Figure 80. A method by which the initially pruned date column can be re-appended to the newly obtained 1D tensors is shown in Figure 81.

Algorithm 10a: CSV Time Series Conversion to 1D Tensors

```
Input:     CSV Timeseries Data without Schema
Output:    1D Tensors
Data:      New York Stock Exchange Timeseries (1997-2021)
begin

        Input CSV timeseries data
        Convert the timeseries data into a data frame. /* Use Python Pandas */
        Prune all columns with the date type or string type data set./* This
        algorithm works best the 'int' and 'float' type tensors.
        If columns containing string types are available and if they are needed in
        the current numeric type computation, cast those columns as 'int' or 'float'
        type tensors.
        Cast all the remaining columns as 1D tensors. Use the Python 'torch' library
        /* If there is a need, append the pruned date/string columns back to the
        resulting tensors */

end
```

Figure 78. Python implementation of algorithm 10: Conversion of CSV time series to 1D tensors

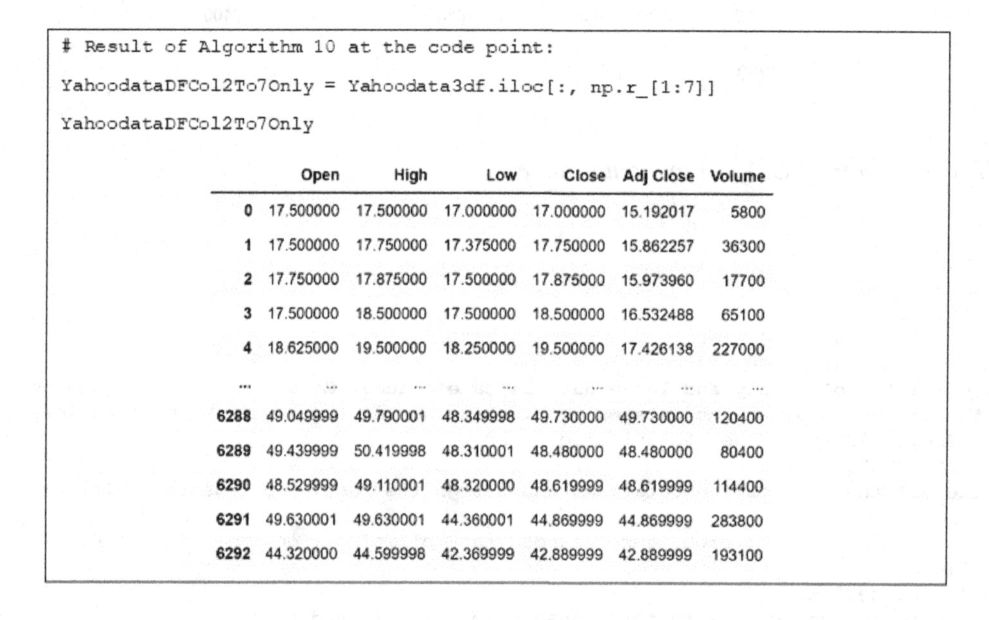

```
# Result of Algorithm 10 at the code point:
YahoodataDFCol2To7Only = Yahoodata3df.iloc[:, np.r_[1:7]]
YahoodataDFCol2To7Only
```

	Open	High	Low	Close	Adj Close	Volume
0	17.500000	17.500000	17.000000	17.000000	15.192017	5800
1	17.500000	17.750000	17.375000	17.750000	15.862257	36300
2	17.750000	17.875000	17.500000	17.875000	15.973960	17700
3	17.500000	18.500000	17.500000	18.500000	16.532488	65100
4	18.625000	19.500000	18.250000	19.500000	17.426138	227000
...	...	...	...	...	...	...
6288	49.049999	49.790001	48.349998	49.730000	49.730000	120400
6289	49.439999	50.419998	48.310001	48.480000	48.480000	80400
6290	48.529999	49.110001	48.320000	48.619999	48.619999	114400
6291	49.630001	49.630001	44.360001	44.869999	44.869999	283800
6292	44.320000	44.599998	42.369999	42.889999	42.889999	193100

Some Python methods of working and interacting with the 1D tensors obtained from the of the original CSV time series of Figure 10 are shown in Figure 82.

Pandas dataframe of the 1D tensors of the original CSV time series of Figure 10 is shown in Figure 83. The transpose of the data frame is shown in Figure 84. It is challenging to generate the transpose of the individual tensors. The Python Pandas method made working with the complete tensors dataframe transpose possible as shown in Figure 84.

As an extension of 1D tensors, we hereby present a Python based algorithm that can be used to comfortably create multi-dimensional (N-dimensional) tensors. The method is shown in Algorithm 10b. As opposed to Algorithm 10a where a Python 'for loop' is used to concatenate column of tensorized

Figure 79. CSV data frame with date column (string) pruned

```
# Result of Algorithm 10 at the code point:
YahoodataTensorizedCols = [torch.tensor(YahoodataDFCol2To7Only[column]) for column in Yahoo-
dataDFCol2To7Only]
YahoodataTensorizedCols

[tensor([17.5000, 17.5000, 17.7500,  ..., 48.5300, 49.6300, 44.3200],
        dtype=torch.float64),
 tensor([17.5000, 17.7500, 17.8750,  ..., 49.1100, 49.6300, 44.6000],
        dtype=torch.float64),
 tensor([17.0000, 17.3750, 17.5000,  ..., 48.3200, 44.3600, 42.3700],
        dtype=torch.float64),
 tensor([17.0000, 17.7500, 17.8750,  ..., 48.6200, 44.8700, 42.8900],
        dtype=torch.float64),
 tensor([15.1920, 15.8623, 15.9740,  ..., 48.6200, 44.8700, 42.8900],
        dtype=torch.float64),
 tensor([  5800,  36300,  17700,  ..., 114400, 283800, 193100])]
```

Figure 80. CSV data tensors

```
# Append the tensorized cols with the pruned date cols if need be
YahoodataTensorizedColsWithDates = [YahoodataTensorizedCols, Yahoodata3dfCol0]
YahoodataTensorizedColsWithDates
[tensor([17.5000, 17.5000, 17.7500,  ..., 48.5300, 49.6300, 44.3200],
        dtype=torch.float64),
 tensor([17.5000, 17.7500, 17.8750,  ..., 49.1100, 49.6300, 44.6000],
        dtype=torch.float64),
............

tensor([  5800,  36300,  17700,  ..., 114400, 283800, 193100])],
 0       4/30/1997
 1        5/1/1997
 2        5/2/1997
 3        5/5/1997
 4        5/6/1997
            ...
 6288    4/25/2022
 6289    4/26/2022
 6290    4/27/2022
 6291    4/28/2022
 6292    4/29/2022
 Name: Date, Length: 6293, dtype: object]
```

time series data columns as bases for creating the 1D tensor, Algorithm 10b functions by using linear combination of 1D tensors as listed bases to create 2D tensors. 3D tensors are also created through the linear combination of bases of 2D tensors; and so on.

The Python implementation of this approach is shown in Figure 85; and the resulting 2D tensor is shown in Figure 86. Further, result of 3D tensor created from a linear combination of the bases of 2D tensors is shown in Figure 87 while a 4D tensor created from the field of 3D bases tensors is shown in Figure 88.

Figure 81. CSV data tensors with the added date (string) column

```
# Access each cols of the tensors

YahoodataTensorizedCols1 - YahoodataTensorizedCols[0]
YahoodataTensorizedCols1

tensor([17.5000, 17.5000, 17.7500,  ..., 48.5300, 49.6300, 44.3200], dtype=torch.float64)

YahoodataTensorizedCols2 = YahoodataTensorizedCols[2]
YahoodataTensorizedCols2

tensor([17.0000, 17.3750, 17.5000,  ..., 48.3200, 44.3600, 42.3700], dtype=torch.float64)

# Access Each col of the combined tensors
#To access a col in the tensor:
print(f”Col 0: {YahoodataTensorizedCols[0]}”)
print(f”Col 1: {YahoodataTensorizedCols[1]}”)
print(f”Col 2: {YahoodataTensorizedCols[2]}”)
print(f”Col 3: {YahoodataTensorizedCols[3]}”)
print(f”Col 4: {YahoodataTensorizedCols[4]}”)
print(f”Col 5: {YahoodataTensorizedCols[5]}”)

Col 0: tensor([17.5000, 17.5000, 17.7500,  ..., 48.5300, 49.6300, 44.3200],
        dtype=torch.float64)
Col 1: tensor([17.5000, 17.7500, 17.8750,  ..., 49.1100, 49.6300, 44.6000],
        dtype=torch.float64)
Col 2: tensor([17.0000, 17.3750, 17.5000,  ..., 48.3200, 44.3600, 42.3700],
        dtype=torch.float64)
Col 3: tensor([17.0000, 17.7500, 17.8750,  ..., 48.6200, 44.8700, 42.8900],
        dtype=torch.float64)
Col 4: tensor([15.1920, 15.8623, 15.9740,  ..., 48.6200, 44.8700, 42.8900],
        dtype=torch.float64)
Col 5: tensor([  5800,  36300,  17700,  ..., 114400, 283800, 193100])
```

Figure 82. Working with 1D tensors from the CSV time series

```
ConvertTensorToNumpyArrayDF = pd.DataFrame(YahoodataTensorizedCols)

ConvertTensorToNumpyArrayDF
```

```
ConvertTensorToNumpyArrayDF = pd.DataFrame(YahoodataTensorizedCols)
ConvertTensorToNumpyArrayDF
```

	0	1	2	3	4	5	6	7	
0	tensor(17.5000, dtype=torch.float64)	tensor(17.5000, dtype=torch.float64)	tensor(17.7500, dtype=torch.float64)	tensor(17.5000, dtype=torch.float64)	tensor(18.6250, dtype=torch.float64)	tensor(20., dtype=torch.float64)	tensor(20., dtype=torch.float64)	tensor(19., dtype=torch.float64)	dty
1	tensor(17.5000, dtype=torch.float64)	tensor(17.7500, dtype=torch.float64)	tensor(17.8750, dtype=torch.float64)	tensor(18.5000, dtype=torch.float64)	tensor(19.5000, dtype=torch.float64)	tensor(20.2500, dtype=torch.float64)	tensor(20., dtype=torch.float64)	tensor(19.6250, dtype=torch.float64)	dty
2	tensor(17., dtype=torch.float64)	tensor(17.3750, dtype=torch.float64)	tensor(17.5000, dtype=torch.float64)	tensor(17.5000, dtype=torch.float64)	tensor(18.2500, dtype=torch.float64)	tensor(19.5000, dtype=torch.float64)	tensor(19.1250, dtype=torch.float64)	tensor(18.8125, dtype=torch.float64)	dty
3	tensor(17., dtype=torch.float64)	tensor(17.7500, dtype=torch.float64)	tensor(17.8750, dtype=torch.float64)	tensor(18.5000, dtype=torch.float64)	tensor(19.5000, dtype=torch.float64)	tensor(20., dtype=torch.float64)	tensor(19.1250, dtype=torch.float64)	tensor(19., dtype=torch.float64)	dty
4	tensor(15.1920, dtype=torch.float64)	tensor(15.8623, dtype=torch.float64)	tensor(15.9740, dtype=torch.float64)	tensor(16.5325, dtype=torch.float64)	tensor(17.4261, dtype=torch.float64)	tensor(17.8730, dtype=torch.float64)	tensor(17.0910, dtype=torch.float64)	tensor(16.9793, dtype=torch.float64)	dty
5	tensor(5800)	tensor(36300)	tensor(17700)	tensor(65100)	tensor(227000)	tensor(75600)	tensor(5600)	tensor(56400)	

6 rows × 6293 columns

Figure 83. Top part of the 1D Tensor dataframe of the CSV time series

```
ConvertTensorToNumpyArrayDFInverted = pd.DataFrame(YahoodataTensorizedCols).transpose()
ConvertTensorToNumpyArrayDFInverted
```


```
ConvertTensorToNumpyArrayDFInverted = pd.DataFrame(YahoodataTensorizedCols).transpose()
ConvertTensorToNumpyArrayDFInverted
```


	0	1	2	3	4	5
0	tensor(17.5000, dtype=torch.float64)	tensor(17.5000, dtype=torch.float64)	tensor(17., dtype=torch.float64)	tensor(17., dtype=torch.float64)	tensor(15.1920, dtype=torch.float64)	tensor(5800)
1	tensor(17.5000, dtype=torch.float64)	tensor(17.7500, dtype=torch.float64)	tensor(17.3750, dtype=torch.float64)	tensor(17.7500, dtype=torch.float64)	tensor(15.8623, dtype=torch.float64)	tensor(36300)
2	tensor(17.7500, dtype=torch.float64)	tensor(17.8750, dtype=torch.float64)	tensor(17.5000, dtype=torch.float64)	tensor(17.8750, dtype=torch.float64)	tensor(15.9740, dtype=torch.float64)	tensor(17700)
3	tensor(17.5000, dtype=torch.float64)	tensor(18.5000, dtype=torch.float64)	tensor(17.5000, dtype=torch.float64)	tensor(18.5000, dtype=torch.float64)	tensor(16.5325, dtype=torch.float64)	tensor(65100)
4	tensor(18.6250, dtype=torch.float64)	tensor(19.5000, dtype=torch.float64)	tensor(18.2500, dtype=torch.float64)	tensor(19.5000, dtype=torch.float64)	tensor(17.4261, dtype=torch.float64)	tensor(227000)
...	...	...	...	...	...	...
6288	tensor(49.0500, dtype=torch.float64)	tensor(49.7900, dtype=torch.float64)	tensor(48.3500, dtype=torch.float64)	tensor(49.7300, dtype=torch.float64)	tensor(49.7300, dtype=torch.float64)	tensor(120400)
6289	tensor(49.4400, dtype=torch.float64)	tensor(50.4200, dtype=torch.float64)	tensor(48.3100, dtype=torch.float64)	tensor(48.4800, dtype=torch.float64)	tensor(48.4800, dtype=torch.float64)	tensor(80400)
6290	tensor(48.5300, dtype=torch.float64)	tensor(49.1100, dtype=torch.float64)	tensor(48.3200, dtype=torch.float64)	tensor(48.6200, dtype=torch.float64)	tensor(48.6200, dtype=torch.float64)	tensor(114400)
6291	tensor(49.6300, dtype=torch.float64)	tensor(49.6300, dtype=torch.float64)	tensor(44.3600, dtype=torch.float64)	tensor(44.8700, dtype=torch.float64)	tensor(44.8700, dtype=torch.float64)	tensor(283800)
6292	tensor(44.3200, dtype=torch.float64)	tensor(44.6000, dtype=torch.float64)	tensor(42.3700, dtype=torch.float64)	tensor(42.8900, dtype=torch.float64)	tensor(42.8900, dtype=torch.float64)	tensor(193100)

6293 rows × 6 columns

Algorithm 10b: CSV Time series Conversion to N-Dimensional Tensors

```
Input:     CSV Timeseries Data without Schema
Output:    N-dimensional Tensors
Data:      New York Stock Exchange Timeseries (1997-2021)
begin

        Input CSV timeseries data
        Convert the timeseries data into a data frame. /* Use Python Pandas */

        Prune all columns with the date type or string type data set./* This
        algorithm works best the 'int' and 'float' type tensors.

        If columns containing string types are available and if they are needed in
        the current numeric-type computation, cast those columns as 'int' or 'float'
        type tensors.

        Cast all the remaining columns as 1D tensors. Use the Python 'torch' library
        Create 2D tensors form the as a linear combination of 1D tensor bases. Use
        Python list

        Create 3D tensors form the as a linear combination of 2D tensor bases. Use
        Python list

        Create N-Dim tensors form the as a linear combination of (N-1)D tensor bases.
        Use Python list

end
```

Figure 84. Transpose of the 1D Tensor dataframe of the CSV time series

```
import csv
import torch
import pandas as pd
import numpy as np
Yahoodata3 = pd.read_csv('/Users/……./YahooData.csv')
Yahoodata3df = pd.DataFrame(Yahoodata3)
#Select several Columns of data from the data frame.
Yahoodata3dfCol1 = Yahoodata3df.iloc[:, 1]
Yahoodata3dfCol2 = Yahoodata3df.iloc[:, 2]
Yahoodata3dfCol3 = Yahoodata3df.iloc[:, 3]
Yahoodata3dfCol4 = Yahoodata3df.iloc[:, 4]
Yahoodata3dfCol5 = Yahoodata3df.iloc[:, 5]
Yahoodata3dfCol6 = Yahoodata3df.iloc[:, 6]

# Convert to Tensor
Yahoodata3dfCol1Tensor = torch.tensor(Yahoodata3dfCol1)
Yahoodata3dfCol2Tensor = torch.tensor(Yahoodata3dfCol2)
Yahoodata3dfCol3Tensor = torch.tensor(Yahoodata3dfCol3)
Yahoodata3dfCol4Tensor = torch.tensor(Yahoodata3dfCol4)
Yahoodata3dfCol5Tensor = torch.tensor(Yahoodata3dfCol5)
Yahoodata3dfCol6Tensor = torch.tensor(Yahoodata3dfCol6)

# Combine  individual Tensor to form a field of 1D Tensors as a basis for multidimensional
tensors
CombineTensors = [Yahoodata3dfCol1Tensor, Yahoodata3dfCol2Tensor, Yahoodata3dfCol3Tensor, Ya-
hoodata3dfCol4Tensor, Yahoodata3dfCol5Tensor, Yahoodata3dfCol6Tensor]
CombineTensors

#Create 2D Tensors from a combination of 1D Tensors as a Python data list
CombineTensors2dim = [CombineTensors, CombineTensors]
CombineTensors2dim

# Create 3D Tensors by linear combination of bases from the field of 2D Tensors
CombineTensors3dim = [CombineTensors2dim, CombineTensors2dim]
CombineTensors3dim

# Create 4D Tensors by linear combination of bases from the field of 3D Tensors
CombineTensors4dim = [CombineTensors3dim , CombineTensors3dim]
CombineTensors4dim
```

Figure 85. Python implementation of algorithm 10b: Conversion of CSV time series to multidimensional tensors

```
# Result at the point…

CombineTensors = [Yahoodata3dfCol1Tensor, Yahoodata3dfCol2Tensor, Yahoodata3dfCol3Tensor, Ya-
hoodata3dfCol4Tensor, Yahoodata3dfCol5Tensor, Yahoodata3dfCol6Tensor]
CombineTensors

……. in Fig. 84


[tensor([17.5000, 17.5000, 17.7500,  ..., 48.5300, 49.6300, 44.3200],
        dtype=torch.float64),
 tensor([17.5000, 17.7500, 17.8750,  ..., 49.1100, 49.6300, 44.6000],
        dtype=torch.float64),
 tensor([17.0000, 17.3750, 17.5000,  ..., 48.3200, 44.3600, 42.3700],
        dtype=torch.float64),
 tensor([17.0000, 17.7500, 17.8750,  ..., 48.6200, 44.8700, 42.8900],
        dtype=torch.float64),
 tensor([15.1920, 15.8623, 15.9740,  ..., 48.6200, 44.8700, 42.8900],
        dtype=torch.float64),
 tensor([  5800,  36300,  17700,  ..., 114400, 283800, 193100])]

# Result at the point…
# Create 2D Tensors from a combination of 1D Tensors as a Python data list
CombineTensors2dim = [CombineTensors, CombineTensors]
CombineTensors2dim
……. in Fig. 84

[[tensor([17.5000, 17.5000, 17.7500,  ..., 48.5300, 49.6300, 44.3200],
         dtype=torch.float64),
  tensor([17.5000, 17.7500, 17.8750,  ..., 49.1100, 49.6300, 44.6000],
         dtype=torch.float64),
  tensor([17.0000, 17.3750, 17.5000,  ..., 48.3200, 44.3600, 42.3700],
         dtype=torch.float64),
  tensor([17.0000, 17.7500, 17.8750,  ..., 48.6200, 44.8700, 42.8900],
         dtype=torch.float64),
  tensor([15.1920, 15.8623, 15.9740,  ..., 48.6200, 44.8700, 42.8900],
         dtype=torch.float64),
  tensor([  5800,  36300,  17700,  ..., 114400, 283800, 193100])],
 [tensor([17.5000, 17.5000, 17.7500,  ..., 48.5300, 49.6300, 44.3200],
         dtype=torch.float64),
  tensor([17.5000, 17.7500, 17.8750,  ..., 49.1100, 49.6300, 44.6000],
         dtype=torch.float64),
  tensor([17.0000, 17.3750, 17.5000,  ..., 48.3200, 44.3600, 42.3700],
         dtype=torch.float64),
  tensor([17.0000, 17.7500, 17.8750,  ..., 48.6200, 44.8700, 42.8900],
         dtype=torch.float64),
  tensor([15.1920, 15.8623, 15.9740,  ..., 48.6200, 44.8700, 42.8900],
         dtype=torch.float64),
  tensor([  5800,  36300,  17700,  ..., 114400, 283800, 193100])]]
```

Figure 86. Python implementation of algorithm 10b: Conversion of CSV time series to multidimensional tensors (1D & 2D Tensors)

```
# Create 3D Tensors from the field of 2D Tensor
CombineTensors3dim = [CombineTensors2dim, CombineTensors2dim]
CombineTensors3dim

[[[tensor([17.5000, 17.5000, 17.7500,  ..., 48.5300, 49.6300, 44.3200],
          dtype=torch.float64),
   tensor([17.5000, 17.7500, 17.8750,  ..., 49.1100, 49.6300, 44.6000],
          dtype=torch.float64),
   tensor([17.0000, 17.3750, 17.5000,  ..., 48.3200, 44.3600, 42.3700],
          dtype=torch.float64),
   tensor([17.0000, 17.7500, 17.8750,  ..., 48.6200, 44.8700, 42.8900],
          dtype=torch.float64),
   tensor([15.1920, 15.8623, 15.9740,  ..., 48.6200, 44.8700, 42.8900],
          dtype=torch.float64),
   tensor([  5800,  36300,  17700,  ..., 114400, 283800, 193100])],
  [tensor([17.5000, 17.5000, 17.7500,  ..., 48.5300, 49.6300, 44.3200],
          dtype=torch.float64),
   tensor([17.5000, 17.7500, 17.8750,  ..., 49.1100, 49.6300, 44.6000],
          dtype=torch.float64),
   tensor([17.0000, 17.3750, 17.5000,  ..., 48.3200, 44.3600, 42.3700],
          dtype=torch.float64),
   tensor([17.0000, 17.7500, 17.8750,  ..., 48.6200, 44.8700, 42.8900],
          dtype=torch.float64),
   tensor([15.1920, 15.8623, 15.9740,  ..., 48.6200, 44.8700, 42.8900],
          dtype=torch.float64),
   tensor([  5800,  36300,  17700,  ..., 114400, 283800, 193100])]],
 [[tensor([17.5000, 17.5000, 17.7500,  ..., 48.5300, 49.6300, 44.3200],
          dtype=torch.float64),
   tensor([17.5000, 17.7500, 17.8750,  ..., 49.1100, 49.6300, 44.6000],
          dtype=torch.float64),
   tensor([17.0000, 17.3750, 17.5000,  ..., 48.3200, 44.3600, 42.3700],
          dtype=torch.float64),
   tensor([17.0000, 17.7500, 17.8750,  ..., 48.6200, 44.8700, 42.8900],
          dtype=torch.float64),
   tensor([15.1920, 15.8623, 15.9740,  ..., 48.6200, 44.8700, 42.8900],
          dtype=torch.float64),
   tensor([  5800,  36300,  17700,  ..., 114400, 283800, 193100])],
  [tensor([17.5000, 17.5000, 17.7500,  ..., 48.5300, 49.6300, 44.3200],
          dtype=torch.float64),
   tensor([17.5000, 17.7500, 17.8750,  ..., 49.1100, 49.6300, 44.6000],
          dtype=torch.float64),
   tensor([17.0000, 17.3750, 17.5000,  ..., 48.3200, 44.3600, 42.3700],
          dtype=torch.float64),
   tensor([17.0000, 17.7500, 17.8750,  ..., 48.6200, 44.8700, 42.8900],
          dtype=torch.float64),
   tensor([15.1920, 15.8623, 15.9740,  ..., 48.6200, 44.8700, 42.8900],
          dtype=torch.float64),
   tensor([  5800,  36300,  17700,  ..., 114400, 283800, 193100])]]]
```

Figure 87. Python implementation of algorithm 10b: Conversion of CSV time series to multidimensional tensors (3D tensors)

```
# Create 4D Tensors by linear combination of bases from the field of 3D Tensors
CombineTensors4dim = [CombineTensors3dim , CombineTensors3dim]
CombineTensors4dim

[[[[tensor([17.5000, 17.5000, 17.7500,  ..., 48.5300, 49.6300, 44.3200],
         dtype=torch.float64),
    tensor([17.5000, 17.7500, 17.8750,  ..., 49.1100, 49.6300, 44.6000],
         dtype=torch.float64),
    tensor([17.0000, 17.3750, 17.5000,  ..., 48.3200, 44.3600, 42.3700],
         dtype=torch.float64),
    tensor([17.0000, 17.7500, 17.8750,  ..., 48.6200, 44.8700, 42.8900],
         dtype=torch.float64),
    tensor([15.1920, 15.8623, 15.9740,  ..., 48.6200, 44.8700, 42.8900],
         dtype=torch.float64),
    tensor([  5800,  36300,  17700,  ..., 114400, 283800, 193100])],
   [tensor([17.5000, 17.5000, 17.7500,  ..., 48.5300, 49.6300, 44.3200],
         dtype=torch.float64),
    .........................
    tensor([15.1920, 15.8623, 15.9740,  ..., 48.6200, 44.8700, 42.8900],
         dtype=torch.float64),
    tensor([  5800,  36300,  17700,  ..., 114400, 283800, 193100])]],
  [[tensor([17.5000, 17.5000, 17.7500,  ..., 48.5300, 49.6300, 44.3200],
         dtype=torch.float64),
    tensor([17.5000, 17.7500, 17.8750,  ..., 49.1100, 49.6300, 44.6000],
         dtype=torch.float64),
    tensor([17.0000, 17.3750, 17.5000,  ..., 48.3200, 44.3600, 42.3700],
         dtype=torch.float64),
    tensor([17.0000, 17.7500, 17.8750,  ..., 48.6200, 44.8700, 42.8900],
         dtype=torch.float64),
    tensor([15.1920, 15.8623, 15.9740,  ..., 48.6200, 44.8700, 42.8900],
         dtype=torch.float64),
    tensor([  5800,  36300,  17700,  ..., 114400, 283800, 193100])],
   [tensor([17.5000, 17.5000, 17.7500,  ..., 48.5300, 49.6300, 44.3200],
         dtype=torch.float64),
    tensor([17.5000, 17.7500, 17.8750,  ..., 49.1100, 49.6300, 44.6000],
         dtype=torch.float64),
    tensor([17.0000, 17.3750, 17.5000,  ..., 48.3200, 44.3600, 42.3700],
         dtype=torch.float64),
    tensor([17.0000, 17.7500, 17.8750,  ..., 48.6200, 44.8700, 42.8900],
         dtype=torch.float64),
    tensor([15.1920, 15.8623, 15.9740,  ..., 48.6200, 44.8700, 42.8900],
         dtype=torch.float64),
    tensor([  5800,  36300,  17700,  ..., 114400, 283800, 193100])]]],
 [[[tensor([17.5000, 17.5000, 17.7500,  ..., 48.5300, 49.6300, 44.3200],
         dtype=torch.float64),
    tensor([17.5000, 17.7500, 17.8750,  ..., 49.1100, 49.6300, 44.6000],
         dtype=torch.float64),
    tensor([17.0000, 17.3750, 17.5000,  ..., 48.3200, 44.3600, 42.3700],
         dtype=torch.float64),
    tensor([17.0000, 17.7500, 17.8750,  ..., 48.6200, 44.8700, 42.8900],
         dtype=torch.float64),
    tensor([15.1920, 15.8623, 15.9740,  ..., 48.6200, 44.8700, 42.8900],
         dtype=torch.float64),
    tensor([  5800,  36300,  17700,  ..., 114400, 283800, 193100])],
.................................
    tensor([15.1920, 15.8623, 15.9740,  ..., 48.6200, 44.8700, 42.8900],
         dtype=torch.float64),
    tensor([  5800,  36300,  17700,  ..., 114400, 283800, 193100])]]]]
```

Figure 88. Python implementation of algorithm 10b: Conversion of CSV time series to multidimensional tensors (4D tensors)

```
import pandas as pd
import numpy as np
import pickle
import json
import csv

# Algorithm to read from csv and convert to Pickle and then Filter.
# This algorithm have no need of conversion to dataframe
Yahoodata = pd.read_csv('/Users/............../YahooData.csv')
# Original Yahoodata Pickle with binary representation
with open('YahooDataBinaryPickleOnly.pickle', 'wb') as f:
    pickle.dump(Yahoodata , f, -1)
#  Load the Binary Pickled file
with open('YahooDataBinaryPickleOnly.pickle', 'rb') as f:
    YahooDataBinaryPickledData = pickle.load(f)
YahooDataBinaryPickledData
# Convert from Binary Pickle to json and read the data in JSON format
pickleFromCSV = pd.read_pickle('YahooDataBinaryPickleOnly.pickle')
pickleFromCSV.to_json('YahoodataConvertedFromBinaryPickleToJSON.json')
# Test to see if the convert JSON is in proper JSON format
with open("YahoodataConvertedFromBinaryPickleToJSON.json", "r") as read_file:
    datajsonPandasMethod = json.load(read_file)
ReadPickledFromPandasDatafameToJSON = pd.read_json('/Users/........./YahoodataConvertedFromBina-
ryPickleToJSON.json')
print(ReadPickledFromPandasDatafameToJSON)
```

ALGORITHM AND PYTHON IMPLEMENTATION FOR CONVERTING CSV DATA FORMATS TO PICKLE BINARIES

The Python Pickle library is very useful for serializing and deserializing Python objects and data files. Serializing is a process of converting data into byte or character streams. The character stream contains all the information needed to reconstruct the data object back to its original object format.

For Data Scientists that are working with huge data sets in CPS, after developing fully functional machine learning models for such CPS, the model can be Pickled so that the Data Scientist will not need to rewrite and retrain the model when there is a need to reuse it in the future.

The process of serializing, i.e., converting Python objects or data into a byte stream is also known as delating or marshalling (Kong et al., 2021; Pathak, 2018). When compared to other near-universal data formats such as JSON, it is instructive to note that JSON is a human-readable text serialization format while Pickle is a non-human-readable binary serialization format. JSON can also be used outside of Python, and it is interoperable with most other development and programming software, whereas Pickle data stream cannot be used in its raw form outside of the Python environment. On the other hand, Pickle can represent a huge amount of Python data types more than the amount that JSON can represent.

Apart from not being useful outside the Python environment, serializing data using Pickle can be potentially dangerous since such serialization can lead to serializing and creating arbitrary byte streams from an unknown source in a computing environment. Hence, unpickling a data set using the Python Pickle module is inherently unsecure.

Due to the challenges associated with using Pickle, in this chapter, we have provided algorithms (Algorithm 11a and Algorithm 11b) and methods by which user can convert time series from CSV to

Pickle and onward into Feather, MessagePack and Parquet data formats. Other algorithms that we have provided in this chapter can also be easily used to navigate through the Pickle-to-other-data formats conversion processes and vice versa. Our algorithms will enable users to be able to use Pickled data outside of Python environments by easily converting the Pickled data sets to other formats. Data sets from other data formats can also be easily Pickled by our series of algorithms. Results of using Algorithm 11a and 11b to convert CSV to Pickle and to other data formats are shown in Figure 89 to Figure 95. Data frames that are resulting from Pickle-to-Feather and Pickle-to-Parquet data formats are similar in attributes to other Feather and Parquet data frames that have been presented earlier.

Algorithm 11a: CSV To Pickle (Binary Data Format) and To JSON (CSV to Pickle to JSON)

```
Input:     CSV Timeseries Data without Schema
Output:    JSON time series from Pickle Binaries
Data:      New York Stock Exchange Timeseries (1997-2021)
begin

           Input CSV timeseries data

           Convert the CSV time series into Pickle binaries. Use 'pickle.dump()' method

           Save onto local memory.

           Use 'pickle.load()' method to load the binaries from local memory.

           Cast the Pickle binaries as JSON using the 'to_json()' method. /* Test to
           see if the resulting data is in proper JSON format. If not, use other JSON
           conversion method such as Algorithm 2 (DictReader() & Sniffer() method)*/

           Convert to a dataframe if needed. Use Python Pandas.

end
```

Algorithm 11b: Pickle Binary Data Conversion to MessagePack, Feather & Parquet Data Formats

```
Input:     CSV Time Series Data
Output:    Pickled binaries, MessagePack, Feather & Parquet Data Formats.
Data:      New York Stock Exchange Timeseries (1997-2021)
begin
           Input Pickled binary timeseries data
           Convert the timeseries data into a JSON time series. Use the 'to_json()'
           method.
           Save onto local memory.
           Read in the JSON file from local memory. Use json.load() or any other
           appropriate method that ensures that the file is read-in as JSON.
           Cast or write the JSON file as Messagefile data. Use Python 'msgpack'
           library.
           Render as MessagePack data
           /* for Parquet, convert the inputted pickle data into a Pandas dataframe,
           then use 'to_parquet()' method with 'pyarrow' engine */
           /* for Feather, use 'pickle.load()' method to read-in the pickle binaries.
           Do not convert into Pandas dataframe. Use 'to_fether()' method to
           the pickle binaries into Feather data format */

end
```

Figure 89. Conversion of CSV to pickle binary format and conversion to JSON

```
# Test to see if the convert JSON is in proper JSON format
with open("YahoodataConvertedFromBinaryPickleToJSON.json", "r") as read_file:
    datajsonPandasMethod = json.load(read_file)

{'Date': {'0': '4/30/1997',
  '1': '5/1/1997',
  '2': '5/2/1997',
  '3': '5/5/1997',
..........
'998': '4/16/2001',
'999': '4/17/2001',
  ...},

'Open': {'0': 17.5,
  '1': 17.5,
  '2': 17.75,
  '3': 17.5,
..............
'997': 3.0,
'998': 3.0,
'999': 3.12,
  ...},
..........
```

Figure 90. Conversion of CSV to pickle binary format and conversion to JSON

```
ReadPickledFromPandasDatafameToJSON = pd.read_json('/Users/............./YahoodataConvertedFromBinaryPickleToJSON.json')
print(ReadPickledFromPandasDatafameToJSON)


            Date       Open       High        Low      Close  Adj Close  Volume
0     1997-04-30  17.500000  17.500000  17.000000  17.000000  15.192017    5800
1     1997-05-01  17.500000  17.750000  17.375000  17.750000  15.862257   36300
2     1997-05-02  17.750000  17.875000  17.500000  17.875000  15.973960   17700
3     1997-05-05  17.500000  18.500000  17.500000  18.500000  16.532488   65100
4     1997-05-06  18.625000  19.500000  18.250000  19.500000  17.426138  227000
...          ...        ...        ...        ...        ...        ...     ...
6288  2022-04-25  49.049999  49.790001  48.349998  49.730000  49.730000  120400
6289  2022-04-26  49.439999  50.419998  48.310001  48.480000  48.480000   80400
6290  2022-04-27  48.529999  49.110001  48.320000  48.619999  48.619999  114400
6291  2022-04-28  49.630001  49.630001  44.360001  44.869999  44.869999  283800
6292  2022-04-29  44.320000  44.599998  42.369999  42.889999  42.889999  193100

[6293 rows x 7 columns]
```

Figure 91. Conversion of CSV to pickle binary format and conversion to JSON

```
# An approach from Pickle to MessagePack. Convert from Pickle to JSON, then to MsgPack
import msgpack
# Define data
pickleFromPandasDatafame = pd.read_pickle('YahooDataBinaryPickleOnly.pickle')
pickleFromPandasDatafame.to_json('YahoodataConvertedFromBinaryPickleToJSON.json')
with open("YahoodataConvertedFromBinaryPickleToJSON.json", "r") as read_file:
    datajsonPandasMethod = json.load(read_file)
MsgPackdata = datajsonPandasMethod
# Write msgpack file
with open('MsgPackdata.msgpack', 'wb') as inputMsgPackdata:
    msgpack.pack(MsgPackdata, inputMsgPackdata)
# Read MsgPack in from storage location
with open('MsgPackdata.msgpack', "rb") as MsgPackdata_file:
    # MsgPackdata_loaded = json.load(MsgPackdata_file)
    MsgPackdata_loaded = msgpack.unpack(MsgPackdata_file)
MsgPackdata_loaded
MsgPackdata_loadedDF = pd.DataFrame(MsgPackdata_loaded)
MsgPackdata_loadedDF
```

Figure 92. Conversion of pickle binary data to messagepack data format


```
{'Date': {'0': '4/30/1997',
  '1': '5/1/1997',
  '2': '5/2/1997',
  '3': '5/5/1997',
  '4': '5/6/1997',
..............

  '997': '4/12/2001',
  '998': '4/16/2001',
  '999': '4/17/2001',
  ...},
 'Open': {'0': 17.5,
  '1': 17.5,
  '2': 17.75,
  '3': 17.5,
.................
  '997': 3.0,
  '998': 3.0,
  '999': 3.12,
  ...},
```


Figure 93. Some result of algorithm 11b: Conversion of pickle binary data to messagepack data format

```
# Read in a pickle file and convert it to Parquet Format
pickleFromPandasDatafame = pd.read_pickle('YahooDataBinaryPickleOnly.pickle')
YahoodataDF = pd.DataFrame(pickleFromPandasDatafame)
YahoodataDF.to_parquet('YahoodataDFParquet.parquet', engine="pyarrow")
YahoodataDFParquet = pd.read_parquet('YahoodataDFParquet.parquet')
YahoodataDFParquet
```

Figure 94. Conversion of pickle binary data to parquet file format

```
# Read in a pickle file and convert it to Feather data format
with open('YahooDataBinaryPickleOnly.pickle', 'rb') as f:
    YahooDataBinaryPickledData = pickle.load(f)
YahooDataBinaryPickledData
YahooDataBinaryPickledData.to_feather('PickleYahoodataToFeather.feather')
# Read in the newly created Feather data
PickleYahoodataFeather = pd.read_feather('PickleYahoodataToFeather.feather')
PickleYahoodataFeather
```

Figure 95. Conversion of pickle binary data to feather file format

```
%%time
df = pd.read_csv('/Users/.............................../YahooData.csv')
with open('YahoodataYAMLformat.yml', 'w') as outfile:
    yaml.dump(
        df.to_dict(orient='records'),
        outfile,
        canonical=True,
        sort_keys=True,
        default_flow_style=False,
        indent=6,
        width=22,
    )
with open('/Users/........./YahoodataYAMLformat.yml') as f:
    yamldata = yaml.safe_load(f)
YahoodataYAMLDF = pd.DataFrame( yamldata)
YahoodataYAMLDF
```

ALGORITHMS' EVALUATION BASED ON DATAFRAME CREATION, SORTING, FINDING UNIQUE VALUES, GROUPING AND TIMES SERIES DATA FILTERING

Results of designed algorithms vis-à-vis the analytics speeds and performance of data frames that are created from the algorithms when the data frames are used for data analytics are discussed in this section.

MessagePack Dataframe (Algorithm 1)

Similar to the JSON (DictReader() / Sniffer() & to_json() methods dataframes) in Figure 17 and Figure 21, the created MessagePack dataframe in Figure 13 has a serial number column automatically appended. The serial number column was not present in the original time series of Figure 10. Resulting data frame columns IDs were arranged in the same order as the original time series data that was presented in Figure 10. However, across all columns, the dataframe created by the MessagePack conversion algorithm

(Algorithm 1) seems not to have the same level of multiple order data precision when compared to the dataframe generated by the 'to_json()' data conversion (Algorithm 3) method.

JSON Dataframe (Algorithm 2)

The DictReader() & Sniffer() Methods dataframe in Figure 17 and the to_json() dataframe in Figure 21 also has a serial number column automatically appended. The column IDs were arranged in the same order as the original data was presented in Figure 10

Feather Dataframe (Algorithm 4)

The created Feather dataframe shown Figure 24 is also automatically appended with a serial number column as its first column. However, the Feather dataframe also has its columns IDs a little bit misaligned. For example, the serial number column (column 1) is now incorrectly identified as the date column while all other IDs are also shifted left by one column ID each. The result ultimately leaves the last dataframe column (previously 'Volume' column in Figure 10) without an ID. Hence, Data Scientists that desires to use Algorithm 4 must be cognizant of this fact and they may have to write additional routine to readjust the column IDs accordingly.

YAML Dataframes (Algorithm 5)

An interesting observation regarding the dataframes generated with YAML data conversion algorithms (Figure 28, Figure 31 and Figure 37) is that the Column IDs are now arranged alphabetically. The 'Adj Close' is now followed by the 'Close' and 'Date' columns respectively. YAML dataframes also have data that have the same level of data precision order as those of the 'to_json()' method in Algorithm 3. In the dataframe (Figure 34) generated when 'index' orientation (Algorithm 5c) is used in the df.to_dict() method, there is lesser precision when compared to the 'List', 'Dict' and 'Record' orientation (Algorithms 5a, 5b and 5d) methods. However, in all cases, IDs are also still arranged alphabetically.

PERFORMANCE EVALUATION IN TERMS OF

Data Analytics Speed

To measure the speed and performance of major algorithms presented in this chapter, Python magic commands for measuring CPU time (%%timeit) and the Wall times (%%time) are used.

Important data analytics processes including sorting, finding unique values, data filtering (querying) and data grouping are also used to evaluate the performance of the algorithms.

It is important to note that, it is not all algorithms that are presented in this chapter that are evaluated in terms of speed and data analytics performances. Algorithms that are evaluated for speed and data analytics performance include those that are used to convert time series data from CSV data formats to other major data formats such as MessagePack, JSON (DictReader() and to_json() methods) , YAML, Feather, Parquet, TSV and HDF5.

The procedure for using Python's %%timeit (CPU time) and %%time (wall time) magic commands to measure the performance of Algorithm 5a for dataframe creation, sorting, finding unique values in data, filtering (querying) and data grouping are shown in Figure 96 to Figure 100. Procedures for all other evaluated algorithms are similar to that of Algorithm 5a even though the logic and lines of codes that are used to implement each algorithm are different.

CPU time basically measures how much time a CPU spent on executing a program. The wall time can also be referred to as elapsed time, wall-clock time or running time. Wall time measures the total time used to execute a program in a computer. Compared to the CPU time, the wall time is often longer because the CPU executing the measured program may also be executing other programs' instructions at the same time. Basically, wall time measures how much time has passed, as if the user is looking at the clock on the wall. CPU time is how many seconds the CPU was busy (Turner-Trauring, 2021), (Knowledgebase, 2005) executing instructions in an algorithm.

Figure 96. Wall time measurement for data filtering (querying) process for Algorithm 5a

```
df = pd.read_csv('/Users/........................../YahooData.csv')
with open('YahoodataYAMLformat.yml', 'w') as outfile:
    yaml.dump(
        df.to_dict(orient='records'),
        outfile,
        canonical=True,
        sort_keys=True,
        default_flow_style=False,
        indent=6,
        width=22,
    )
with open('/Users/........./YahoodataYAMLformat.yml') as f:
    yamldata = yaml.safe_load(f)
YahoodataYAMLDF = pd.DataFrame( yamldata)
YahoodataYAMLDF.query('Open == "17.5"')
```

Figure 97. Wall time measurement for finding time series unique values in 'Open' and 'Close' columns for Algorithm 5a

```
df = pd.read_csv('/Users/........................../YahooData.csv')
with open('YahoodataYAMLformat.yml', 'w') as outfile:
    yaml.dump(
        df.to_dict(orient='records'),
        outfile,
        canonical=True,
        sort_keys=True,
        default_flow_style=False,
        indent=6,
        width=22,
    )
with open('/Users/........./YahoodataYAMLformat.yml') as f:
    yamldata = yaml.safe_load(f)
YahoodataYAMLDF = pd.DataFrame( yamldata)
pd.unique(YahoodataJSONDataFrame[['Open', 'Close']].values.ravel('K'))
```

Figure 98. CPU time measurement for sorting by 'Volume' column for Algorithm 5a

```
%%timeit
df = pd.read_csv('/Users/................................/YahooData.csv')
with open('YahoodataYAMLformat.yml', 'w') as outfile:
    yaml.dump(
        df.to_dict(orient='records'),
        outfile,
        canonical=True,
        sort_keys=True,
        default_flow_style=False,
        indent=6,
        width=22,
    )
with open('/Users/........./YahoodataYAMLformat.yml') as f:
    yamldata = yaml.safe_load(f)
YahoodataYAMLDF = pd.DataFrame( yamldata)
YahoodataYAMLDF.sort_values(by=['Volume'])
```

Figure 99. CPU time measurement for grouping by 'Open' column for Algorithm 5a

```
%%timeit
df = pd.read_csv('/Users/................................/YahooData.csv')
with open('YahoodataYAMLformat.yml', 'w') as outfile:
    yaml.dump(
        df.to_dict(orient='records'),
        outfile,
        canonical=True,
        sort_keys=True,
        default_flow_style=False,
        indent=6,
        width=22,
    )
with open('/Users/........./YahoodataYAMLformat.yml') as f:
    yamldata = yaml.safe_load(f)
YahoodataYAMLDF = pd.DataFrame( yamldata)
YahoodataYAMLDFGroupBy = YahoodataYAMLDF.groupby('Open').mean()
```

For IIoT and CPS processes analytics, CPU time measurement is very important because it is one of the foremost resources that cloud providers such as Google and AWS are selling. Hence algorithms that uses more CPU time will be in need of more virtual machines if such algorithms are included with cloud layer applications. Essentially, the more CPU time a process consumes, the more the cores that will be needed to be provisioned so as to maintain acceptable time latencies.

Full codes for all other algorithms and the CPU times and wall times for each algorithm for data frame creation and other data analytics processes are presented in Appendix A-i to A-xi.

Evaluation of each algorithm enable users to have a basic idea regarding how each algorithm perform in practice using a CPU that is similar to off-the-shelf CPUs that are commonly available on IoT devices. All the algorithms are evaluated in Anaconda environment on a Window 10 Enterprise HP computer with Intel(R) Xeon(R) E-2176M CPU @ 2.71 GHz and a 32 GB installed RAM. Since all the algorithms are evaluated at the point of dataframe creation, data sorting, grouping, filtering and finding unique values, the practitioner will have an idea of the performance of each evaluated data format conversion algorithm with regards to using them for real-time analytics and other time-constrained activities.

The practitioner will also have some idea with regards to the selection of appropriate data formats when speed of data analyses and data format conversion are of importance.

Matlab graphs and results of the performance of each evaluated algorithm with regards to sorting, filtering, finding unique values, dataframe creation and grouping are shown in Figure 100 to Figure 104. Figure 100 to Figure 104 are graphs of wall and CPU times averages for all evaluations shown in Appendix A-i to A-xi. Minimum and maximum CPU and wall times for all the algorithms evaluated are also reported.

Figure 100. Clockwise from top left: Average CPU & Wall times, Maximum CPU & Wall times, Minimum CPU and Wall times for all evaluated algorithms for dataframe creation.

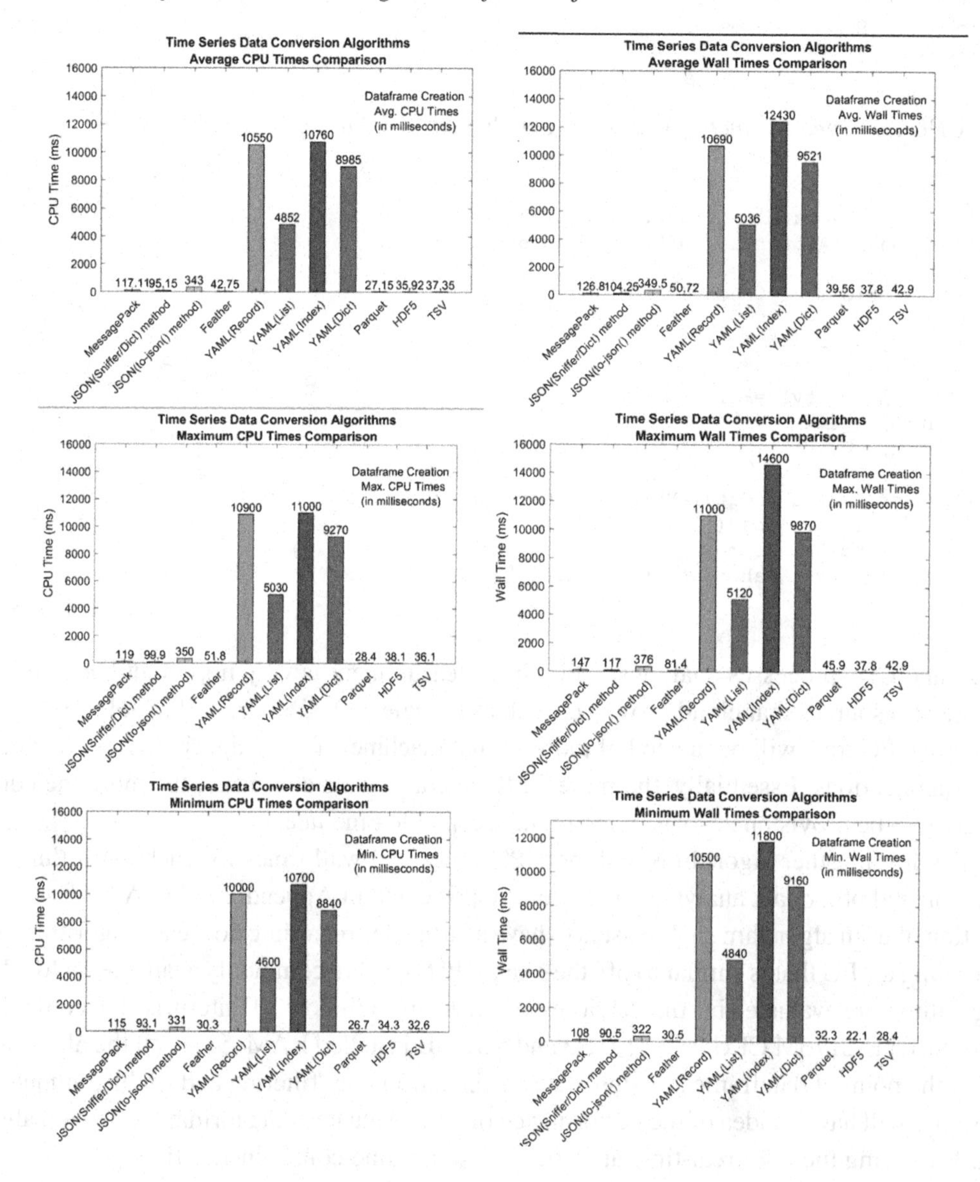

Figure 101. From left to right: Average CPU & Wall times, Maximum CPU & Wall times, Minimum CPU and Wall times for all evaluated algorithms for data analytics: filtering i.e., data querying.

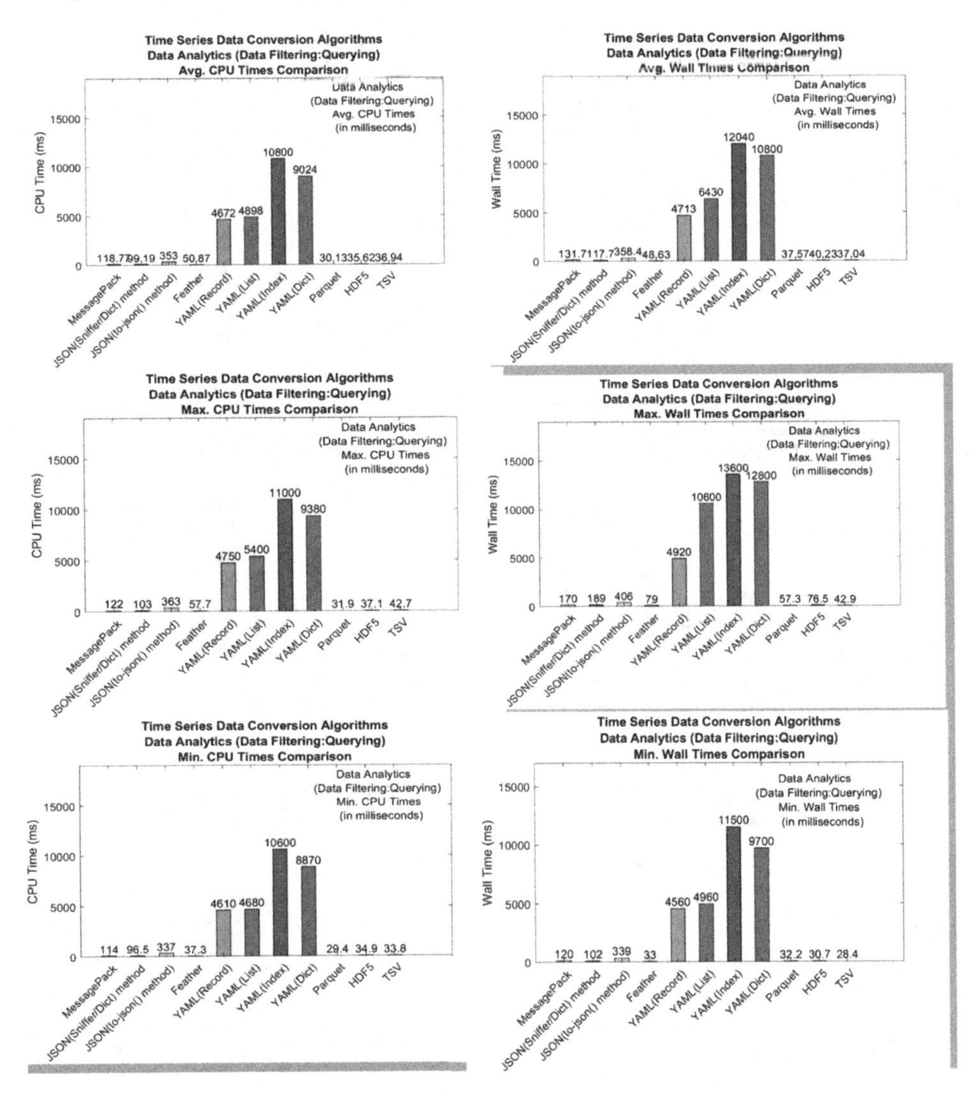

It is important though to point out that Algorithm 1, Algorithm 2 and Algorithm 5c does not support data grouping. Generally, though, when the averages of ten readings per algorithm reported in Appendix A-i to A-xi are considered, it is found from Figure 100 to Figure 100 that the Parquet data format have the lowest speed requirement when dataframe creation, finding unique values, data filtering (querying), data sorting and data grouping are considered.

Results of the analytics shown from Figure 100 to Figure 104 reveals that the Parquet data format conversion algorithm (Algorithm 6) shown in Appendix A-ix have the lowest average CPU times of 27.15 ms, 28.54 ms, 30.13 ms, 30.73 ms, 28.70 ms for dataframe creation, finding unique values, data querying (filtering), data grouping and data sorting respectively. Algorithm 6 also have the lowest average wall times for dataframe creation, finding unique values and data querying. Average wall times recorded for these data analytics activities are 39.56 ms, 28.54 ms and 37.57 ms respectively. Algorithm

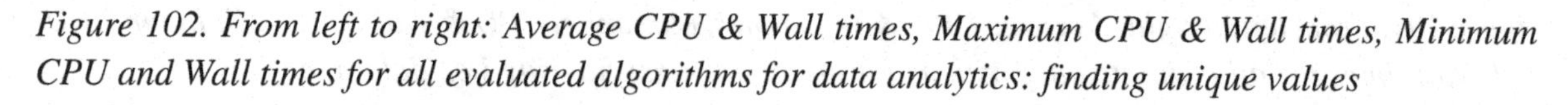

Figure 102. From left to right: Average CPU & Wall times, Maximum CPU & Wall times, Minimum CPU and Wall times for all evaluated algorithms for data analytics: finding unique values

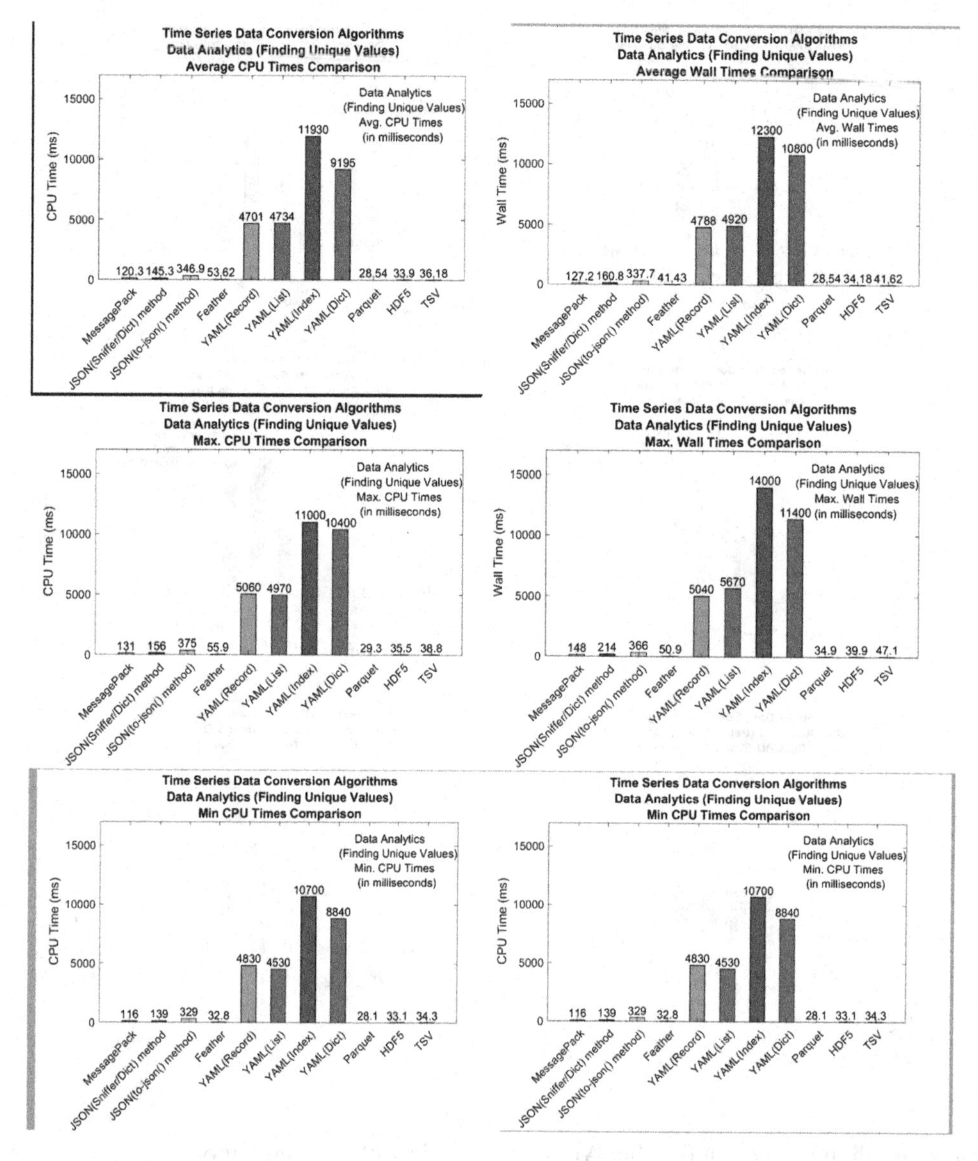

7 (HDF5 data conversion) and Algorithm 8 (TSV data conversion) have the lowest average wall times for data grouping and data sorting respectively.

Across all evaluated algorithms, Algorithm 6 generally have the lowest average CPU and wall times for dataframe creation, finding unique values, data filtering, data grouping and data sorting. The lowest average CPU and wall times for Parquet data format conversion algorithm is followed by HDF5, TSV and Feather data format conversion algorithms respectively.

The YAML data conversion algorithms generally have the highest average CPU and wall times. Algorithm 5c (YAML data conversion using index orientation method) have the highest average wall time (12480 ms). As stated in this chapter and in Part I, some IIoT and CPS applications are best served

Figure 103. From left to right: Average CPU & Wall times, Maximum CPU & Wall times, Minimum CPU and Wall times for all evaluated algorithms for data analytics: grouping

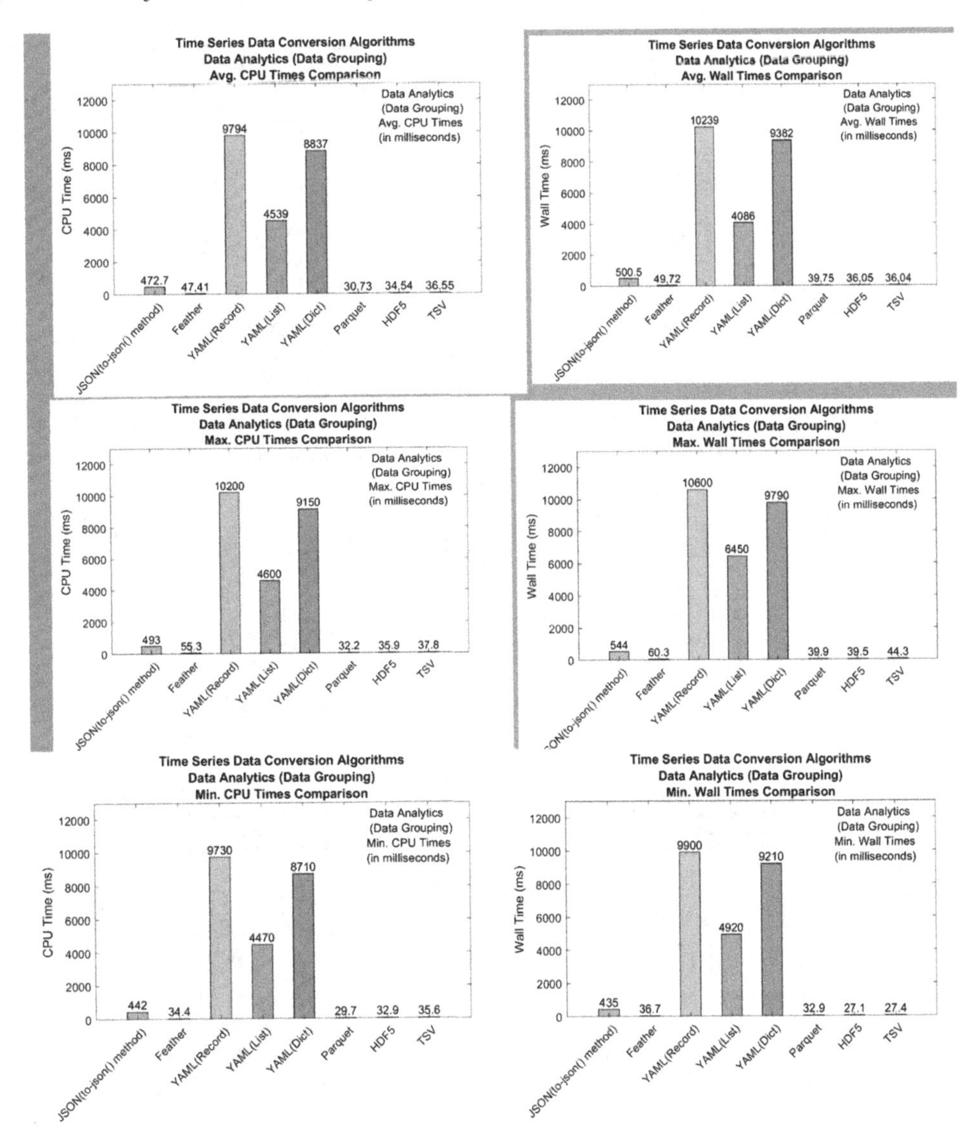

by using YAML data format. If this is the case for any application a user is considering, based on the results graphed in Figure 100 to Figure 104, user can make an informed decision with regards to the selection of YAML data format orientation method. It could be observed from Figure 100 to Figure 104 that the YAML list orientation method (Algorithm 5b) generally has the lowest CPU and wall times for all data analytics activities considered; except for CPU maximum times recorded in Figure 104 where the YAML list orientation method has a slightly higher maximum CPU and wall times than the YAML record orientation method (Algorithm 5c).

Figure 104. From left to right: Average CPU and wall times, maximum CPU and wall times, minimum CPU and wall times for all evaluated algorithms for data analytics: sorting

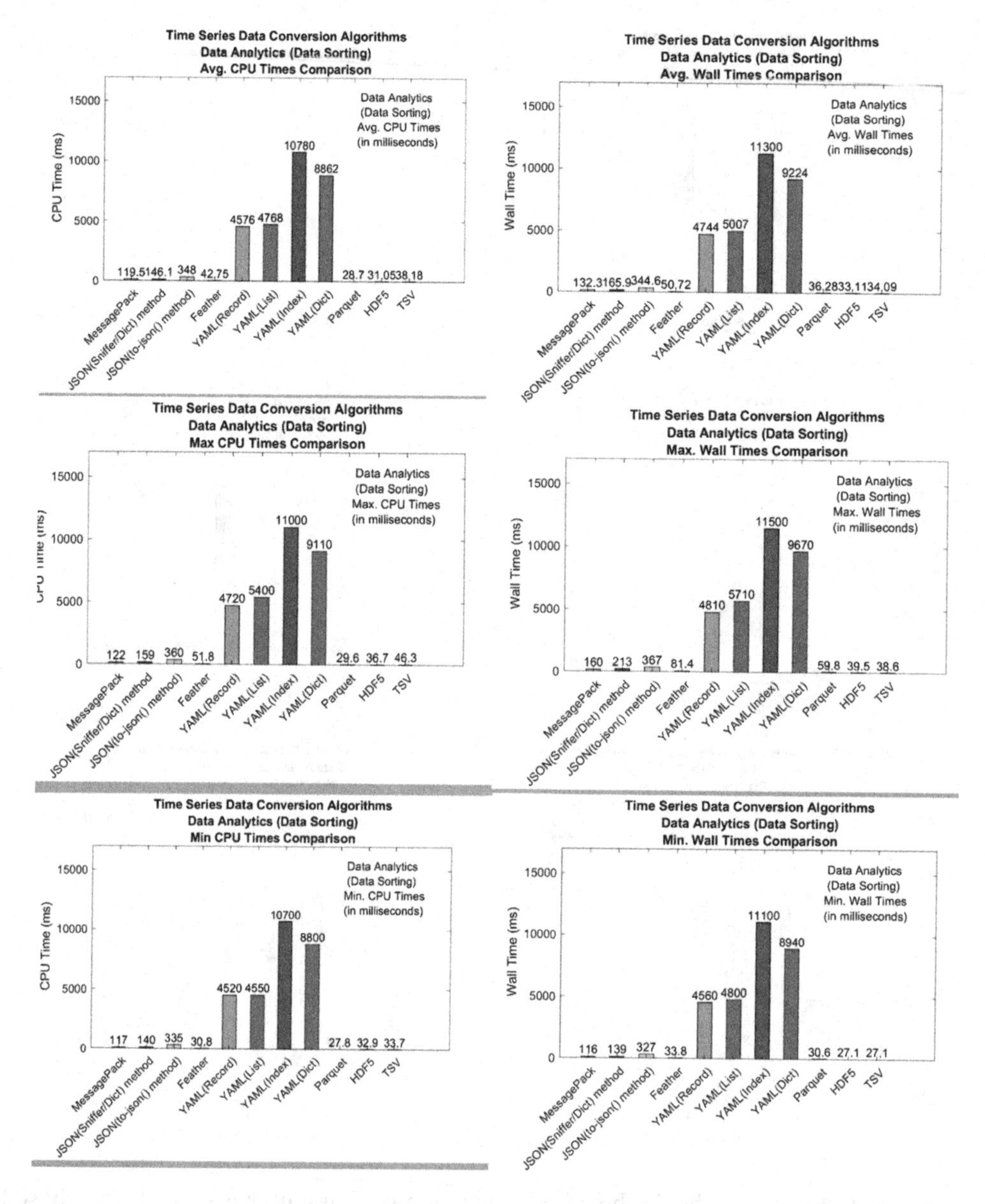

Further, it could be observed from Figure 100 to Figure 104 that the YAML index orientation method (Algorithm 5c) generally has the highest CPU and wall times for all the data analytics activities considered.

If the user decides to use JSON dataframes for any analytics activities, it could be observed form Figure 100 to Figure 104 that dataframes generated from Algorithm 2 consistently have lower CPU and wall times for all analytics activities than the JSON dataframes created with Algorithm 3.

Further, dataframes created with Algorithm 1 (MessagePack data format conversion) have consistently lower CPU and wall averages than dataframes created with JSON and YAML algorithms.

CONCLUSION

In this chapter, a comprehensive discussion regarding the advantages and otherwise of using different types of data formats available including Parquet, CSV, Avro, JSON, HDF5, Tensors, Numpy Arrays, Pickle, YAML, etc. for AI and CPS initiatives are comprehensively discussed. Importance of new methods that make use of data from various machine learning data modalities including sound, images, text and videos and data semantics for ML model training are discussed.

New algorithms that can facilitate robust data conversion among the listed data formats and their Python implementations are provided. Algorithms that facilitate schema integration and metadata generation and their Python implementations are also provided.

From extensively literature searches, and to the best of knowledge of the authors, our algorithms and Python implementations that are freely provided in this chapter constitute the most comprehensive treatise on data format conversion, schema integration and metadata generation for various type of data formats that are available.

It is our most profound believe that Data Engineers, Data Scientists, IoT Engineers, CPS practitioners and DevOps Engineers will derive full benefits from using algorithms and Python implementations that are provided in this chapter.

REFERENCES

Ahmed, Ferzund, Rehman, Usman Ali, Sarwar, & Mehmood. (2017). Modern Data Formats for Big Bioinformatics Data Analytics. *Int'l Journal of Advanced Computer Sc. & Applications, 8*(4).

Apache Arrow. (n.d.). *Feather File Format*. Apache Arrow White Paper. Available: https://arrow.apache.org/docs/python/feather.html

Armbrust, M., Das, T., Sun, L., Yavuz, B., Zhu, S., Murthy, M., Torres, J., van Hovell, H., Ionescu, A., Łuszczak, A., Świtakowski, M., Szafrański, M., Li, X., Ueshin, T., Mokhtar, M., Boncz, P., Ghodsi, A., Paranjpye, S., Senster, P., ... Zaharia, M. (2020). Delta Lake: High Performance ACID Table Storage over Cloud Object Stores. *PVLDB, 13*(12), 3411–3424. doi:10.14778/3415478.3415560

Baheti, R., & Gill, H. (2011). Cyber-Physical Systems. *The Impact of Control Technology, 12*(1), 161–166. Available: http://ieeecss.org/sites/ieeecss/files/2019-07/IoCT-Part3-02 CyberphysicalSystems.pdf

Baltrusaitis, T., Ahuja, C., & Morency, L. (2017). Multimodal Machine Learning: A Survey and Taxonomy. arXiv:1705.09406v2 [cs.LG]

Belov, Tatarintsev, & Nikulchev. (2021). Choosing a Data Storage Format in the Apache Hadoop System. *Symmetry, 13*(195). 13020195 doi:10.3390/sym

Brandi, G., & Matteo, T. (2021). Predicting Multidimensional Data via Tensor Learning. arXiv:2002.04328v3 [stat.ML]

Bynum, A. (2022). *Parsing Improperly Formatted JSON Objects in the Databricks Lakehouse*. White Paper, Databricks. Available: https://www.databricks.com/blog/2022/09/07/parsing-improperly-formatted-json-objects-databricks-lakehouse.html

Chen, J., Adebomi, O. E., Olusayo, O. S., & Kulesza, W. (2010). The Evaluation of the Gaussian Mixture Probability Hypothesis Density approach for Multi-target Tracking. *2010 IEEE International Conference on Imaging Systems and Techniques*, 182-185. 10.1109/IST.2010.5548541

Dowling, J. (2019). *Guide to File Formats for Machine Learning: Columnar, Training, Inferencing, and the Feature Store*. Towards Data Science. Available: https://towardsdatascience.com/guide-to-file-formats-for-machine-learning-columnar-training-inferencing-and-the-feature-store-2e0c3d18d4f9

Duff, I., & Lewis, J. (n.d.). The Rutherford-Boeing Sparse Matrix Collection. White Paper, Rutherford Appleton Laboratory.

Duniam, G., Kitaeff, S., Wicenec, A., German, G., & Shen, A. (2021). Source finding with SoFiA-2 and very large source files. White Paper, University of Western Australia.

Ezeelive Technologies. (n.d.). *JSON – Its Advantages and Disadvantages*. Available: https://ezeelive.com/json-advantages-disadvantages/

Firth, S. (2022). *CSV on the Web: Creating Descriptive Metadata Files*. Available: https://www.stevenfirth.com/csv-on-the-web-creating-descriptive-metadata-files/

FROSTT. (n.d.). *Formidable Repository of Open Sparse Tensor and Tools*. Available: http://frostt.io/tensors/file-formats.html

Gabriel, E. (n.d.). *Big Data Analytics Data Formats – HDF5 and Parquet files*. COSC 6339, University of Houston. Available: http://cs.uh.edu/~gabriel/courses/cosc6339_f18/BDA_16_DataFormats.pdf

GeeksforGeeks. (n.d.). *Numpy Arrays in Python*. Available: https://www.geeksforgeeks.org/numpy-array-in-python/

Hulsebos, M. (2019). Sherlock: A Deep Learning Approach to Semantic Data Type Detection. arXiv:1905.10688v1 [cs.LG

IBM. (n.d.a). *Node Side Network Configuration*. IBM Cloud Pak for Data System, Documentation. Available: https://www.ibm.com/docs/en/cloud-paks/cloudpak-data-system/1.0?topic=configuration-node-side-network

IBM. (n.d.b). *What Are IBM Cloud Paks?* IBM Cloud Pak for Data System, Documentation. Available: https://www.ibm.com/docs/en/cloud-paks

Ji, Y., Wang, Q., Li, X., & Liu, J. (2019). A Survey on Tensor Techniques and Applications in Machine Learning. *IEEE Access : Practical Innovations, Open Solutions*, *7*, 162950–162990. doi:10.1109/ACCESS.2019.2949814

John, S. M., & Oyekanlu, E. (2010). Impact of Packet Losses on the Quality of Video Streaming. *Technical Report*. Blekinge Institute of Technology.

Knowledgebase. (2005). *Difference Between CPU Time and Wall Time*. Available: https://service.futurequest.net/index.php?/Knowledgebase/Article/View/407

Koidan, K. (2019). *Array vs. List in Python – What's the Difference?* LearnPython. Available: https://learnpython.com/blog/python-array-vs-list/

Kong, Q., Siauw, T., & Bayen, A. (2021). *Python Programming and Numerical Methods – A Guide for Engineers and Scientists*. Academy Press.

Kumar, A. (2022). *Tensor Explained with Python Numpy Examples*. Available: https://vitalflux.com/tensor-explained-with-python-numpy-examples/

Levy, E. (2022). *What is the Parquet File Format? Use Cases and Benefits*. Upsolver. Available: https://www.upsolver.com/blog/apache-parquet-why-use

Library of Congress. (n.d.). *TSV, Tab Separated Values*. Available: https://www.loc.gov/preservation/digital/formats/fdd/fdd000533.shtml

Litvin, Y. (2019). *Petastorm: A Light-Weight Approach to Building ML Pipelines @Uber*. Uber White Paper. Available: https://qcon.ai/system/files/presentation-slides/yevgeni_-_petastorm_16th_apr_2019_.pdf

Megagon Labs. (n.d.). *Learning to Detect Semantic Types from Large Table Corpora*. White Paper. Available: https://megagonlabs.medium.com/learning-to-detect-semantic-types-from-large-table-corpora-fe22fcd97060

MessagePack. (n.d.). *It's Like JSON but Fast and Small*. MessagePack White Paper. Available: https://msgpack.org/index.html

Morency, L., & Baltrusaitis, T. (n.d.). *Tutorial on Multimodal Machine Learning*. White Paper, Language Technologies Institute, Carnegie Mellon University. Available: https://www.cs.cmu.edu/~morency/MMML-Tutorial-ACL2017.pdf

Nagowah, S. D., Ben Sta, H., & Gobin-Rahimbux, B. A. (2018). An Overview of Semantic Interoperability Ontologies and Frameworks for IoT. *2018 Sixth International Conference on Enterprise Systems (ES)*, 82-89. 10.1109/ES.2018.00020

NIST. (n.d.). *Framework for Cyber-Physical Systems: Volume 1, Overview*. NIST Special Publication 1500-201. Available: https://nvlpubs.nist.gov/nistpubs/SpecialPublications/NIST.SP.1500-201.pdf

Oliveiros, S. (2016). *How to Choose a Data Format*. KDnuggets White Paper. Available: https://www.kdnuggets.com/2016/11/how-to-choose-data-format.html

Oyekanlu, E. (2017). Predictive Edge Computing for Time-series of Industrial IoT and Large-scale Critical Infrastructure Based on Open-source Software Analytic of Big Data. *2017 IEEE International Conference on Big Data (Big Data)*, 1663-1669. 10.1109/BigData.2017.8258103

Oyekanlu, E. (2018a). Distributed Osmotic Computing Approach to Implementation of Explainable Predictive Deep Learning at Industrial IoT Network Edges with Real-Time Adaptive Wavelet Graphs. *2018 IEEE First International Conference on Artificial Intelligence and Knowledge Engineering (AIKE)*, 179-188. 10.1109/AIKE.2018.00042

Oyekanlu, E. (2018b). Osmotic Collaborative Computing for Machine Learning and Cybersecurity Applications in Industrial IoT Networks and Cyber Physical Systems with Gaussian Mixture Models. *2018 IEEE 4th International Conference on Collaboration and Internet Computing (CIC)*, 326-335. 10.1109/CIC.2018.00051

Oyekanlu, E. (2018c). *Powerline Communication for the Smart Grid and Internet of Things - Powerline Narrowband Frequency Channel Characterization Based on the TMS320C2000 C28x Digital Signal Processor*. Drexel University ProQuest Dissertations Publishing.

Oyekanlu, E. (2018d). Fault-Tolerant Real-Time Collaborative Network Edge Analytics for Industrial IoT and Cyber Physical Systems with Communication Network Diversity. *2018 IEEE 4th International Conference on Collaboration and Internet Computing (CIC)*, 336-345. 10.1109/CIC.2018.00052

Oyekanlu, E., Mulroy, G., & Kuhn, D. (2022). Data Engineering for Factory of the Future: from Factory Floor to the Cloud – Part I: Performance Evaluation of State-of-the-Art Data Formats for Time Series Applications. In *Applied AI and Multimedia Technologies for Smart Manufacturing and CPS Applications. IGI Global.*

Oyekanlu, E., Nelatury, C., Fatade, A. O., Alaba, O., & Abass, O. (2017). Edge computing for industrial IoT and the smart grid: Channel capacity for M2M communication over the power line. *2017 IEEE 3rd International Conference on Electro-Technology for National Development (NIGERCON)*, 1-11, 10.1109/NIGERCON.2017.8281938

Oyekanlu, E., Onidare, S., & Oladele, P. (2018). Towards statistical machine learning for edge analytics in large scale networks: Real-time Gaussian function generation with generic DSP. *2018 First International Colloquium on Smart Grid Metrology (SmaGriMet)*, 1-6. 10.23919/SMAGRIMET.2018.8369850

Oyekanlu, E., & Scoles, K. (2018a). Real-Time Distributed Computing at Network Edges for Large Scale Industrial IoT Networks. *2018 IEEE World Congress on Services (SERVICES)*, 63-64. 10.1109/SERVICES.2018.00045

Oyekanlu, E., & Scoles, K. (2018b). Towards Low-Cost, Real-Time, Distributed Signal and Data Processing for Artificial Intelligence Applications at Edges of Large Industrial and Internet Networks. *2018 IEEE First International Conference on Artificial Intelligence and Knowledge Engineering (AIKE)*, 166-167, 10.1109/AIKE.2018.00037

Oyekanlu, E., Scoles, K., & Oladele, P. O. (2018). Advanced Signal Processing for Communication Networks and Industrial IoT Machines Using Low-Cost Fixed-Point Digital Signal Processor. *2018 10th International Conference on Advanced Infocomm Technology (ICAIT)*, 93-101. 10.1109/ICAIT.2018.8686577

Oyekanlu, E. A., Smith, A. C., Thomas, W. P., Mulroy, G., Hitesh, D., Ramsey, M., Kuhn, D. J., Mcghinnis, J. D., Buonavita, S. C., Looper, N. A., Ng, M., Ng'oma, A., Liu, W., Mcbride, P. G., Shultz, M. G., Cerasi, C., & Sun, D. (2020). A Review of Recent Advances in Automated Guided Vehicle Technologies: Integration Challenges and Research Areas for 5G-Based Smart Manufacturing Applications. *IEEE Access : Practical Innovations, Open Solutions*, *8*, 202312–202353. doi:10.1109/ACCESS.2020.3035729

Pathak, O. (2018). *Understanding Python Pickling with Example*. GeeksforGeeks Publication. Available: https://www.geeksforgeeks.org/understanding-python-pickling-example/

Peternel, G. (2021). *The Fundamentals of Data Warehouse + Data Lake = Lake House*. Towards Data Science. Available: https://towardsdatascience.com/the-fundamentals-of-data-ware house-data-lake-lake-house-ff640851c832

Pivarski, J., Osborne, I., Das, P., Biswas, A., & Elmer, P. (2020). Awkward Array: JSON-like Data, Numpy-like Idioms. *Proc. of the 19th Python in Science Conference*. 10.25080/Majora-342d178e-00b

Pool Party. (n.d.). *Introducing Semantic AI Ingredients for a sustainable Enterprise AI Strategy*. White Paper. Available: http://www.baonenterprises.com/uploads/1/0/1/6/101618342/sem antic-ai-white-paper_en6.7.18.pdf

Progress. (2017). *Benefits of JSON*. Available: https:// docs.progress.com/en-US/bundle/openedge-abl-use-json -117/page/Benefits-of-JSON.html

Pykes, K. (2021). *Tensors and Arrays – What's the Difference?* Towards Data Science. Available: https://towardsdatascience.com/tensors-and-arrays-2611d48676 d5

Python. (n.d.). *CSV File Reading and Writing*. White Paper. Available: https://docs.python.org/3/library/ csv.html

PyYAML 6.0. (n.d.). *Pip Install YAML – Project Description*. Available: https://pypi.org/project/PyYAML/

Rabanser, S., Shchur, O., & Gunnemann, S. (2017). Introduction to Tensor Decomposition and Their Applications in Machine Learning. arXiv:1711.10781v1 [stat.ML]

Ramm, J. (2021). *Feather Documentation*. White Paper, Release 0.1.0. Available: https://buildmedia.readthedocs.org/media/pdf/plume/stable/pl ume.pdf

Rashidibajgan, S., Hupperich, T., Doss, R., & Pan, L. (2020). Opportunistic Tracking in Cyber-Physical Systems. *2020 IEEE 19th International Conference on Trust, Security and Privacy in Computing and Communications (TrustCom)*, 1672-1679. 10.1109/TrustCom50675.2020.00230

Retter, A., Underdown, D., & Walpole, R. (n.d.). *CSV Schema Language 1.2 - A Language for Defining and Validating CSV Data*. W3C White Paper. Available: https://digital-preservation.github.io/csv-schema/csv-schema -1.2.html

Sharma, S. (2021). *Ansible for Beginners – Overview, Architecture and Use Cases*. White Paper, K21 Academy. Available: https://k21academy.com/ansible/ansible-for-beginners/

Stanford University. (n.d.). *Parsing JSON with jq*. Available: http://www.compciv.org/recipes/cli/jq-for-parsing-json/

Sztipanovits. (2013). Strategic R&D Opportunities for 21st Century Cyber-Physical Systems – Connecting Computer and Information Systems with the Physical World. NIST White Paper. Available: https://www.nist.gov/system/files/documents/el/12-Cyber-Physical-Systems020113_final.pdf

TACCO. (n.d.). *Defining Tensors*. TACCO White Paper. Available: http://tensor-compiler.org/docs/tensors

TensorFlow. (n.d.). *TFRecords and tf.train.Example*. Available: https://www.tensorflow.org/tutorials/load_data/tfrecord

Tokcan, N., Gryak, J., Najarian, K., & Deriksen, H. (2020). Algebraic Methods for Tensor Data. arXiv:2005.12988v1 [math.RT]

Turner-Trauring, I. (2021). *Where's Your Bottleneck? CPU Time vs Wallclock Time*. Available: https://pythonspeed.com/articles/blocking-cpu-or-io/

Tutorialspoint. (n.d.). *JSON Quick Guide*. Available: https://www.tutorialspoint.com/json/pdf/json_quick_guide.pdf

Tyson, M. (n.d.). *What is JSON? The Universal Data Format*. White Paper, Infoworld. Available: https://www.infoworld.com/article/3222851/what-is-json-a-better-format-for-data-exchange.html

Uber. (n.d.). *Introducing Petastorm: Uber ATG's Data Access Library for Deep Learning*. Uber White Paper. Available: https://www.uber.com/blog/petastorm/

W3C. (2012). *SPARQL 1.1 Query Results CSV and TSV Formats*. White Paper. Available: https://www.w3.org/2009/sparql/docs/csv-tsv-results/results-csv-tsv.html

Ye, X., Song, W. S., Hong, S. H., Kim, Y. C., & Yoo, N. H. (2022). Toward Data Interoperability of Enterprise and Control Applications via the Industry 4.0 Asset Administration Shell. *IEEE Access : Practical Innovations, Open Solutions*, *10*, 35795–35803. doi:10.1109/ACCESS.2022.3163738

Zaczyński. (n.d.). *YAML: The Missing Battery in Python*. Available: https://realpython.com/python-yaml/

Zhang. (n.d.). Sato: Contextual Semantic Type Detection in Tables. *Proceedings of the VLDB Endowment, 13*(11). doi:10.14778/3407790.3407793

APPENDIX I

Results

Algorithm 1: Convert CSV Time series into MessagePack DataFrame

Dataframe Creation

```
%%time
CSVFileInput = r'C:\Users\........................\Yahoodata.csv'
JSONFileOutput = r'C:\Users\..................\Yahoodata2.json'
def FileInput_to_json(CSVFileInput, JSONFileOutput):
    fileToStoreConvertedArray = []
    with open(CSVFileInput,  newline='') as inputfile:
        dialect = csv.Sniffer().sniff(inputfile.read(1024))
        inputfile.seek(0)
        csvReader = csv.DictReader(inputfile)
        #convert each csv row into python dict
        for row in csvReader:
            fileToStoreConvertedArray.append(row)
    with open(JSONFileOutput, 'w', newline='') as jsonfile:
        jsonString = json.dumps(fileToStoreConvertedArray, indent=4)
        jsonfile.write(jsonString)
start = time.perf_counter()
FileInput_to_json(CSVFileInput ,JSONFileOutput)
finish = time.perf_counter()
with open("Yahoodata2.json", "r") as read_file:
    datajson = json.load(read_file)
MsgPackdata = datajson
# Write msgpack file
with open('MsgPackdata.msgpack', 'wb') as inputMsgPackdata:
    msgpack.pack(MsgPackdata, inputMsgPackdata)
# Read MsgPack in from storage location
with open('MsgPackdata.msgpack', "rb") as MsgPackdata_file:
    # MsgPackdata_loaded = json.load(MsgPackdata_file)
    MsgPackdata_loaded = msgpack.unpack(MsgPackdata_file)
newloadedMsgPack = {"convertMSGPckToDF": MsgPackdata_loaded}
MsgPackDF = pd.DataFrame(newloadedMsgPack['convertMSGPckToDF'])
MsgPackDF
```

Wall Times (ms)

```
Wall time: 108 ms
Wall time: 120 ms
```

```
Wall time: 140 ms
Wall time: 131 ms
Wall time: 119 ms
Wall time: 133 ms
Wall time: 147 ms
Wall time: 129 ms
Wall time: 120 ms
Wall time: 121 ms
```

CPU Times (ms)

```
117 ms ± 5.34 ms per loop (mean ± std. dev. of 7 runs, 10 loops each)
115 ms ± 2.45 ms per loop (mean ± std. dev. of 7 runs, 10 loops each)
119 ms ± 3.07 ms per loop (mean ± std. dev. of 7 runs, 10 loops each)
119 ms ± 3.43 ms per loop (mean ± std. dev. of 7 runs, 10 loops each)
116 ms ± 1.63 ms per loop (mean ± std. dev. of 7 runs, 10 loops each)
118 ms ± 3.3 ms per loop (mean ± std. dev. of 7 runs, 10 loops each)
116 ms ± 2.15 ms per loop (mean ± std. dev. of 7 runs, 10 loops each)
115 ms ± 1.59 ms per loop (mean ± std. dev. of 7 runs, 10 loops each)
119 ms ± 1.94 ms per loop (mean ± std. dev. of 7 runs, 10 loops each)
119 ms ± 3.13 ms per loop (mean ± std. dev. of 7 runs, 10 loops each)
```

Filtering (Data Querying)

Wall Times (ms)

```
Wall time: 120 ms
Wall time: 135 ms
Wall time: 123 ms
Wall time: 126 ms
Wall time: 129 ms
Wall time: 123 ms
Wall time: 170 ms
Wall time: 126 ms
Wall time: 136 ms
Wall time: 129 ms
```

CPU Times (ms)

```
114 ms ± 3.64 ms per loop (mean ± std. dev. of 7 runs, 10 loops each)
118 ms ± 2.09 ms per loop (mean ± std. dev. of 7 runs, 10 loops each)
120 ms ± 3.87 ms per loop (mean ± std. dev. of 7 runs, 10 loops each)
121 ms ± 7.18 ms per loop (mean ± std. dev. of 7 runs, 10 loops each)
118 ms ± 4.36 ms per loop (mean ± std. dev. of 7 runs, 10 loops each)
117 ms ± 2.86 ms per loop (mean ± std. dev. of 7 runs, 10 loops each)
119 ms ± 3.86 ms per loop (mean ± std. dev. of 7 runs, 10 loops each)
```

```
122 ms ± 4.5 ms per loop (mean ± std. dev. of 7 runs, 10 loops each)
120 ms ± 3.43 ms per loop (mean ± std. dev. of 7 runs, 10 loops each)
117 ms ± 2.16 ms per loop (mean ± std. dev. of 7 runs, 10 loops each)

# CPU time of filtering MsgPack
CSVFileInput = r'C:\Users\..................\Yahoodata.csv'
JSONFileOutput = r'C:\Users\..............\Yahoodata2.json'
def FileInput_to_json(CSVFileInput, JSONFileOutput):
    fileToStoreConvertedArray = []
    with open(CSVFileInput,  newline='') as inputfile:
        dialect = csv.Sniffer().sniff(inputfile.read(1024))
        inputfile.seek(0)
        csvReader = csv.DictReader(inputfile)
        #convert each csv row into python dict
        for row in csvReader:
            fileToStoreConvertedArray.append(row)
    with open(JSONFileOutput, 'w', newline='') as jsonfile:
        jsonString = json.dumps(fileToStoreConvertedArray, indent=4)
        jsonfile.write(jsonString)
start = time.perf_counter()
FileInput_to_json(CSVFileInput ,JSONFileOutput)
finish = time.perf_counter()
with open("Yahoodata2.json", "r") as read_file:
    datajson = json.load(read_file)
MsgPackdata = datajson
# Write msgpack file
with open('MsgPackdata.msgpack', 'wb') as inputMsgPackdata:
    msgpack.pack(MsgPackdata, inputMsgPackdata)
# Read MsgPack in from storage location
with open('MsgPackdata.msgpack', "rb") as MsgPackdata_file:
    # MsgPackdata_loaded = json.load(MsgPackdata_file)
    MsgPackdata_loaded = msgpack.unpack(MsgPackdata_file)
newloadedMsgPack = {"convertMSGPckToDF": MsgPackdata_loaded}
MsgPackDF = pd.DataFrame(newloadedMsgPack['convertMSGPckToDF'])
MsgPackDF.query('Open == "17.5"')
```

	Date	Open	High	Low	Close	Adj Close	Volume
0	4/30/1997	17.5	17.5	17	17	15.192017	5800
1	5/1/1997	17.5	17.75	17.375	17.75	15.862257	36300
3	5/5/1997	17.5	18.5	17.5	18.5	16.532488	65100
105	9/29/1997	17.5	18.25	17.5	17.5	15.638843	4300
106	9/30/1997	17.5	17.75	17.5	17.5	15.638843	21000
108	10/2/1997	17.5	17.5	17.25	17.4375	15.582988	6800
115	10/13/1997	17.5	17.625	17.25	17.4375	15.582988	9200
135	11/10/1997	17.5	17.5	17	17	15.192017	3100
168	12/29/1997	17.5	17.625	17	17	15.192017	3300
498	4/22/1999	17.5	19	17.5	19	16.979313	160800
529	6/7/1999	17.5	17.625	17.375	17.625	15.750552	53200
565	7/28/1999	17.5	17.5	17.25	17.375	15.527136	27400
4039	5/20/2013	17.5	17.74	17.4	17.469999	16.068762	78200
4051	6/6/2013	17.5	18.24	17.5	18.23	16.767805	101500

Finding Unique Values

....

```
newloadedMsgPack = {"convertMSGPckToDF": MsgPackdata_loaded}
MsgPackDF = pd.DataFrame(newloadedMsgPack['convertMSGPckToDF'])
pd.unique(MsgPackDF[['Open', 'Close']].values.ravel('K'))
array(['17.5', '17.75', '18.625', ..., '48.48', '44.869999', '42.889999'],
      dtype=object)
```

Wall Times (ms): 107 ms, 128, 131, 136, 121, 133, 121, 148, 128, 119

CPU Times (ms)

```
131 ms ± 6.54 ms per loop (mean ± std. dev. of 7 runs, 10 loops each)
122 ms ± 9.75 ms per loop (mean ± std. dev. of 7 runs, 10 loops each)
122 ms ± 5.07 ms per loop (mean ± std. dev. of 7 runs, 10 loops each)
120 ms ± 2.58 ms per loop (mean ± std. dev. of 7 runs, 10 loops each)
122 ms ± 2.93 ms per loop (mean ± std. dev. of 7 runs, 10 loops each)
117 ms ± 2.88 ms per loop (mean ± std. dev. of 7 runs, 10 loops each)
119 ms ± 1.86 ms per loop (mean ± std. dev. of 7 runs, 10 loops each)
119 ms ± 4.23 ms per loop (mean ± std. dev. of 7 runs, 10 loops each)
115 ms ± 3.41 ms per loop (mean ± std. dev. of 7 runs, 10 loops each)
```

```
116 ms ± 3.77 ms per loop (mean ± std. dev. of 7 runs, 10 loops each)
```

Data Sorting

....

```
newloadedMsgPack = {"convertMSGPckToDF": MsgPackdata_loaded}
MsgPackDF = pd.DataFrame(newloadedMsgPack['convertMSGPckToDF'])
MsgPackDF.sort_values(by=['Volume'])
```

```
Wall time: 140 ms
```

	Date	Open	High	Low	Close	Adj Close	Volume
1275	5/28/2002	4.39	4.39	4.39	4.39	3.923114	0
3216	2/10/2010	3.79	3.79	3.79	3.79	3.386926	0
166	12/24/1997	17.625	17.625	17.625	17.625	15.750552	100
81	8/25/1997	22	22	22	22	19.660265	1000
949	2/2/2001	2.19	2.22	2.14	2.22	1.983899	1000
...	...	...	...	...	...	...	...
5250	3/12/2018	28.290001	28.42	28.120001	28.290001	26.830349	99600
1730	3/17/2004	4.74	4.9	4.74	4.89	4.369939	99800
4495	3/12/2015	24.110001	24.75	24.01	24.1	22.375002	99800
4509	4/1/2015	23.85	23.950001	23.35	23.549999	21.864365	99900
5845	7/22/2020	18.43	18.82	18.32	18.75	18.372919	99900

6293 rows × 7 columns

Wall Times (ms): 140 ms, 148, 121, 122, 160, 116, 126, 128, 141, 121

CPU Times (ms)

```
122 ms ± 4.23 ms per loop (mean ± std. dev. of 7 runs, 10 loops each)
118 ms ± 3.28 ms per loop (mean ± std. dev. of 7 runs, 10 loops each)
119 ms ± 2.89 ms per loop (mean ± std. dev. of 7 runs, 10 loops each)
117 ms ± 2.9 ms per loop (mean ± std. dev. of 7 runs, 10 loops each)
120 ms ± 2.93 ms per loop (mean ± std. dev. of 7 runs, 10 loops each)
120 ms ± 3.3 ms per loop (mean ± std. dev. of 7 runs, 10 loops each)
121 ms ± 3.23 ms per loop (mean ± std. dev. of 7 runs, 10 loops each)
119 ms ± 2.65 ms per loop (mean ± std. dev. of 7 runs, 10 loops each)
122 ms ± 1.84 ms per loop (mean ± std. dev. of 7 runs, 10 loops each)
117 ms ± 2.41 ms per loop (mean ± std. dev. of 7 runs, 10 loops each)
```

APPENDIX II

Results

Algorithm 2: CSV Timeseries Conversion to JSON Timeseries Format – DictReader() & Sniffer() Methods

Dataframe Creation

```
# Wall Time
CSVFileInput = r'C:\Users\.........\Yahoodata.csv'
JSONFileOutput = r'C:\Users\.......\Yahoodata2.json'
def FileInput_to_json(CSVFileInput, JSONFileOutput):
    fileToStoreConvertedArray = []
    with open(CSVFileInput,  newline='') as inputfile:
        dialect = csv.Sniffer().sniff(inputfile.read(1024))
        inputfile.seek(0)
        csvReader = csv.DictReader(inputfile)
        #convert each csv row into python dict
        for row in csvReader:
            fileToStoreConvertedArray.append(row)
      with open(JSONFileOutput, 'w', newline='') as jsonfile:
        jsonString = json.dumps(fileToStoreConvertedArray, indent=4)
        jsonfile.write(jsonString)
start = time.perf_counter()
FileInput_to_json(CSVFileInput ,JSONFileOutput)
finish = time.perf_counter()
print(f"Conversion of all TimeSeries file rows completed successfully in {fin-
ish - start:0.4f} seconds")
# Read in JSON data
with open("Yahoodata2.json", "r") as read_file:
    datajson = json.load(read_file)
datajsonPandasDataFrame = pd.DataFrame(datajson)
```

	Date	Open	High	Low	Close	Adj Close	Volume
0	4/30/1997	17.5	17.5	17	17	15.192017	5800
1	5/1/1997	17.5	17.75	17.375	17.75	15.862257	36300
2	5/2/1997	17.75	17.875	17.5	17.875	15.97396	17700
3	5/5/1997	17.5	18.5	17.5	18.5	16.532488	65100
4	5/6/1997	18.625	19.5	18.25	19.5	17.426138	227000
...	...	...	...	...	...	...	...
6288	4/25/2022	49.049999	49.790001	48.349998	49.73	49.73	120400
6289	4/26/2022	49.439999	50.419998	48.310001	48.48	48.48	80400
6290	4/27/2022	48.529999	49.110001	48.32	48.619999	48.619999	114400
6291	4/28/2022	49.630001	49.630001	44.360001	44.869999	44.869999	283800
6292	4/29/2022	44.32	44.599998	42.369999	42.889999	42.889999	193100

6293 rows × 7 columns

Wall Times (ms): 106 ms, 117, 105, 104, 104, 90.5, 101, 103, 109, 103

```
CPU Times (ms)
94.5 ms ± 1.31 ms per loop (mean ± std. dev. of 7 runs, 10 loops each)
93.1 ms ± 621 µs per loop (mean ± std. dev. of 7 runs, 10 loops each)
93.6 ms ± 1.4 ms per loop (mean ± std. dev. of 7 runs, 10 loops each)
99.9 ms ± 4.14 ms per loop (mean ± std. dev. of 7 runs, 10 loops each)
96.1 ms ± 2.61 ms per loop (mean ± std. dev. of 7 runs, 10 loops each
94.3 ms ± 1.04 ms per loop (mean ± std. dev. of 7 runs, 10 loops each)
95.1 ms ± 2.12 ms per loop (mean ± std. dev. of 7 runs, 10 loops each)
94.7 ms ± 1.89 ms per loop (mean ± std. dev. of 7 runs, 10 loops each)
95.7 ms ± 1.32 ms per loop (mean ± std. dev. of 7 runs, 10 loops each)
94.5 ms ± 2.55 ms per loop (mean ± std. dev. of 7 runs, 10 loops each)
```

Data Filtering (Querying)

....

```
with open("Yahoodata2.json", "r") as read_file:
    datajson = json.load(read_file)
datajsonPandasDataFrame = pd.DataFrame(datajson)
datajsonPandasDataFrame.query('Open == "17.5"')
```

	Date	Open	High	Low	Close	Adj Close	Volume
0	4/30/1997	17.5	17.5	17	17	15.192017	5800
1	5/1/1997	17.5	17.75	17.375	17.75	15.862257	36300
3	5/5/1997	17.5	18.5	17.5	18.5	16.532488	65100
105	9/29/1997	17.5	18.25	17.5	17.5	15.638843	4300
106	9/30/1997	17.5	17.75	17.5	17.5	15.638843	21000
108	10/2/1997	17.5	17.5	17.25	17.4375	15.582988	6800
115	10/13/1997	17.5	17.625	17.25	17.4375	15.582988	9200
135	11/10/1997	17.5	17.5	17	17	15.192017	3100
168	12/29/1997	17.5	17.625	17	17	15.192017	3300
498	4/22/1999	17.5	19	17.5	19	16.979313	160800
529	6/7/1999	17.5	17.625	17.375	17.625	15.750552	53200
565	7/28/1999	17.5	17.5	17.25	17.375	15.527136	27400
4039	5/20/2013	17.5	17.74	17.4	17.469999	16.068762	78200
4051	6/6/2013	17.5	18.24	17.5	18.23	16.767805	101500

Wall Times (ms): 107 ms, 189, 107, 111, 110, 105, 103, 114, 114, 117, 102

CPU Times (ms)

```
96.5 ms ± 1.47 ms per loop (mean ± std. dev. of 7 runs, 10 loops each)
98.4 ms ± 1.79 ms per loop (mean ± std. dev. of 7 runs, 10 loops each)
102 ms ± 3.7 ms per loop (mean ± std. dev. of 7 runs, 10 loops each)
101 ms ± 3.43 ms per loop (mean ± std. dev. of 7 runs, 10 loops each)
103 ms ± 4.15 ms per loop (mean ± std. dev. of 7 runs, 10 loops each)
98.5 ms ± 2.27 ms per loop (mean ± std. dev. of 7 runs, 10 loops each)
97.9 ms ± 1.6 ms per loop (mean ± std. dev. of 7 runs, 10 loops each)
98.3 ms ± 2.71 ms per loop (mean ± std. dev. of 7 runs, 10 loops each)
97.6 ms ± 1.57 ms per loop (mean ± std. dev. of 7 runs, 10 loops each)
98.7 ms ± 1.7 ms per loop (mean ± std. dev. of 7 runs, 10 loops each)
```

Finding Unique Values

```
datajsonPandasDataFrame = pd.DataFrame(datajson)
datajsonPandasDataFrame.query('Open == "17.5"')
pd.unique(datajsonPandasDataFrame [['Open', 'Close']].values.ravel('K'))
array(['17.5', '17.75', '18.625', ..., '48.48', '44.869999', '42.889999'],
```

```
    dtype=object)
```

Wall Times (ms): 127 ms, 177, 163, 214, 167, 140, 146, 160, 151, 163

CPU Times (ms)

```
148 ms ± 12.8 ms per loop (mean ± std. dev. of 7 runs, 10 loops each)
156 ms ± 10.2 ms per loop (mean ± std. dev. of 7 runs, 10 loops each)
153 ms ± 9.27 ms per loop (mean ± std. dev. of 7 runs, 10 loops each)
147 ms ± 5.74 ms per loop (mean ± std. dev. of 7 runs, 10 loops each)
139 ms ± 8.2 ms per loop (mean ± std. dev. of 7 runs, 10 loops each)
139 ms ± 9.19 ms per loop (mean ± std. dev. of 7 runs, 10 loops each)
142 ms ± 8.99 ms per loop (mean ± std. dev. of 7 runs, 10 loops each)
146 ms ± 6.78 ms per loop (mean ± std. dev. of 7 runs, 10 loops each)
139 ms ± 8.09 ms per loop (mean ± std. dev. of 7 runs, 10 loops each)
144 ms ± 7 ms per loop (mean ± std. dev. of 7 runs, 10 loops each)
```

Data Sorting

Wall Time (ms): 175 ms, 148, 166, 168, 164, 213, 160, 140, 186, 139

CPU Times (ms)

```
159 ms ± 5.95 ms per loop (mean ± std. dev. of 7 runs, 10 loops each)
143 ms ± 9.06 ms per loop (mean ± std. dev. of 7 runs, 10 loops each)
146 ms ± 12.6 ms per loop (mean ± std. dev. of 7 runs, 10 loops each)
150 ms ± 9.07 ms per loop (mean ± std. dev. of 7 runs, 10 loops each)
154 ms ± 6.12 ms per loop (mean ± std. dev. of 7 runs, 10 loops each)
140 ms ± 5.99 ms per loop (mean ± std. dev. of 7 runs, 10 loops each)
143 ms ± 5.86 ms per loop (mean ± std. dev. of 7 runs, 10 loops each)
141 ms ± 6.15 ms per loop (mean ± std. dev. of 7 runs, 10 loops each)
145 ms ± 11.5 ms per loop (mean ± std. dev. of 7 runs, 10 loops each)
140 ms ± 9.94 ms per loop (mean ± std. dev. of 7 runs, 10 loops each)
```

APPENDIX III

Results

Algorithm 3: CSV Timeseries to JSON Timeseries Format Conversion – to_json() Method

Dataframe Creation

```
YahooPricedata = pd.read_csv('/Users/............./YahooData.csv')
YahooPriceDataFrame = pd.DataFrame(YahooPricedata)
YahooPriceDataFrame.to_json('YahooPriceDataFrameToJSON.json')
YahoodataJSONDataFrame = pd.read_json('/Users/............................/YahooPriceDataFrame-
ToJSON.json')
YahoodataJSONDataFrame
```

Wall time (ms): 343 ms, 371, 353, 354, 335, 338, 376, 352, 341, 332

CPU Times (ms)

```
341 ms ± 14.5 ms per loop (mean ± std. dev. of 7 runs, 1 loop each)
347 ms ± 14.3 ms per loop (mean ± std. dev. of 7 runs, 1 loop each)
348 ms ± 25.6 ms per loop (mean ± std. dev. of 7 runs, 1 loop each)
350 ms ± 15.7 ms per loop (mean ± std. dev. of 7 runs, 1 loop each)
344 ms ± 12.7 ms per loop (mean ± std. dev. of 7 runs, 1 loop each)
346 ms ± 14.4 ms per loop (mean ± std. dev. of 7 runs, 1 loop each)
342 ms ± 20.6 ms per loop (mean ± std. dev. of 7 runs, 1 loop each)
342 ms ± 9.51 ms per loop (mean ± std. dev. of 7 runs, 1 loop each)
331 ms ± 7.39 ms per loop (mean ± std. dev. of 7 runs, 1 loop each)
339 ms ± 15.2 ms per loop (mean ± std. dev. of 7 runs, 1 loop each)
```

Filtering (Data Querying)

```
%%time
YahooPricedata = pd.read_csv('/Users/..................../YahooData.csv')
YahooPriceDataFrame = pd.DataFrame(YahooPricedata)
YahooPriceDataFrame.to_json('YahooPriceDataFrameToJSON.json')
YahoodataJSONDataFrame = pd.read_json('/Users/....................../YahooPriceDataFrameTo-
JSON.json')
YahoodataJSONDataFrame.query('Open == "17.5"')
```

Wall time (ms): 406 ms, 387, 339, 367, 356, 349, 345, 348, 347, 340

CPU Time (ms)

```
348 ms ± 16.9 ms per loop (mean ± std. dev. of 7 runs, 1 loop each)
356 ms ± 14.5 ms per loop (mean ± std. dev. of 7 runs, 1 loop each)
356 ms ± 17.7 ms per loop (mean ± std. dev. of 7 runs, 1 loop each)
354 ms ± 14.6 ms per loop (mean ± std. dev. of 7 runs, 1 loop each)
345 ms ± 9.39 ms per loop (mean ± std. dev. of 7 runs, 1 loop each)
363 ms ± 18.2 ms per loop (mean ± std. dev. of 7 runs, 1 loop each)
337 ms ± 12.7 ms per loop (mean ± std. dev. of 7 runs, 1 loop each)
364 ms ± 20.2 ms per loop (mean ± std. dev. of 7 runs, 1 loop each)
344 ms ± 8.32 ms per loop (mean ± std. dev. of 7 runs, 1 loop each)
363 ms ± 18.2 ms per loop (mean ± std. dev. of 7 runs, 1 loop each)
```

Finding Unique Values

```
%%time
YahooPricedata = pd.read_csv('/Users/......................./YahooData.csv')
YahooPriceDataFrame = pd.DataFrame(YahooPricedata)
YahooPriceDataFrame.to_json('YahooPriceDataFrameToJSON.json')
YahoodataJSONDataFrame = pd.read_json('/Users/..........  ........./YahooPriceDataFrameTo-
JSON.json')
pd.unique(YahoodataJSONDataFrame[['Open', 'Close']].values.ravel('K'))
array(())
```

Wall times (ms): 353 ms, 331, 326, 344, 332, 341, 345, 330, 334, 328

CPU Times (ms)

```
329 ms ± 8.36 ms per loop (mean ± std. dev. of 7 runs, 1 loop each)
339 ms ± 11 ms per loop (mean ± std. dev. of 7 runs, 1 loop each)
342 ms ± 10.9 ms per loop (mean ± std. dev. of 7 runs, 1 loop each)
355 ms ± 13.7 ms per loop (mean ± std. dev. of 7 runs, 1 loop each)
334 ms ± 9.12 ms per loop (mean ± std. dev. of 7 runs, 1 loop each)
339 ms ± 10 ms per loop (mean ± std. dev. of 7 runs, 1 loop each)
355 ms ± 17.5 ms per loop (mean ± std. dev. of 7 runs, 1 loop each)
343 ms ± 12.3 ms per loop (mean ± std. dev. of 7 runs, 1 loop each)
375 ms ± 13.8 ms per loop (mean ± std. dev. of 7 runs, 1 loop each)
358 ms ± 32 ms per loop (mean ± std. dev. of 7 runs, 1 loop each)
```

Data Sorting

```
%%time
YahooPricedata = pd.read_csv('/Users/................................/YahooData.csv')
YahooPriceDataFrame = pd.DataFrame(YahooPricedata)
YahooPriceDataFrame.to_json('YahooPriceDataFrameToJSON.json')
YahoodataJSONDataFrame = pd.read_json('/Users/........................ ...../YahooPriceDataFrame-
ToJSON.json')
YahoodataJSONDataFrame.sort_values(by=['Volume'])
```

Wall times (ms): 340 ms, 345, 367, 353, 345, 351, 340, 337, 341, 357

CPU times (ms):

```
347 ms ± 17.8 ms per loop (mean ± std. dev. of 7 runs, 1 loop each)
342 ms ± 20.3 ms per loop (mean ± std. dev. of 7 runs, 1 loop each)
347 ms ± 7.22 ms per loop (mean ± std. dev. of 7 runs, 1 loop each)
360 ms ± 18.3 ms per loop (mean ± std. dev. of 7 runs, 1 loop each)
353 ms ± 16.9 ms per loop (mean ± std. dev. of 7 runs, 1 loop each)
354 ms ± 17.9 ms per loop (mean ± std. dev. of 7 runs, 1 loop each)
336 ms ± 6.32 ms per loop (mean ± std. dev. of 7 runs, 1 loop each)
335 ms ± 9.43 ms per loop (mean ± std. dev. of 7 runs, 1 loop each)
360 ms ± 12.9 ms per loop (mean ± std. dev. of 7 runs, 1 loop each)
346 ms ± 9.29 ms per loop (mean ± std. dev. of 7 runs, 1 loop each)
```

Grouping

```
%%time
YahooPricedata = pd.read_csv('/Users/......................................./YahooData.csv')
YahooPriceDataFrame = pd.DataFrame(YahooPricedata)
YahooPriceDataFrame.to_json('YahooPriceDataFrameToJSON.json')
YahoodataJSONDataFrame = pd.read_json('/Users/................................./YahooPriceDataFrame-
ToJSON.json')
YahoodataJSONDataFrameGroupBy = YahoodataJSONDataFrame.groupby('High').mean()
YahoodataJSONDataFrameGroupBy
```

Wall time (ms): 544 ms, 510, 482, 511, 543, 471, 435, 524, 459, 526

CPU Times (ms):

```
442 ms ± 25.2 ms per loop (mean ± std. dev. of 7 runs, 1 loop each)
482 ms ± 55.2 ms per loop (mean ± std. dev. of 7 runs, 1 loop each)
466 ms ± 21.5 ms per loop (mean ± std. dev. of 7 runs, 1 loop each)
457 ms ± 29.2 ms per loop (mean ± std. dev. of 7 runs, 1 loop each)
468 ms ± 24.6 ms per loop (mean ± std. dev. of 7 runs, 1 loop each)
478 ms ± 33.6 ms per loop (mean ± std. dev. of 7 runs, 1 loop each)
487 ms ± 40.4 ms per loop (mean ± std. dev. of 7 runs, 1 loop each)
493 ms ± 29.8 ms per loop (mean ± std. dev. of 7 runs, 1 loop each)
479 ms ± 53.9 ms per loop (mean ± std. dev. of 7 runs, 1 loop each)
475 ms ± 43.5 ms per loop (mean ± std. dev. of 7 runs, 1 loop each)
```

	Open	Low	Close	Adj Close	Volume
High					
1.312500	1.187500	1.125000	1.187500	1.061207	87300.0
1.375000	1.375000	1.093750	1.281250	1.144987	149550.0
1.437500	1.337500	1.287500	1.412500	1.262278	66400.0
1.470000	1.410000	1.400000	1.450000	1.295790	76200.0
1.480000	1.430000	1.370000	1.410000	1.260044	189300.0
...	...	...	...	...	...
64.110001	63.090000	62.099998	63.799999	63.652538	189400.0
65.160004	64.639999	61.049999	61.740002	61.597301	180800.0
65.379997	64.870003	64.419998	64.440002	64.291061	204200.0
66.290001	64.400002	64.129997	64.989998	64.839790	142900.0
66.330002	65.555001	64.620003	65.200001	65.049305	132000.0

2296 rows × 5 columns

APPENDIX IV

Results

Algorithm 4: Conversion of CSV Time Series into Feather DataFrame

Dataframe Creation

```
%%time
Yahoodata3 = pd.read_csv('/Users/............/YahooData.csv')
Yahoodata3df = pd.DataFrame(Yahoodata3)
Yahoodata3df.to_feather('YahoodataFeather.feather')
YahoodataFeather = pd.read_feather('YahoodataFeather.feather')
YahoodataFeather
```

Wall Times (ms): 33.8 ms, 81.4, 33.9, 55.1, 46.1, 36.7, 57.2, 45.8, 30.5, 45.4

CPU Times (ms):

```
44.9 ms ± 7.29 ms per loop (mean ± std. dev. of 7 runs, 10 loops each)
33.8 ms ± 6.52 ms per loop (mean ± std. dev. of 7 runs, 10 loops each)
39.5 ms ± 7.49 ms per loop (mean ± std. dev. of 7 runs, 10 loops each)
45.8 ms ± 2.73 ms per loop (mean ± std. dev. of 7 runs, 10 loops each)
30.3 ms ± 3.33 ms per loop (mean ± std. dev. of 7 runs, 10 loops each)
48.6 ms ± 2.1 ms per loop (mean ± std. dev. of 7 runs, 10 loops each)
51.8 ms ± 2.14 ms per loop (mean ± std. dev. of 7 runs, 10 loops each)
46.5 ms ± 2.49 ms per loop (mean ± std. dev. of 7 runs, 10 loops each)
38 ms ± 8.73 ms per loop (mean ± std. dev. of 7 runs, 10 loops each)
48.3 ms ± 4.53 ms per loop (mean ± std. dev. of 7 runs, 10 loops each)
```

Data Filtering (Querying)

```
%%timeit
Yahoodata3 = pd.read_csv('/Users/........................./YahooData.csv')
Yahoodata3df = pd.DataFrame(Yahoodata3)
Yahoodata3df.to_feather('YahoodataFeather.feather')
YahoodataFeather = pd.read_feather('YahoodataFeather.feather')
YahoodataFeatherFiltered = YahoodataFeather.query('Open == "17.5"')
YahoodataFeatherFiltered
```

Wall Times (ms): 79 ms, 39.5, 71.3, 37.2, 39.9, 36.7, 61.1, 38.5, 33, 50.1

CPU Times (ms):

```
52.1 ms ± 6.41 ms per loop (mean ± std. dev. of 7 runs, 10 loops each)
```

```
42.6 ms ± 11.7 ms per loop (mean ± std. dev. of 7 runs, 10 loops each)
54.2 ms ± 4.82 ms per loop (mean ± std. dev. of 7 runs, 10 loops each)
37.3 ms ± 6.93 ms per loop (mean ± std. dev. of 7 runs, 10 loops each)
57.7 ms ± 2.19 ms per loop (mean ± std. dev. of 7 runs, 10 loops each)
47.2 ms ± 11.2 ms per loop (mean ± std. dev. of 7 runs, 10 loops eac
56.7 ms ± 1.71 ms per loop (mean ± std. dev. of 7 runs, 10 loops each)
53.9 ms ± 4.79 ms per loop (mean ± std. dev. of 7 runs, 10 loops each)
53.2 ms ± 3.66 ms per loop (mean ± std. dev. of 7 runs, 10 loops each)
53.8 ms ± 2.41 ms per loop (mean ± std. dev. of 7 runs, 10 loops each)
```

Finding Unique Values

```
%%time
# Algorithms to read unique values in 'Open' and 'Close' columns. The 'K' ar-
gument reads them out as they occur in memory
Yahoodata3 = pd.read_csv('/Users/................................../YahooData.csv')
Yahoodata3df = pd.DataFrame(Yahoodata3)
Yahoodata3df.to_feather('YahoodataFeather.feather')
YahoodataFeather = pd.read_feather('YahoodataFeather.feather')
pd.unique(YahoodataFeather[['Open']].values.ravel('K'))
```

Wall Times (ms): 50.9 ms, 25, 20.9, 30.5, 43.8, 47.2, 33.8, 56.2, 38.1, 34

CPU Times (ms)

```
55.9 ms ± 1.55 ms per loop (mean ± std. dev. of 7 runs, 10 loops each)
45.2 ms ± 8.23 ms per loop (mean ± std. dev. of 7 runs, 10 loops each)
53.7 ms ± 1.95 ms per loop (mean ± std. dev. of 7 runs, 10 loops each)
50.2 ms ± 1.65 ms per loop (mean ± std. dev. of 7 runs, 10 loops each)
32.8 ms ± 2.37 ms per loop (mean ± std. dev. of 7 runs, 10 loops each)
45.4 ms ± 3.28 ms per loop (mean ± std. dev. of 7 runs, 10 loops each)
47.3 ms ± 4.87 ms per loop (mean ± std. dev. of 7 runs, 10 loops each)
52.4 ms ± 2.64 ms per loop (mean ± std. dev. of 7 runs, 10 loops each)
52.2 ms ± 3.2 ms per loop (mean ± std. dev. of 7 runs, 10 loops each)
50.8 ms ± 3.35 ms per loop (mean ± std. dev. of 7 runs, 10 loops each)
50.3 ms ± 2.85 ms per loop (mean ± std. dev. of 7 runs, 10 loops each)
```

Data Sorting

```
%%time
YahooPricedata = pd.read_csv('/Users/........./YahooData.csv')
YahooPriceDataFrame = pd.DataFrame(YahooPricedata)
YahooPriceDataFrame.to_json('YahooPriceDataFrameToJSON.json')
YahoodataJSONDataFrame = pd.read_json('/Users/............/YahooPriceDataFrameToJSON.
```

```
json')
YahoodataJSONDataFrame.sort_values(by=['Volume'])
```

Wall Times (ms): 33.5 ms, 34.6,37.2, 45.3, 43.6, 36.2, 30.4, 32.7, 54.4, 44.1

CPU Times (ms):

```
32.1 ms ± 3.82 ms per loop (mean ± std. dev. of 7 runs, 10 loops each)
41.2 ms ± 5.34 ms per loop (mean ± std. dev. of 7 runs, 10 loops each)
49 ms ± 4.23 ms per loop (mean ± std. dev. of 7 runs, 10 loops each)
48.9 ms ± 5.41 ms per loop (mean ± std. dev. of 7 runs, 10 loops each)
41.6 ms ± 8.61 ms per loop (mean ± std. dev. of 7 runs, 10 loops each)
47.2 ms ± 5.93 ms per loop (mean ± std. dev. of 7 runs, 10 loops each)
39.3 ms ± 8.55 ms per loop (mean ± std. dev. of 7 runs, 10 loops each)
37.2 ms ± 6.45 ms per loop (mean ± std. dev. of 7 runs, 10 loops each)
50.9 ms ± 3.76 ms per loop (mean ± std. dev. of 7 runs, 10 loops each)
50.6 ms ± 3.78 ms per loop (mean ± std. dev. of 7 runs, 10 loops each)
```

Data Grouping

```
%%time
Yahoodata3 = pd.read_csv('/Users/................................/YahooData.csv')
Yahoodata3df = pd.DataFrame(Yahoodata3)
Yahoodata3df.to_feather('YahoodataFeather.feather')
YahoodataFeather = pd.read_feather('YahoodataFeather.feather')
YahoodataFeatherFrameGroupBy = YahoodataFeather.groupby('High').mean()
YahoodataFeatherFrameGroupBy
```

Wall Times (ms): 55.4 ms, 60.3, 44.7, 52.3, 53.3, 36.7, 37.7, 39.4, 57.1, 60.3, 55.2

CPU Times (ms):

```
55.3 ms ± 2.39 ms per loop (mean ± std. dev. of 7 runs, 10 loops each)
51.5 ms ± 3.5 ms per loop (mean ± std. dev. of 7 runs, 10 loops each)
37.2 ms ± 7.14 ms per loop (mean ± std. dev. of 7 runs, 10 loops each)
44.6 ms ± 7.71 ms per loop (mean ± std. dev. of 7 runs, 10 loops each)
54.2 ms ± 3.46 ms per loop (mean ± std. dev. of 7 runs, 10 loops each)
52.9 ms ± 6.16 ms per loop (mean ± std. dev. of 7 runs, 10 loops each)
48.7 ms ± 2.81 ms per loop (mean ± std. dev. of 7 runs, 10 loops each)
50 ms ± 4.59 ms per loop (mean ± std. dev. of 7 runs, 10 loops each)
34.4 ms ± 4.69 ms per loop (mean ± std. dev. of 7 runs, 10 loops each)
45.3 ms ± 7.78 ms per loop (mean ± std. dev. of 7 runs, 10 loops each)
```

High	Open	Low	Close	Adj Close	Volume
1.312500	1.187500	1.125000	1.187500	1.061207	87300.0
1.375000	1.375000	1.093750	1.281250	1.144987	149550.0
1.437500	1.337500	1.287500	1.412500	1.262278	66400.0
1.470000	1.410000	1.400000	1.450000	1.295790	76200.0
1.480000	1.430000	1.370000	1.410000	1.260044	189300.0
...	...	...	...	...	...
64.110001	63.090000	62.099998	63.799999	63.652538	189400.0
65.160004	64.639999	61.049999	61.740002	61.597301	180800.0
65.379997	64.870003	64.419998	64.440002	64.291061	204200.0
66.290001	64.400002	64.129997	64.989998	64.839790	142900.0
66.330002	65.555001	64.620003	65.200001	65.049305	132000.0

2296 rows × 5 columns

APPENDIX V

Results

Algorithm 5a: CSV Timeseries to YAML Data Format Conversion (Dictionary Orientation = Records)

Dataframe Creation

```
%%time
df = pd.read_csv('/Users/............................../YahooData.csv')
with open('YahoodataYAMLformat.yml', 'w') as outfile:
    yaml.dump(
        df.to_dict(orient='records'),
        outfile,
        canonical=True,
        sort_keys=True,
        default_flow_style=False,
        indent=6,
        width=22,
    )
with open('/Users/........./YahoodataYAMLformat.yml') as f:
    yamldata = yaml.safe_load(f)
YahoodataYAMLDF = pd.DataFrame( yamldata)
YahoodataYAMLDF
```

Important - Note that the set of Algorithms 5a - 5d have wall and CPU times in order of seconds. Not milliseconds like other algorithms designed in this chapter.

Wall time (seconds): Wall time: 10.5 s, 10.6 s, 10.7, 10.6, 10.6, 10.5, 10.6, 10.7, 11 s

CPU Times (seconds)

```
10 s ± 193 ms per loop (mean ± std. dev. of 7 runs, 1 loop each)
10.9 s ± 1.25 s per loop (mean ± std. dev. of 7 runs, 1 loop each)
10.6 s ± 222 ms per loop (mean ± std. dev. of 7 runs, 1 loop each)
10.5 s ± 103 ms per loop (mean ± std. dev. of 7 runs, 1 loop each)
10.4 s ± 1.91 s per loop (mean ± std. dev. of 7 runs, 1 loop each)
10.2 s ± 301 ms per loop (mean ± std. dev. of 7 runs, 1 loop each)
10.8 s ± 88 ms per loop (mean ± std. dev. of 7 runs, 1 loop each)
10.8 s ± 203 s per loop (mean ± std. dev. of 7 runs, 1 loop each)
10.4 s ± 178 ms per loop (mean ± std. dev. of 7 runs, 1 loop each)
```

```
10.9 s ± 117 ms per loop (mean ± std. dev. of 7 runs, 1 loop each)
```

Data Filtering (Querying)

```
%%time
df = pd.read_csv('/Users/............................/YahooData.csv')
with open('YahoodataYAMLformat.yml', 'w') as outfile:
    yaml.dump(
        df.to_dict(orient='records'),
        outfile,
        canonical=True,
        sort_keys=True,
        default_flow_style=False,
        indent=6,
        width=22,
    )
with open('/Users/......................…../YahoodataYAMLformat.yml') as f:
    yamldata = yaml.safe_load(f)
YahoodataYAMLDF = pd.DataFrame( yamldata)
YahoodataYAMLDF.query('Open == "17.5"')
```

Wall Times (seconds): 10.1 s, 10.3, 10.3, 10.1, 10.3, 10.5, 10.2, 10.5, 11.2, 10.3

CPU Times (seconds)

```
9.85 s ± 306 ms per loop (mean ± std. dev. of 7 runs, 1 loop each)
9.98 s ± 326 ms per loop (mean ± std. dev. of 7 runs, 1 loop each)
9.8 s ± 164 ms per loop (mean ± std. dev. of 7 runs, 1 loop each)
9.8 s ± 84.3 ms per loop (mean ± std. dev. of 7 runs, 1 loop each)
9.79 s ± 88.2 ms per loop (mean ± std. dev. of 7 runs, 1 loop each)
10.3 s ± 434 ms per loop (mean ± std. dev. of 7 runs, 1 loop each)
9.84 s ± 207 ms per loop (mean ± std. dev. of 7 runs, 1 loop each)
9.82 s ± 261 ms per loop (mean ± std. dev. of 7 runs, 1 loop each)
9.87 s ± 152 ms per loop (mean ± std. dev. of 7 runs, 1 loop each)
9.82 s ± 261 ms per loop (mean ± std. dev. of 7 runs, 1 loop each)
```

	Adj Close	Close	Date	High	Low	Open	Volume
0	15.192017	17.000000	4/30/1997	17.500	17.000	17.5	5800
1	15.862257	17.750000	5/1/1997	17.750	17.375	17.5	36300
3	16.532488	18.500000	5/5/1997	18.500	17.500	17.5	65100
105	15.638843	17.500000	9/29/1997	18.250	17.500	17.5	4300
106	15.638843	17.500000	9/30/1997	17.750	17.500	17.5	21000
108	15.582988	17.437500	10/2/1997	17.500	17.250	17.5	6800
115	15.582988	17.437500	10/13/1997	17.625	17.250	17.5	9200
135	15.192017	17.000000	11/10/1997	17.500	17.000	17.5	3100
168	15.192017	17.000000	12/29/1997	17.625	17.000	17.5	3300
498	16.979313	19.000000	4/22/1999	19.000	17.500	17.5	160800
529	15.750552	17.625000	6/7/1999	17.625	17.375	17.5	53200
565	15.527136	17.375000	7/28/1999	17.500	17.250	17.5	27400
4039	16.068762	17.469999	5/20/2013	17.740	17.400	17.5	78200
4051	16.767805	18.230000	6/6/2013	18.240	17.500	17.5	101500

Finding Unique Values

```
%%time
df = pd.read_csv('/Users/...........................YahooData.csv')
with open('YahoodataYAMLformat.yml', 'w') as outfile:
    yaml.dump(
        df.to_dict(orient='records'),
        outfile,
        canonical=True,
        sort_keys=True,
        default_flow_style=False,
        indent=6,
        width=22,
    )
with open('/Users/........./YahoodataYAMLformat.yml') as f:
    yamldata = yaml.safe_load(f)
YahoodataYAMLDF = pd.DataFrame( yamldata)
pd.unique(YahoodataJSONDataFrame[['Open', 'Close']].values.ravel('K'))
array([17.5     , 17.75    , 18.625    , ..., 48.48  , 44.869999, 42.889999])
```

Wall Times (seconds): 10.4 s, 10.8, 11.2, 11.1, 11.1, 11, 10.4, 10.9, 10.6, 10.9s

CPU Times (seconds):

```
9.74 s ± 97.8 ms per loop (mean ± std. dev. of 7 runs, 1 loop each)
9.73 s ± 127 ms per loop (mean ± std. dev. of 7 runs, 1 loop each)
9.75 s ± 156 ms per loop (mean ± std. dev. of 7 runs, 1 loop each)
9.68 s ± 73.2 ms per loop (mean ± std. dev. of 7 runs, 1 loop each)
9.72 s ± 140 ms per loop (mean ± std. dev. of 7 runs, 1 loop each)
9.7 s ± 64.9 ms per loop (mean ± std. dev. of 7 runs, 1 loop each)
9.75 s ± 241 ms per loop (mean ± std. dev. of 7 runs, 1 loop each)
9.83 s ± 334 ms per loop (mean ± std. dev. of 7 runs, 1 loop each)
10.2 s ± 308 ms per loop (mean ± std. dev. of 7 runs, 1 loop each)
9.68 s ± 68.1 ms per loop (mean ± std. dev. of 7 runs, 1 loop each)
```

Data Sorting

Wall Times (seconds): 10.2 s, 10.3, 10.3, 10.6, 10.2, 10.1, 10.2, 10.2, 10.3, 10.3

CPU Times (seconds):

```
9.7 s ± 108 ms per loop (mean ± std. dev. of 7 runs, 1 loop each)
9.69 s ± 28.9 ms per loop (mean ± std. dev. of 7 runs, 1 loop each)
9.71 s ± 120 ms per loop (mean ± std. dev. of 7 runs, 1 loop each)
9.79 s ± 129 ms per loop (mean ± std. dev. of 7 runs, 1 loop each)
9.76 s ± 117 ms per loop (mean ± std. dev. of 7 runs, 1 loop each)
9.72 s ± 71 ms per loop (mean ± std. dev. of 7 runs, 1 loop each)
9.73 s ± 111 ms per loop (mean ± std. dev. of 7 runs, 1 loop each)
9.61 s ± 86.8 ms per loop (mean ± std. dev. of 7 runs, 1 loop each)
9.73 s ± 173 ms per loop (mean ± std. dev. of 7 runs, 1 loop each)
9.65 s ± 76 ms per loop (mean ± std. dev. of 7 runs, 1 loop each)
```

```
%%timeit
df = pd.read_csv('/Users/............................/YahooData.csv')
with open('YahoodataYAMLformat.yml', 'w') as outfile:
    yaml.dump(
        df.to_dict(orient='records'),
        outfile,
        canonical=True,
        sort_keys=True,
        default_flow_style=False,
        indent=6,
```

```
        width=22,
    )
with open('/Users/......../YahoodataYAMLformat.yml') as f:
    yamldata = yaml.safe_load(f)
YahoodataYAMLDF = pd.DataFrame( yamldata)
YahoodataYAMLDF.sort_values(by=['Volume'])
```

Data Grouping

```
%%timeit
df = pd.read_csv('/Users/.........................../YahooData.csv')
with open('YahoodataYAMLformat.yml', 'w') as outfile:
    yaml.dump(
        df.to_dict(orient='records'),
        outfile,
        canonical=True,
        sort_keys=True,
        default_flow_style=False,
        indent=6,
        width=22,
    )
with open('/Users/......../YahoodataYAMLformat.yml') as f:
    yamldata = yaml.safe_load(f)
YahoodataYAMLDF = pd.DataFrame( yamldata)
YahoodataYAMLDFGroupBy = YahoodataYAMLDF.groupby('Open').mean()
```

Wall Times (seconds): 10.2 s, 10.1, 10.3, 9.9, 9.99, 10.3, 10.4, 10.1, 10.5, 10.6

CPU Times (seconds):

```
9.75 s ± 110 ms per loop (mean ± std. dev. of 7 runs, 1 loop each)
9.74 s ± 129 ms per loop (mean ± std. dev. of 7 runs, 1 loop each)
9.8 s ± 56 ms per loop (mean ± std. dev. of 7 runs, 1 loop each)
9.74 s ± 96.3 ms per loop (mean ± std. dev. of 7 runs, 1 loop each)
9.74 s ± 90.5 ms per loop (mean ± std. dev. of 7 runs, 1 loop each)
9.73 s ± 78.7 ms per loop (mean ± std. dev. of 7 runs, 1 loop each)
9.75 s ± 90.5 ms per loop (mean ± std. dev. of 7 runs, 1 loop each)
9.75 s ± 52.7 ms per loop (mean ± std. dev. of 7 runs, 1 loop each)
9.74 s ± 104 ms per loop (mean ± std. dev. of 7 runs, 1 loop each)
10.2 s ± 117 ms per loop (mean ± std. dev. of 7 runs, 1 loop each)
```

APPENDIX VI

Results

Algorithm 5b: CSV Timeseries to YAML Data Format Conversion (Dictionary Orientation = List)

Dataframe Creation

```
%%time
df = pd.read_csv('/Users/............................/YahooData.csv')
with open('YahoodataYAMLformat.yml', 'w') as outfile:
    yaml.dump(
        df.to_dict(orient='list'),
        outfile,
        canonical=True,
        sort_keys=True,
        default_flow_style=False,
        indent=6,
        width=22,
    )
with open('/Users/........................./YahoodataYAMLformat.yml') as f:
    yamldata = yaml.safe_load(f)
YahoodataYAMLDF = pd.DataFrame( yamldata)
YahoodataYAMLDF
```

Important - Note that the set of Algorithms 5a - 5d have wall and CPU times in order of seconds. Not milliseconds like other algorithms designed in this chapter.

Wall Times (seconds): 5.2 s, 5.14, 4.84, 4.85, 5.05, 5.06, 5.12, 5.09, 5.06, 4.95

CPU Times (seconds)

```
4.8 s ± 294 ms per loop (mean ± std. dev. of 7 runs, 1 loop each)
5.13 s ± 159 ms per loop (mean ± std. dev. of 7 runs, 1 loop each)
5.03 s ± 294 ms per loop (mean ± std. dev. of 7 runs, 1 loop each)
4.85 s ± 167 ms per loop (mean ± std. dev. of 7 runs, 1 loop each)
4.55 s ± 55.1 ms per loop (mean ± std. dev. of 7 runs, 1 loop each)
4.6 s ± 63.6 ms per loop (mean ± std. dev. of 7 runs, 1 loop each)
4.91 s ± 76.2 ms per loop (mean ± std. dev. of 7 runs, 1 loop each)
5.03 s ± 140 ms per loop (mean ± std. dev. of 7 runs, 1 loop each)
5.03 s ± 172 ms per loop (mean ± std. dev. of 7 runs, 1 loop each)
```

4.59 s ± 110 ms per loop (mean ± std. dev. of 7 runs, 1 loop each)

Data Filtering (Querying)

```
%%time
df = pd.read_csv('/Users/............................/YahooData.csv')
with open('YahoodataYAMLformat.yml', 'w') as outfile:
    yaml.dump(
        df.to_dict(orient='list'),
        outfile,
        canonical=True,
        sort_keys=True,
        default_flow_style=False,
        indent=6,
        width=22,
    )
with open('/Users/........./YahoodataYAMLformat.yml') as f:
    yamldata = yaml.safe_load(f)
YahoodataYAMLDF = pd.DataFrame( yamldata)
YahoodataYAMLDF.query('Open == "17.5"')
```

Wall Times (seconds): 4.96 s, 10.6, 7.08, 6.49, 5.92, 5.74, 5.8, 5.95, 5.89, 5.87

CPU Times (seconds)

5 s ± 309 ms per loop (mean ± std. dev. of 7 runs, 1 loop each)
4.82 s ± 253 ms per loop (mean ± std. dev. of 7 runs, 1 loop each)
4.75 s ± 134 ms per loop (mean ± std. dev. of 7 runs, 1 loop each)
4.68 s ± 122 ms per loop (mean ± std. dev. of 7 runs, 1 loop each)
4.8 s ± 114 ms per loop (mean ± std. dev. of 7 runs, 1 loop each)
4.77 s ± 305 ms per loop (mean ± std. dev. of 7 runs, 1 loop each)
4.98 s ± 165 ms per loop (mean ± std. dev. of 7 runs, 1 loop each)
4.8 s ± 157 ms per loop (mean ± std. dev. of 7 runs, 1 loop each)
4.98 s ± 321 ms per loop (mean ± std. dev. of 7 runs, 1 loop each)
5.4 s ± 606 ms per loop (mean ± std. dev. of 7 runs, 1 loop each)

	Adj Close	Close	Date	High	Low	Open	Volume
0	15.192017	17.000000	4/30/1997	17.500	17.000	17.5	5800
1	15.862257	17.750000	5/1/1997	17.750	17.375	17.5	36300
3	16.532488	18.500000	5/5/1997	18.500	17.500	17.5	65100
105	15.638843	17.500000	9/29/1997	18.250	17.500	17.5	4300
106	15.638843	17.500000	9/30/1997	17.750	17.500	17.5	21000
108	15.582988	17.437500	10/2/1997	17.500	17.250	17.5	6800
115	15.582988	17.437500	10/13/1997	17.625	17.250	17.5	9200
135	15.192017	17.000000	11/10/1997	17.500	17.000	17.5	3100
168	15.192017	17.000000	12/29/1997	17.625	17.000	17.5	3300
498	16.979313	19.000000	4/22/1999	19.000	17.500	17.5	160800
529	15.750552	17.625000	6/7/1999	17.625	17.375	17.5	53200
565	15.527136	17.375000	7/28/1999	17.500	17.250	17.5	27400
4039	16.068762	17.469999	5/20/2013	17.740	17.400	17.5	78200
4051	16.767805	18.230000	6/6/2013	18.240	17.500	17.5	101500

Finding Unique Values

```
%%time
df = pd.read_csv('/Users/............................/YahooData.csv')
with open('YahoodataYAMLformat.yml', 'w') as outfile:
    yaml.dump(
        df.to_dict(orient='records'),
        outfile,
        canonical=True,
        sort_keys=True,
        default_flow_style=False,
        indent=6,
        width=22,
    )
with open('/Users/........./YahoodataYAMLformat.yml') as f:
    yamldata = yaml.safe_load(f)
YahoodataYAMLDF = pd.DataFrame( yamldata)
pd.unique(YahoodataYAMLDF [['Open', 'Close']].values.ravel('K'))
array([17.5     , 17.75     , 18.625     , ..., 48.48     , 44.869999, 42.889999])
```

Wall Times (seconds): 4.71 s, 4.72, 5.67, 4.72, 4.64, 5.42, 4.67, 4.95, 5.03, 4.67

CPU Times (seconds):

```
4.95 s ± 358 ms per loop (mean ± std. dev. of 7 runs, 1 loop each)
4.68 s ± 168 ms per loop (mean ± std. dev. of 7 runs, 1 loop each)
4.85 s ± 292 ms per loop (mean ± std. dev. of 7 runs, 1 loop each)
4.97 s ± 220 ms per loop (mean ± std. dev. of 7 runs, 1 loop each)
4.62 s ± 82.6 ms per loop (mean ± std. dev. of 7 runs, 1 loop each)
4.79 s ± 192 ms per loop (mean ± std. dev. of 7 runs, 1 loop each)
4.53 s ± 102 ms per loop (mean ± std. dev. of 7 runs, 1 loop each)
4.78 s ± 151 ms per loop (mean ± std. dev. of 7 runs, 1 loop each)
4.62 s ± 62 ms per loop (mean ± std. dev. of 7 runs, 1 loop each)
4.55 s ± 96.8 ms per loop (mean ± std. dev. of 7 runs, 1 loop each)
```

Data Sorting

Wall Times (seconds): 5.12 s, 5.21, 4.82, 4.99, 4.75, 4.86, 4.95, 5.71, 4.8, 4.86

CPU Times (seconds):

```
4.75 s ± 63.1 ms per loop (mean ± std. dev. of 7 runs, 1 loop each)
5.4 s ± 272 ms per loop (mean ± std. dev. of 7 runs, 1 loop each)
4.77 s ± 258 ms per loop (mean ± std. dev. of 7 runs, 1 loop each)
4.74 s ± 303 ms per loop (mean ± std. dev. of 7 runs, 1 loop each)
4.73 s ± 177 ms per loop (mean ± std. dev. of 7 runs, 1 loop each)
4.61 s ± 75.7 ms per loop (mean ± std. dev. of 7 runs, 1 loop each)
4.78 s ± 84.4 ms per loop (mean ± std. dev. of 7 runs, 1 loop each)
4.55 s ± 85.6 ms per loop (mean ± std. dev. of 7 runs, 1 loop each)
4.78 s ± 229 ms per loop (mean ± std. dev. of 7 runs, 1 loop each)
4.57 s ± 93.5 ms per loop (mean ± std. dev. of 7 runs, 1 loop each)
```

```
%%time
df = pd.read_csv('/Users/………………………../YahooData.csv')
with open('YahoodataYAMLformat.yml', 'w') as outfile:
    yaml.dump(
        df.to_dict(orient='records'),
        outfile,
        canonical=True,
        sort_keys=True,
        default_flow_style=False,
        indent=6,
```

```
        width=22,
    )
with open('/Users/......../YahoodataYAMLformat.yml') as f:
    yamldata = yaml.safe_load(f)
YahoodataYAMLDF = pd.DataFrame( yamldata)
YahoodataYAMLDF.sort_values(by=['Volume'])
```

	Adj Close	Close	Date	High	Low	Open	Volume
1275	3.923114	4.390000	5/28/2002	4.3900	4.390000	4.3900	0
3216	3.386926	3.790000	2/10/2010	3.7900	3.790000	3.7900	0
166	15.750552	17.625000	12/24/1997	17.6250	17.625000	17.6250	100
1302	3.717577	4.160000	7/5/2002	4.1600	4.160000	4.1600	200
997	2.680944	3.000000	4/12/2001	3.0000	3.000000	3.0000	400
...	...	...	...	...	...	...	...
577	10.221101	11.437500	8/13/1999	14.5000	10.500000	14.5000	1111300
4245	15.817896	17.129999	3/14/2014	17.4100	16.709999	17.0200	1165400
892	1.675590	1.875000	11/9/2000	1.9375	1.500000	1.9375	1379700
3813	7.443938	8.160000	6/22/2012	8.5900	7.850000	8.3500	1857100
4244	15.725554	17.030001	3/13/2014	18.3400	16.799999	18.1700	2073400

```
%%time
df = pd.read_csv('/Users/...........................YahooData.csv')
with open('YahoodataYAMLformat.yml', 'w') as outfile:
    yaml.dump(
        df.to_dict(orient='list'),
        outfile,
        canonical=True,
        sort_keys=True,
        default_flow_style=False,
        indent=6,
        width=22,
    )
with open('/Users/......../YahoodataYAMLformat.yml') as f:
    yamldata = yaml.safe_load(f)
YahoodataYAMLDF = pd.DataFrame( yamldata)
YahoodataYAMLDFGroupBy = YahoodataYAMLDF.groupby('Open').mean()
```

Data Grouping

```
Wall Times (seconds): 5.24 s, 5.96, 5.37, 5.16, 5.38, 4.92, 5.5, 5.43, 5.42,
6.45,

CPU Times (seconds):
4.54 s ± 109 ms per loop (mean ± std. dev. of 7 runs, 1 loop each)
4.49 s ± 67.5 ms per loop (mean ± std. dev. of 7 runs, 1 loop each)
4.49 s ± 74.4 ms per loop (mean ± std. dev. of 7 runs, 1 loop each)
4.6 s ± 123 ms per loop (mean ± std. dev. of 7 runs, 1 loop each)
4.5 s ± 78.5 ms per loop (mean ± std. dev. of 7 runs, 1 loop each)
4.5 s ± 78.5 ms per loop (mean ± std. dev. of 7 runs, 1 loop each)
4.53 s ± 64.3 ms per loop (mean ± std. dev. of 7 runs, 1 loop each)
4.53 s ± 101 ms per loop (mean ± std. dev. of 7 runs, 1 loop each)
4.47 s ± 54.1 ms per loop (mean ± std. dev. of 7 runs, 1 loop each)
4.74 s ± 81.1 ms per loop (mean ± std. dev. of 7 runs, 1 loop each)
```

	Adj Close	Close	High	Low	Volume
Open					
1.187500	1.172913	1.312500	1.375000	1.156250	69150.000000
1.312500	1.340472	1.500000	1.541667	1.270833	106033.333333
1.375000	1.186876	1.328125	1.406250	1.203125	100550.000000
1.380000	1.340472	1.500000	1.510000	1.380000	39300.000000
1.410000	1.295790	1.450000	1.470000	1.400000	76200.000000
...	...	...	...	...	...
64.400002	64.839790	64.989998	66.290001	64.129997	142900.000000
64.639999	61.597301	61.740002	65.160004	61.049999	180800.000000
64.870003	64.291061	64.440002	65.379997	64.419998	204200.000000
65.199997	65.757668	65.910004	66.330002	64.830002	134300.000000
65.910004	64.340942	64.489998	66.330002	64.410004	129700.000000

APPENDIX VII

Results

Algorithm 5c: CSV Timeseries to YAML Data Format Conversion (Dictionary Orientation = Index)

Dataframe Creation

```
%%time
df = pd.read_csv('/Users/...........................//YahooData.csv')
with open('YahoodataYAMLformat.yml', 'w') as outfile:
    yaml.dump(
        df.to_dict(orient='index'),
        outfile,
        canonical=True,
        sort_keys=True,
        default_flow_style=False,
        indent=6,
        width=22,
    )
with open('/Users/........./YahoodataYAMLformat.yml') as f:
    yamldata = yaml.safe_load(f)
YahoodataYAMLDF = pd.DataFrame( yamldata)
YahoodataYAMLDF
```

Important - Note that the set of Algorithms 5a - 5d have wall and CPU times in order of seconds. Not milliseconds like other algorithms designed in this chapter.

Dataframe Generation

Wall Times (seconds): 12.1 s, 13.6, 14.6, 12.1, 12.2, 12, 11.8, 11.9, 11.6,12.4

CPU Times (seconds)

10.7 s ± 181 ms per loop (mean ± std. dev. of 7 runs, 1 loop each)
10.7 s ± 128 ms per loop (mean ± std. dev. of 7 runs, 1 loop each)
10.7 s ± 112 ms per loop (mean ± std. dev. of 7 runs, 1 loop each)
10.7 s ± 132 ms per loop (mean ± std. dev. of 7 runs, 1 loop each)
11 s ± 433 ms per loop (mean ± std. dev. of 7 runs, 1 loop each)
10.7 s ± 92 ms per loop (mean ± std. dev. of 7 runs, 1 loop each)
10.8 s ± 130 ms per loop (mean ± std. dev. of 7 runs, 1 loop each)

```
10.7 s ± 79.7 ms per loop (mean ± std. dev. of 7 runs, 1 loop each)
10.8 s ± 119 ms per loop (mean ± std. dev. of 7 runs, 1 loop each)
10.6 s ± 89 ms per loop (mean ± std. dev. of 7 runs, 1 loop each)
```

Data Filtering

```
%%time
df = pd.read_csv('/Users/...................................../YahooData.csv')
with open('YahoodataYAMLformat3b.yml', 'w') as outfile:
    yaml.dump(
        df.to_dict(orient='index'),
        outfile,
        canonical=True,
        sort_keys=True,
        default_flow_style=False,
        indent=6,
        width=22,
    )
with open('/Users/................................./YahoodataYAMLformat3b.yml') as f:
    yamldata = yaml.safe_load(f)
YahoodataYAMLDF = pd.DataFrame( yamldata).transpose()
YahoodataYAMLDF.query('Open == "17.5"')
```

Wall Times (seconds): 12 s, 12.2, 11.9, 11.8, 12, 12.3, 13.6, 11.5, 11.6, 11.5

CPU Times (seconds)

```
11.2 s ± 160 ms per loop (mean ± std. dev. of 7 runs, 1 loop each)
11 s ± 231 ms per loop (mean ± std. dev. of 7 runs, 1 loop each)
10.6 s ± 155 ms per loop (mean ± std. dev. of 7 runs, 1 loop each)
10.8 s ± 391 ms per loop (mean ± std. dev. of 7 runs, 1 loop each)
10.7 s ± 79.7 ms per loop (mean ± std. dev. of 7 runs, 1 loop each)
10.8 s ± 98.9 ms per loop (mean ± std. dev. of 7 runs, 1 loop each)
10.8 s ± 131 ms per loop (mean ± std. dev. of 7 runs, 1 loop each)
10.7 s ± 123 ms per loop (mean ± std. dev. of 7 runs, 1 loop each)
10.7 s ± 56.3 ms per loop (mean ± std. dev. of 7 runs, 1 loop each)
10.7 s ± 117 ms per loop (mean ± std. dev. of 7 runs, 1 loop each)
```

	Adj Close	Close	Date	High	Low	Open	Volume
0	15.192017	17.000000	4/30/1997	17.500	17.000	17.5	5800
1	15.862257	17.750000	5/1/1997	17.750	17.375	17.5	36300
3	16.532488	18.500000	5/5/1997	18.500	17.500	17.5	65100
105	15.638843	17.500000	9/29/1997	18.250	17.500	17.5	4300
106	15.638843	17.500000	9/30/1997	17.750	17.500	17.5	21000
108	15.582988	17.437500	10/2/1997	17.500	17.250	17.5	6800
115	15.582988	17.437500	10/13/1997	17.625	17.250	17.5	9200
135	15.192017	17.000000	11/10/1997	17.500	17.000	17.5	3100
168	15.192017	17.000000	12/29/1997	17.625	17.000	17.5	3300
498	16.979313	19.000000	4/22/1999	19.000	17.500	17.5	160800
529	15.750552	17.625000	6/7/1999	17.625	17.375	17.5	53200
565	15.527136	17.375000	7/28/1999	17.500	17.250	17.5	27400
4039	16.068762	17.469999	5/20/2013	17.740	17.400	17.5	78200
4051	16.767805	18.230000	6/6/2013	18.240	17.500	17.5	101500

Finding Unique Values

```
%%time
df = pd.read_csv('/Users/.........................../YahooData.csv')
with open('YahoodataYAMLformat.yml', 'w') as outfile:
    yaml.dump(
        df.to_dict(orient='index'),
        outfile,
        canonical=True,
        sort_keys=True,
        default_flow_style=False,
        indent=6,
        width=22,
    )
with open('/Users/......../YahoodataYAMLformat.yml') as f:
    yamldata = yaml.safe_load(f)
YahoodataYAMLDF = pd.DataFrame( yamldata)
pd.unique(YahoodataYAMLDF [['Open', 'Close']].values.ravel('K'))
array([17.5, 17.75, 18.625, ..., 48.48, 44.869999, 42.889998999999996],
```

```
dtype=object)
```

Wall Times (seconds): 12.4 s, 12.7, 12.3, 12.1, 14, 12.4, 11.7, 11.5, 12.2, 11.7

CPU Times (seconds):

```
10.8 s ± 143 ms per loop (mean ± std. dev. of 7 runs, 1 loop each)
10.8 s ± 117 ms per loop (mean ± std. dev. of 7 runs, 1 loop each)
10.8 s ± 103 ms per loop (mean ± std. dev. of 7 runs, 1 loop each)
11 s ± 214 ms per loop (mean ± std. dev. of 7 runs, 1 loop each)
10.8 s ± 123 ms per loop (mean ± std. dev. of 7 runs, 1 loop each)
10.8 s ± 221 ms per loop (mean ± std. dev. of 7 runs, 1 loop each)
10.9 s ± 206 ms per loop (mean ± std. dev. of 7 runs, 1 loop each)
10.9 s ± 261 ms per loop (mean ± std. dev. of 7 runs, 1 loop each)
10.7 s ± 47.6 ms per loop (mean ± std. dev. of 7 runs, 1 loop each)
11 s ± 179 ms per loop (mean ± std. dev. of 7 runs, 1 loop each)
```

Data Sorting

```
%%time
df = pd.read_csv('/Users/………………………../YahooData.csv')
with open('YahoodataYAMLformat.yml', 'w') as outfile:
    yaml.dump(
        df.to_dict(orient='index'),
        outfile,
        canonical=True,
        sort_keys=True,
        default_flow_style=False,
        indent=6,
        width=22,
    )
with open('/Users/………/YahoodataYAMLformat.yml') as f:
    yamldata = yaml.safe_load(f)
YahoodataYAMLDF = pd.DataFrame( yamldata)
YahoodataYAMLDF.sort_values(by=['Volume'])
```

Wall Times (seconds): 11.5 s, 11.5, 11.4, 11.3, 11.3, 11.2, 11.1, 11.1, 11.4, 11.2

	Adj Close	Close	Date	High	Low	Open	Volume
1275	3.923114	4.390000	5/28/2002	4.3900	4.390000	4.3900	0
3216	3.386926	3.790000	2/10/2010	3.7900	3.790000	3.7900	0
166	15.750552	17.625000	12/24/1997	17.6250	17.625000	17.6250	100
1302	3.717577	4.160000	7/5/2002	4.1600	4.160000	4.1600	200
997	2.680944	3.000000	4/12/2001	3.0000	3.000000	3.0000	400
...	...	...	...	...	...	...	...
577	10.221101	11.437500	8/13/1999	14.5000	10.500000	14.5000	1111300
4245	15.817896	17.129999	3/14/2014	17.4100	16.709999	17.0200	1165400
892	1.675590	1.875000	11/9/2000	1.9375	1.500000	1.9375	1379700
3813	7.443938	8.160000	6/22/2012	8.5900	7.850000	8.3500	1857100
4244	15.725554	17.030001	3/13/2014	18.3400	16.799999	18.1700	2073400

CPU Times (seconds):

```
10.7 s ± 87.8 ms per loop (mean ± std. dev. of 7 runs, 1 loop each)
10.9 s ± 118 ms per loop (mean ± std. dev. of 7 runs, 1 loop each)
10.8 s ± 115 ms per loop (mean ± std. dev. of 7 runs, 1 loop each)
10.7 s ± 114 ms per loop (mean ± std. dev. of 7 runs, 1 loop each)
10.7 s ± 98.8 ms per loop (mean ± std. dev. of 7 runs, 1 loop each)
10.7 s ± 95.7 ms per loop (mean ± std. dev. of 7 runs, 1 loop each)
10.8 s ± 89.4 ms per loop (mean ± std. dev. of 7 runs, 1 loop each)
10.8 s ± 65.2 ms per loop (mean ± std. dev. of 7 runs, 1 loop each)
10.7 s ± 121 ms per loop (mean ± std. dev. of 7 runs, 1 loop each)
11 s ± 165 ms per loop (mean ± std. dev. of 7 runs, 1 loop each)
```

Data Grouping

The Index orientation method does not support data grouping

APPENDIX VIII

Results

Algorithm 5d: CSV Timeseries to YAML Data Format Conversion (Dictionary Orientation = Dict)

Dataframe Creation

```
%%time
df = pd.read_csv('/Users/.......................... ./YahooData.csv')
with open('YahoodataYAMLformat.yml', 'w') as outfile:
    yaml.dump(
        df.to_dict(orient='dict'),
        outfile,
        canonical=True,
        sort_keys=True,
        default_flow_style=False,
        indent=6,
        width=22,
    )
with open('/Users/........./YahoodataYAMLformat.yml') as f:
    yamldata = yaml.safe_load(f)
YahoodataYAMLDF = pd.DataFrame( yamldata)
YahoodataYAMLDF
```

Important - Note that the set of Algorithms 5a - 5d have wall and CPU times in order of seconds. Not milliseconds like other algorithms designed in this chapter.

Wall Times (seconds): 9.48 s, 9.16, 9.36, 9.23, 9.65, 9.49, 9.64, 9.87, 9.57, 9.76,

CPU Times (seconds)

8.86 s ± 92.4 ms per loop (mean ± std. dev. of 7 runs, 1 loop each)
9.27 s ± 294 ms per loop (mean ± std. dev. of 7 runs, 1 loop each)
9.24 s ± 387 ms per loop (mean ± std. dev. of 7 runs, 1 loop each)
8.86 s ± 97.8 ms per loop (mean ± std. dev. of 7 runs, 1 loop each)
8.84 s ± 108 ms per loop (mean ± std. dev. of 7 runs, 1 loop each)
8.84 s ± 70.6 ms per loop (mean ± std. dev. of 7 runs, 1 loop each)
8.94 s ± 133 ms per loop (mean ± std. dev. of 7 runs, 1 loop each)
8.92 s ± 103 ms per loop (mean ± std. dev. of 7 runs, 1 loop each)

```
8.9 s ± 109 ms per loop (mean ± std. dev. of 7 runs, 1 loop each)
9.17 s ± 233 ms per loop (mean ± std. dev. of 7 runs, 1 loop each)
```

Data Filtering

```
%%time
df = pd.read_csv('/Users/oyekanlea2/Desktop/TimeSeriesCodes/YahooData.csv')
with open('YahoodataYAMLformat3b.yml', 'w') as outfile:
    yaml.dump(
        df.to_dict(orient='dict'),
        outfile,
        canonical=True,
        sort_keys=True,
        default_flow_style=False,
        indent=6,
        width=22,
    )
with open('/Users/oyekanlea2/Desktop/TimeSeriesCodes/YahoodataYAMLformat3b.
yml') as f:
    yamldata = yaml.safe_load(f)
YahoodataYAMLDF = pd.DataFrame( yamldata).transpose()
YahoodataYAMLDF.query('Open == "17.5"')
```

Wall Times (seconds): 10 s, 10.7, 9.7, 10.2, 10.3, 10.8, 10.3, 12.8, 11.9, 11.3,

CPU Times (seconds)

```
8.9 s ± 84.1 ms per loop (mean ± std. dev. of 7 runs, 1 loop each)
9.34 s ± 107 ms per loop (mean ± std. dev. of 7 runs, 1 loop each)
9.38 s ± 72.6 ms per loop (mean ± std. dev. of 7 runs, 1 loop each)
9.26 s ± 186 ms per loop (mean ± std. dev. of 7 runs, 1 loop each)
8.99 s ± 148 ms per loop (mean ± std. dev. of 7 runs, 1 loop each)
8.84 s ± 85.2 ms per loop (mean ± std. dev. of 7 runs, 1 loop each)
8.93 s ± 141 ms per loop (mean ± std. dev. of 7 runs, 1 loop each)
8.87 s ± 113 ms per loop (mean ± std. dev. of 7 runs, 1 loop each)
8.91 s ± 129 ms per loop (mean ± std. dev. of 7 runs, 1 loop each)
8.82 s ± 97.1 ms per loop (mean ± std. dev. of 7 runs, 1 loop each)
```

	Adj Close	Close	Date	High	Low	Open	Volume
0	15.192017	17.000000	4/30/1997	17.500	17.000	17.5	5800
1	15.862257	17.750000	5/1/1997	17.750	17.375	17.5	36300
3	16.532488	18.500000	5/5/1997	18.500	17.500	17.5	65100
105	15.638843	17.500000	9/29/1997	18.250	17.500	17.5	4300
106	15.638843	17.500000	9/30/1997	17.750	17.500	17.5	21000
108	15.582988	17.437500	10/2/1997	17.500	17.250	17.5	6800
115	15.582988	17.437500	10/13/1997	17.625	17.250	17.5	9200
135	15.192017	17.000000	11/10/1997	17.500	17.000	17.5	3100
168	15.192017	17.000000	12/29/1997	17.625	17.000	17.5	3300
498	16.979313	19.000000	4/22/1999	19.000	17.500	17.5	160800
529	15.750552	17.625000	6/7/1999	17.625	17.375	17.5	53200
565	15.527136	17.375000	7/28/1999	17.500	17.250	17.5	27400
4039	16.068762	17.469999	5/20/2013	17.740	17.400	17.5	78200
4051	16.767805	18.230000	6/6/2013	18.240	17.500	17.5	101500

Finding Unique Values

```
%%time
df = pd.read_csv('/Users/.........................../YahooData.csv')
with open('YahoodataYAMLformat.yml', 'w') as outfile:
    yaml.dump(
        df.to_dict(orient='dict'),
        outfile,
        canonical=True,
        sort_keys=True,
        default_flow_style=False,
        indent=6,
        width=22,
    )
with open('/Users/........./YahoodataYAMLformat.yml') as f:
    yamldata = yaml.safe_load(f)
YahoodataYAMLDF = pd.DataFrame( yamldata)
pd.unique(YahoodataYAMLDF [['Open', 'Close']].values.ravel('K'))
array([17.5, 17.75, 18.625, ..., 48.48, 44.869999, 42.889998999999996],
```

```
dtype=object)
```

Wall Times (seconds): 10.4 s, 10.7, 10.4, 10.4, 10.9, 11, 11.4, 11.1, 10.6, 11.1,

CPU Times (seconds):

```
8.85 s ± 81.4 ms per loop (mean ± std. dev. of 7 runs, 1 loop each)
8.84 s ± 58.8 ms per loop (mean ± std. dev. of 7 runs, 1 loop each)
8.88 s ± 81.8 ms per loop (mean ± std. dev. of 7 runs, 1 loop each)
8.91 s ± 104 ms per loop (mean ± std. dev. of 7 runs, 1 loop each)
9.01 s ± 158 ms per loop (mean ± std. dev. of 7 runs, 1 loop each)
8.86 s ± 82.3 ms per loop (mean ± std. dev. of 7 runs, 1 loop each)
8.93 s ± 119 ms per loop (mean ± std. dev. of 7 runs, 1 loop each)
8.97 s ± 104 ms per loop (mean ± std. dev. of 7 runs, 1 loop each)
10.4 s ± 723 ms per loop (mean ± std. dev. of 7 runs, 1 loop each)
10.3 s ± 269 ms per loop (mean ± std. dev. of 7 runs, 1 loop each)
```

Data Sorting

```
%%time
df = pd.read_csv('/Users/.........................../YahooData.csv')
with open('YahoodataYAMLformat.yml', 'w') as outfile:
    yaml.dump(
        df.to_dict(orient='dict'),
        outfile,
        canonical=True,
        sort_keys=True,
        default_flow_style=False,
        indent=6,
        width=22,
    )
with open('/Users/........./YahoodataYAMLformat.yml') as f:
    yamldata = yaml.safe_load(f)
YahoodataYAMLDF = pd.DataFrame( yamldata)
YahoodataYAMLDF.sort_values(by=['Volume'])
```

Grouping

	Adj Close	Close	Date	High	Low	Open	Volume
1275	3.923114	4.390000	5/28/2002	4.3900	4.390000	4.3900	0
3216	3.386926	3.790000	2/10/2010	3.7900	3.790000	3.7900	0
166	15.750552	17.625000	12/24/1997	17.6250	17.625000	17.6250	100
1302	3.717577	4.160000	7/5/2002	4.1600	4.160000	4.1600	200
997	2.680944	3.000000	4/12/2001	3.0000	3.000000	3.0000	400
...	...	...	...	...	...	...	...
577	10.221101	11.437500	8/13/1999	14.5000	10.500000	14.5000	1111300
4245	15.817896	17.129999	3/14/2014	17.4100	16.709999	17.0200	1165400
892	1.675590	1.875000	11/9/2000	1.9375	1.500000	1.9375	1379700
3813	7.443938	8.160000	6/22/2012	8.5900	7.850000	8.3500	1857100
4244	15.725554	17.030001	3/13/2014	18.3400	16.799999	18.1700	2073400

Wall Times (seconds): 9 s, 9.22, 8.94, 9.22, 9.13, 9.32, 9.08, 9.53, 9.67, 9.13,

CPU Times (seconds):

```
8.87 s ± 125 ms per loop (mean ± std. dev. of 7 runs, 1 loop each)
8.8 s ± 55.5 ms per loop (mean ± std. dev. of 7 runs, 1 loop each)
8.81 s ± 67.5 ms per loop (mean ± std. dev. of 7 runs, 1 loop each)
8.85 s ± 107 ms per loop (mean ± std. dev. of 7 runs, 1 loop each)
8.83 s ± 87.7 ms per loop (mean ± std. dev. of 7 runs, 1 loop each)
8.82 s ± 82.1 ms per loop (mean ± std. dev. of 7 runs, 1 loop each)
8.88 s ± 190 ms per loop (mean ± std. dev. of 7 runs, 1 loop each)
8.86 s ± 83.7 ms per loop (mean ± std. dev. of 7 runs, 1 loop each)
9.11 s ± 259 ms per loop (mean ± std. dev. of 7 runs, 1 loop each)
8.79 s ± 122 ms per loop (mean ± std. dev. of 7 runs, 1 loop each)
%%time
df = pd.read_csv('/Users/.........................../YahooData.csv')
with open('YahoodataYAMLformat.yml', 'w') as outfile:
    yaml.dump(
        df.to_dict(orient='dict'),
        outfile,
        canonical=True,
        sort_keys=True,
        default_flow_style=False,
        indent=6,
        width=22,
```

```
    )
with open('/Users/......../YahoodataYAMLformat.yml') as f:
    yamldata = yaml.safe_load(f)
YahoodataYAMLDF = pd.DataFrame( yamldata)
YahoodataYAMLDFGroupBy = YahoodataYAMLDF.groupby('Open').mean()
```

Wall Times (seconds): 9.72 s, 9.39, 9.79, 9.23, 9.37, 9.3, 9.21, 9.43, 9.37, 9.46

CPU Times (seconds):

```
8.86 s ± 206 ms per loop (mean ± std. dev. of 7 runs, 1 loop each)
8.76 s ± 73.1 ms per loop (mean ± std. dev. of 7 runs, 1 loop each)
8.71 s ± 91.8 ms per loop (mean ± std. dev. of 7 runs, 1 loop each)
8.75 s ± 190 ms per loop (mean ± std. dev. of 7 runs, 1 loop each)
8.76 s ± 149 ms per loop (mean ± std. dev. of 7 runs, 1 loop each)
8.74 s ± 62.9 ms per loop (mean ± std. dev. of 7 runs, 1 loop each)
9.15 s ± 191 ms per loop (mean ± std. dev. of 7 runs, 1 loop each)
8.75 s ± 133 ms per loop (mean ± std. dev. of 7 runs, 1 loop each)
9.07 s ± 154 ms per loop (mean ± std. dev. of 7 runs, 1 loop each)
8.82 s ± 134 ms per loop (mean ± std. dev. of 7 runs, 1 loop each)
```

	Adj Close	Close	High	Low	Volume
Open					
1.187500	1.172913	1.312500	1.375000	1.156250	69150.000000
1.312500	1.340472	1.500000	1.541667	1.270833	106033.333333
1.375000	1.186876	1.328125	1.406250	1.203125	100550.000000
1.380000	1.340472	1.500000	1.510000	1.380000	39300.000000
1.410000	1.295790	1.450000	1.470000	1.400000	76200.000000
...	...	...	...	...	...
64.400002	64.839790	64.989998	66.290001	64.129997	142900.000000
64.639999	61.597301	61.740002	65.160004	61.049999	180800.000000
64.870003	64.291061	64.440002	65.379997	64.419998	204200.000000
65.199997	65.757668	65.910004	66.330002	64.830002	134300.000000
65.910004	64.340942	64.489998	66.330002	64.410004	129700.000000

APPENDIX IX

Results

Algorithm 6: CSV Timeseries to Parquet Data Format Conversion & Parquet to Multi-data Format Conversion Methods

Dataframe Creation

```
%%timeit
Yahoodata = pd.read_csv('/Users/............................../YahooData.csv')
YahoodataDF = pd.DataFrame(Yahoodata)
YahoodataDF.to_parquet('YahoodataDFParquet3.parquet', engine="pyarrow")
YahoodataDFParquet = pd.read_parquet('YahoodataDFParquet3.parquet')
YahoodataDFParquet.query('Open == "17.5"')
```

Wall Times (ms): 37.9 ms, 42.8 ms, 35.9, 32.3, 32.6, 45.9, 32.9, 32.9, 32.6, 33.6, 36.2

CPU Times (ms)

```
26.8 ms ± 155 µs per loop (mean ± std. dev. of 7 runs, 10 loops each)
26.9 ms ± 203 µs per loop (mean ± std. dev. of 7 runs, 10 loops each)
26.8 ms ± 197 µs per loop (mean ± std. dev. of 7 runs, 10 loops each)
26.8 ms ± 207 µs per loop (mean ± std. dev. of 7 runs, 10 loops each)
26.7 ms ± 148 µs per loop (mean ± std. dev. of 7 runs, 10 loops each)
27.1 ms ± 528 µs per loop (mean ± std. dev. of 7 runs, 10 loops each)
28.4 ms ± 2.09 ms per loop (mean ± std. dev. of 7 runs, 10 loops each)
27.3 ms ± 502 µs per loop (mean ± std. dev. of 7 runs, 10 loops each)
27.4 ms ± 722 µs per loop (mean ± std. dev. of 7 runs, 10 loops each)
27.3 ms ± 406 µs per loop (mean ± std. dev. of 7 runs, 10 loops each)
```

Data Filtering (Querying)

Wall Times (ms): 32.2 ms, 36.9, 34.3, 34.9, 33.6, 38.9, 32.6, 40.1, 34.9, 57.3

CPU Times (ms)

```
31.8 ms ± 2.28 ms per loop (mean ± std. dev. of 7 runs, 10 loops each)
31.9 ms ± 1.16 ms per loop (mean ± std. dev. of 7 runs, 10 loops each)
29.4 ms ± 227 µs per loop (mean ± std. dev. of 7 runs, 10 loops each)
29.5 ms ± 258 µs per loop (mean ± std. dev. of 7 runs, 10 loops each)
29.5 ms ± 257 µs per loop (mean ± std. dev. of 7 runs, 10 loops each)
```

```
29.7 ms ± 690 µs per loop (mean ± std. dev. of 7 runs, 10 loops each)
29.8 ms ± 286 µs per loop (mean ± std. dev. of 7 runs, 10 loops each)
30.4 ms ± 567 µs per loop (mean ± std. dev. of 7 runs, 10 loops each)
29.9 ms ± 533 µs per loop (mean ± std. dev. of 7 runs, 10 loops each)
29.4 ms ± 264 µs per loop (mean ± std. dev. of 7 runs, 10 loops each)
```

Finding Unique Values

```
array([17.5     , 17.75     , 18.625     , ..., 48.48     , 44.869999, 42.889999])
```

Wall Times (ms): 32.9 ms, 32.9, 33.9, 34.9, 31.9, 30.9, 30.9, 31.9, 29.9, 32.9

CPU Times (ms):

```
28.1 ms ± 253 µs per loop (mean ± std. dev. of 7 runs, 10 loops each)
29.2 ms ± 870 µs per loop (mean ± std. dev. of 7 runs, 10 loops each)
28.7 ms ± 1.03 ms per loop (mean ± std. dev. of 7 runs, 10 loops each)
28.3 ms ± 324 µs per loop (mean ± std. dev. of 7 runs, 10 loops each)
28 ms ± 205 µs per loop (mean ± std. dev. of 7 runs, 10 loops each)
27.9 ms ± 238 µs per loop (mean ± std. dev. of 7 runs, 10 loops each)
28.4 ms ± 547 µs per loop (mean ± std. dev. of 7 runs, 10 loops each)
28.4 ms ± 958 µs per loop (mean ± std. dev. of 7 runs, 10 loops each)
29.3 ms ± 885 µs per loop (mean ± std. dev. of 7 runs, 10 loops each)
29.1 ms ± 338 µs per loop (mean ± std. dev. of 7 runs, 10 loops each)
```

Data Sorting

Wall Times (ms): 38.4 ms, 30.9, 30.6, 31.6, 33.9, 34, 34.9, 32.9, 35.8, 59.8

CPU Times (ms):

```
29.1 ms ± 603 µs per loop (mean ± std. dev. of 7 runs, 10 loops each)
28.6 ms ± 813 µs per loop (mean ± std. dev. of 7 runs, 10 loops each)
28.5 ms ± 983 µs per loop (mean ± std. dev. of 7 runs, 10 loops each)
28.5 ms ± 339 µs per loop (mean ± std. dev. of 7 runs, 10 loops each)
27.8 ms ± 197 µs per loop (mean ± std. dev. of 7 runs, 10 loops each)
28.4 ms ± 898 µs per loop (mean ± std. dev. of 7 runs, 10 loops each)
28.1 ms ± 281 µs per loop (mean ± std. dev. of 7 runs, 10 loops each)
29.2 ms ± 1.26 ms per loop (mean ± std. dev. of 7 runs, 10 loops each)
29.6 ms ± 441 µs per loop (mean ± std. dev. of 7 runs, 10 loops each)
29.2 ms ± 1.05 ms per loop (mean ± std. dev. of 7 runs, 10 loops each)
```

	Date	Open	High	Low	Close	Adj Close	Volume
1275	5/28/2002	4.3900	4.3900	4.390000	4.390000	3.923114	0
3216	2/10/2010	3.7900	3.7900	3.790000	3.790000	3.386926	0
166	12/24/1997	17.6250	17.6250	17.625000	17.625000	15.750552	100
1302	7/5/2002	4.1600	4.1600	4.160000	4.160000	3.717577	200
997	4/12/2001	3.0000	3.0000	3.000000	3.000000	2.680944	400
...	...	...	...	...	...	...	...
577	8/13/1999	14.5000	14.5000	10.500000	11.437500	10.221101	1111300
4245	3/14/2014	17.0200	17.4100	16.709999	17.129999	15.817896	1165400
892	11/9/2000	1.9375	1.9375	1.500000	1.875000	1.675590	1379700
3813	6/22/2012	8.3500	8.5900	7.850000	8.160000	7.443938	1857100
4244	3/13/2014	18.1700	18.3400	16.799999	17.030001	15.725554	2073400

6293 rows × 7 columns

Data Grouping

Wall Times (ms): 33.9 ms, 35.1, 35.9, 36.9, 38.9, 39.9, 37.8, 34.9, 33.9, 32.9

CPU Times (ms):

```
30.1 ms ± 446 µs per loop (mean ± std. dev. of 7 runs, 10 loops each)
30 ms ± 889 µs per loop (mean ± std. dev. of 7 runs, 10 loops each)
29.7 ms ± 348 µs per loop (mean ± std. dev. of 7 runs, 10 loops each)
29.9 ms ± 1.21 ms per loop (mean ± std. dev. of 7 runs, 10 loops each)
31.3 ms ± 1.04 ms per loop (mean ± std. dev. of 7 runs, 10 loops each)
30.5 ms ± 679 µs per loop (mean ± std. dev. of 7 runs, 10 loops each)
31.9 ms ± 691 µs per loop (mean ± std. dev. of 7 runs, 10 loops each)
32.2 ms ± 2.22 ms per loop (mean ± std. dev. of 7 runs, 10 loops each)
30.2 ms ± 364 µs per loop (mean ± std. dev. of 7 runs, 10 loops each)
31.5 ms ± 974 µs per loop (mean ± std. dev. of 7 runs, 10 loops each)
```

Open	High	Low	Close	Adj Close	Volume
1.187500	1.375000	1.156250	1.312500	1.172913	69150.000000
1.312500	1.541667	1.270833	1.500000	1.340472	106033.333333
1.375000	1.406250	1.203125	1.328125	1.186876	100550.000000
1.380000	1.510000	1.380000	1.500000	1.340472	39300.000000
1.410000	1.470000	1.400000	1.450000	1.295790	76200.000000
...	...	...	...	...	...
64.400002	66.290001	64.129997	64.989998	64.839790	142900.000000
64.639999	65.160004	61.049999	61.740002	61.597301	180800.000000
64.870003	65.379997	64.419998	64.440002	64.291061	204200.000000
65.199997	66.330002	64.830002	65.910004	65.757668	134300.000000
65.910004	66.330002	64.410004	64.489998	64.340942	129700.000000

2299 rows × 5 columns

APPENDIX X

Results

Algorithm 7a: Multiformat Data Conversion Algorithm: CSV to HDF5

Dataframe Creation

```
YahooPricedata = pd.read_csv('/Users/.........../YahooData.csv')
#Convert the CSV explicitly to Pandas Dataframe
YahooPriceDataFrame = pd.DataFrame(YahooPricedata)
#Convert the data Pandas Dataframe to HDF5 format
YahooPriceDataFrame.to_hdf('YahooPriceDataFrameToHDF.h5', key='df', mode='w')
YahoodataHDF5DataFrame = pd.read_hdf('/Users/........./YahooPriceDataFrameToHDF.h5')
YahoodataHDF5DataFrame
```

Wall Times (ms): 37.8 ms, 31.3, 22.2, 31.2, 37.8, 31.3, 22.1, 31.1, 22.2, 31.2

CPU Times (ms)

```
34.7 ms ± 456 µs per loop (mean ± std. dev. of 7 runs, 10 loops each)
36.7 ms ± 5.82 ms per loop (mean ± std. dev. of 7 runs, 10 loops each)
35.3 ms ± 2.42 ms per loop (mean ± std. dev. of 7 runs, 10 loops each)
36.7 ms ± 2.76 ms per loop (mean ± std. dev. of 7 runs, 10 loops each)
36 ms ± 1.74 ms per loop (mean ± std. dev. of 7 runs, 10 loops each)
37.9 ms ± 3.68 ms per loop (mean ± std. dev. of 7 runs, 10 loops each)
34.8 ms ± 1.44 ms per loop (mean ± std. dev. of 7 runs, 10 loops each)
34.3 ms ± 948 µs per loop (mean ± std. dev. of 7 runs, 10 loops each)
34.7 ms ± 2.04 ms per loop (mean ± std. dev. of 7 runs, 10 loops each)
38.1 ms ± 3.12 ms per loop (mean ± std. dev. of 7 runs, 10 loops each)
```

Data Filtering (Querying)

Wall Times (ms): 30.7 ms, 38, 31.3, 35.8, 30.8, 34.4, 41.6, 43.2, 40, 76.5

CPU Times (ms)

```
36.3 ms ± 1.85 ms per loop (mean ± std. dev. of 7 runs, 10 loops each)
35 ms ± 190 µs per loop (mean ± std. dev. of 7 runs, 10 loops each)
35.6 ms ± 780 µs per loop (mean ± std. dev. of 7 runs, 10 loops each)
36.1 ms ± 964 µs per loop (mean ± std. dev. of 7 runs, 10 loops each)
35.8 ms ± 799 µs per loop (mean ± std. dev. of 7 runs, 10 loops each)
34.9 ms ± 292 µs per loop (mean ± std. dev. of 7 runs, 10 loops each)
```

35.1 ms ± 470 µs per loop (mean ± std. dev. of 7 runs, 10 loops each)
35.1 ms ± 523 µs per loop (mean ± std. dev. of 7 runs, 10 loops each)
35.2 ms ± 353 µs per loop (mean ± std. dev. of 7 runs, 10 loops each)
37.1 ms ± 680 µs per loop (mean ± std. dev. of 7 runs, 10 loops each)

Finding Unique Values

Wall Times (ms): 34.8, 35.1, 38.5, 39.9, 31.2, 32.6, 37.2, 33.8, 32.7, 26

CPU Times (ms):
33.5 ms ± 338 µs per loop (mean ± std. dev. of 7 runs, 10 loops each)
34 ms ± 1.26 ms per loop (mean ± std. dev. of 7 runs, 10 loops each)
33.9 ms ± 740 µs per loop (mean ± std. dev. of 7 runs, 10 loops each)
34.4 ms ± 1.4 ms per loop (mean ± std. dev. of 7 runs, 10 loops each)
33.7 ms ± 399 µs per loop (mean ± std. dev. of 7 runs, 10 loops each)
33.1 ms ± 378 µs per loop (mean ± std. dev. of 7 runs, 10 loops each)
33.2 ms ± 325 µs per loop (mean ± std. dev. of 7 runs, 10 loops each)
33.3 ms ± 353 µs per loop (mean ± std. dev. of 7 runs, 10 loops each)
34.4 ms ± 1.04 ms per loop (mean ± std. dev. of 7 runs, 10 loops each)
35.5 ms ± 709 µs per loop (mean ± std. dev. of 7 runs, 10 loops each)

Data Sorting

Wall Times (ms): 27.9 ms, 39.3, 32.8, 30.9, 27.1, 36.9, 39.5, 38.1, 27.3, 31.3,

CPU Times (ms):
35.6 ms ± 1.23 ms per loop (mean ± std. dev. of 7 runs, 10 loops each)
36.7 ms ± 545 µs per loop (mean ± std. dev. of 7 runs, 10 loops each)
33.3 ms ± 677 µs per loop (mean ± std. dev. of 7 runs, 10 loops each)
33.8 ms ± 590 µs per loop (mean ± std. dev. of 7 runs, 10 loops each)
35.2 ms ± 2.65 ms per loop (mean ± std. dev. of 7 runs, 10 loops each)
35.9 ms ± 1.68 ms per loop (mean ± std. dev. of 7 runs, 10 loops each)
32.9 ms ± 338 µs per loop (mean ± std. dev. of 7 runs, 10 loops each)
33.3 ms ± 210 µs per loop (mean ± std. dev. of 7 runs, 10 loops each)
33.8 ms ± 484 µs per loop (mean ± std. dev. of 7 runs, 10 loops each)
32.3 ms ± 107 µs per loop (mean ± std. dev. of 7 runs, 10 loops each)

Data Grouping

Wall Times (ms): 31.3 ms, 27.3, 38.1, 39.5, 36.9, 27.1, 30.9, 32.8, 39.3, 29.4, 27.9

CPU Times (ms):

```
32.9 ms ± 443 µs per loop (mean ± std. dev. of 7 runs, 10 loops each)
33.7 ms ± 1.23 ms per loop (mean ± std. dev. of 7 runs, 10 loops each)
34.5 ms ± 679 µs per loop (mean ± std. dev. of 7 runs, 10 loops each)
34.7 ms ± 563 µs per loop (mean ± std. dev. of 7 runs, 10 loops each)
35.5 ms ± 1.66 ms per loop (mean ± std. dev. of 7 runs, 10 loops each)
34.5 ms ± 688 µs per loop (mean ± std. dev. of 7 runs, 10 loops each)
34.6 ms ± 505 µs per loop (mean ± std. dev. of 7 runs, 10 loops each)
33.9 ms ± 101 ms per loop (mean ± std. dev. of 7 runs, 10 loops each)
35.2 ms ± 222 µs per loop (mean ± std. dev. of 7 runs, 10 loops each)
33.6 ms ± 107 µs per loop (mean ± std. dev. of 7 runs, 10 loops each)
```

APPENDIX XI

Results

Algorithm 8a: CSV Timeseries to TSV Data Format Conversion

Dataframe Creation

```
%%timeit
with open('/Users/............../YahooData.csv','r') as inputdata, open('/Users/................/
CSV2TSV.tsv', 'w') as TSVGenerator:
    inputdata = csv.reader(inputdata)
    TSVGenerator = csv.writer(TSVGenerator, delimiter='\t')
    for row in inputdata:
        TSVGenerator.writerow(row)
#Read in converted TSV File as input. Use a CSV Reader. Use a tab (\t) separa-
tor in the pd.read_csv Pandas reader
YahoodataTSV = pd.read_csv('/Users/oyekanlea2/Desktop/TimeSeriesCodes/CSV2TSV.
tsv' ,sep='\t')
YahoodataTSVDataFrame = pd.DataFrame(YahoodataTSV )
YahoodataTSVDataFrame
# YahoodataTSVDataFrame.query('Open == "17.5"')
# pd.unique(YahoodataTSVDataFrame[['Open']].values.ravel('K'))
# YahoodataTSVDataFrameGroupBy = YahoodataTSVDataFrame.groupby('High').mean()
# YahoodataTSVDataFrame.sort_values(by=['Volume'])
```

Dataframe Generation

Wall Times (ms): 42.9 ms, 29.3, 38.2, 35.8, 36.2, 31.7, 37.7, 29.3, 31.7, 28.4

CPU Times (ms)

```
33 ms ± 484 µs per loop (mean ± std. dev. of 7 runs, 10 loops each)
33.2 ms ± 431 µs per loop (mean ± std. dev. of 7 runs, 10 loops each)
36.5 ms ± 898 µs per loop (mean ± std. dev. of 7 runs, 10 loops each)
34.6 ms ± 1.2 ms per loop (mean ± std. dev. of 7 runs, 10 loops each)
34.6 ms ± 1.71 ms per loop (mean ± std. dev. of 7 runs, 10 loops each)
33.2 ms ± 530 µs per loop (mean ± std. dev. of 7 runs, 10 loops each)
33.2 ms ± 530 µs per loop (mean ± std. dev. of 7 runs, 10 loops each)
34.2 ms ± 698 µs per loop (mean ± std. dev. of 7 runs, 10 loops each)
34.2 ms ± 1.15 ms per loop (mean ± std. dev. of 7 runs, 10 loops each)
32.6 ms ± 473 µs per loop (mean ± std. dev. of 7 runs, 10 loops each)
```

Data Filtering (Querying)

Wall Times (ms): 28.4 ms, 31.7, 29.3, 37.7, 31.7, 36.2, 35.8, 38.2, 29.3, 42.9,

CPU Times (ms)

```
37 ms ± 837 µs per loop (mean ± std. dev. of 7 runs, 10 loops each)
36.4 ms ± 1.37 ms per loop (mean ± std. dev. of 7 runs, 10 loops each)
42.7 ms ± 8.09 ms per loop (mean ± std. dev. of 7 runs, 10 loops each)
37.7 ms ± 1.34 ms per loop (mean ± std. dev. of 7 runs, 10 loops each)
38.1 ms ± 1.36 ms per loop (mean ± std. dev. of 7 runs, 10 loops each)
37.9 ms ± 1.19 ms per loop (mean ± std. dev. of 7 runs, 10 loops each)
34.3 ms ± 1.1 ms per loop (mean ± std. dev. of 7 runs, 10 loops each)
36.4 ms ± 2.77 ms per loop (mean ± std. dev. of 7 runs, 10 loops each)
35.1 ms ± 1.56 ms per loop (mean ± std. dev. of 7 runs, 10 loops each)
33.8 ms ± 533 µs per loop (mean ± std. dev. of 7 runs, 10 loops each)
```

Finding Unique Values

```
array([17.5     , 17.75     , 18.625     , ..., 48.529999, 49.630001, 44.32     ])
```

Wall Times (ms): 41.9 ms, 40.6, 43.9, 47.1, 41.6, 45.6, 31.8, 42.2, 45.5, 36

CPU Times (ms):

```
34.8 ms ± 342 µs per loop (mean ± std. dev. of 7 runs, 10 loops each)
35.9 ms ± 1.43 ms per loop (mean ± std. dev. of 7 runs, 10 loops each)
36.3 ms ± 2.22 ms per loop (mean ± std. dev. of 7 runs, 10 loops each)
34.3 ms ± 589 µs per loop (mean ± std. dev. of 7 runs, 10 loops each)
34.3 ms ± 1.13 ms per loop (mean ± std. dev. of 7 runs, 10 loops each)
36.7 ms ± 2.09 ms per loop (mean ± std. dev. of 7 runs, 10 loops each)
35.4 ms ± 1.31 ms per loop (mean ± std. dev. of 7 runs, 10 loops each)
37.4 ms ± 5.08 ms per loop (mean ± std. dev. of 7 runs, 10 loops each)
38.8 ms ± 4.11 ms per loop (mean ± std. dev. of 7 runs, 10 loops each)
37.9 ms ± 2.75 ms per loop (mean ± std. dev. of 7 runs, 10 loops each)
```

Data Sorting

Wall Times (ms): 32.4 ms, 27.1, 36.1, 36.3, 33.1, 35.9, 38.6, 36.9, 32.4, 32.1

CPU Times (ms):

```
36.3 ms ± 1.45 ms per loop (mean ± std. dev. of 7 runs, 10 loops each)
37.5 ms ± 1.56 ms per loop (mean ± std. dev. of 7 runs, 10 loops each)
35.3 ms ± 1.64 ms per loop (mean ± std. dev. of 7 runs, 10 loops each)
```

35.6 ms ± 1.7 ms per loop (mean ± std. dev. of 7 runs, 10 loops each)
46.3 ms ± 8.82 ms per loop (mean ± std. dev. of 7 runs, 10 loops each)
36.3 ms ± 2.72 ms per loop (mean ± std. dev. of 7 runs, 10 loops each)
34.8 ms ± 943 µs per loop (mean ± std. dev. of 7 runs, 10 loops each)
44.5 ms ± 7.67 ms per loop (mean ± std. dev. of 7 runs, 10 loops each)
41.5 ms ± 4.07 ms per loop (mean ± std. dev. of 7 runs, 10 loops each)
33.7 ms ± 428 µs per loop (mean ± std. dev. of 7 runs, 10 loops each)

Data Grouping

Wall Times (ms): 43.1 ms, 34.7, 42.9, 31.1, 27.9, 44.3, 37, 27.4, 41.9, 30.1

CPU Times (ms):
36.6 ms ± 798 µs per loop (mean ± std. dev. of 7 runs, 10 loops each)
37.8 ms ± 1.81 ms per loop (mean ± std. dev. of 7 runs, 10 loops each)
35.6 ms ± 1.03 ms per loop (mean ± std. dev. of 7 runs, 10 loops each)
35.9 ms ± 623 µs per loop (mean ± std. dev. of 7 runs, 10 loops each)
35.8 ms ± 469 µs per loop (mean ± std. dev. of 7 runs, 10 loops each)
36.4 ms ± 891 µs per loop (mean ± std. dev. of 7 runs, 10 loops each)
37.8 ms ± 1.92 ms per loop (mean ± std. dev. of 7 runs, 10 loops each)
37 ms ± 1.61 ms per loop (mean ± std. dev. of 7 runs, 10 loops each)
35.9 ms ± 1.05 ms per loop (mean ± std. dev. of 7 runs, 10 loops each)
36.7 ms ± 740 µs per loop (mean ± std. dev. of 7 runs, 10 loops each)

Chapter 3
Deep Learning-Based Industrial Fault Diagnosis Using Induction Motor Bearing Signals

Saiful Islam
Ahsanullah University of Science and Technology, Bangladesh

Sovon Chakraborty
European University of Bangladesh, Bangladesh

Jannatun Naeem Muna
United International University, Bangladesh

Moumita Kabir
European University of Bangladesh, Bangladesh

Zurana Mehrin Ruhi
Brac University, Bangladesh

Jia Uddin
Woosong University, South Korea

ABSTRACT

Earlier detection of faults in industrial types of machinery can reduce the cost of production. Observing these machines for humans is always a difficult task, for that purpose we need an automated process that can constantly monitor these machines. Without continuous monitoring, a huge downfall can happen that can cost enormous monitory value. In this research, we propose some transfer learning models along with LSTM for earlier detection of faults from vibration signals. Open source Case Western Reserve University (CWRU) dataset has been used to detect four types of signals using transfer learning models. The four classes are Normal, Inner, Ball, Outer. The dataset has divided into three parts namely set1, set2, and set3. VGG19, DenseNet-121, ResNet-50, InceptionV3, and LSTM are applied to that dataset for detecting faults in this signal. The earlier result shows VGG19, LSTM and InceptionV3 can predict the faults in signal with 100% accuracy in the validation set where DenseNet-121, Resnet-50 show an accuracy of 97% and 98% respectively.

DOI: 10.4018/978-1-7998-7852-0.ch003

Copyright © 2023, IGI Global. Copying or distributing in print or electronic forms without written permission of IGI Global is prohibited.

INTRODUCTION

With the advent of the 4th industrial revolution, industry experts along with scientists all over the world have put on a considerable amount of effort to minimize all kinds of errors associated with a specific industry process & maximize efficiency. Keeping this objective in mind our research work illustrates the possible methods & strategies to detect industrial faults. In electromechanical engineering, motion is usually identified by mechanical device structures, which leads to satisfactory records of almost 70% of the gross energy ingestion in the field of modern manufacturing economics (Khan & Kim, 2016; Saidur, 2010).

Statistics have shown that a significant portion of industrial damage accounted for delayed diagnosis or identification of errors in the last one or two decades. The main motive of our research is to implement an effective indicator that can differentiate faulty signals & accurate signals of an induction motor.

Our research primarily focused on induction motors due to their extensive practical applications in the industry. The record has shown that almost 70% of the equipment & machinery in the commercial sector uses a three-phase induction motor. From residential to commercial & industry level, an induction motor is considered to be the most widely used type of machine. The simple yet robust construction, affordability with minimal maintenance & high-reliability characteristics always provide the induction motor a competitive edge in the industry as well as in residence. Induction motors have widespread applications in pumps, wind turbines & generators, where they are accountable for more or less than 70% of the gross energy consumption (Khan & Kim, 2016; Saidur, 2010). From household appliances like pumps, compressors, small fans, mixers, drilling machines to heavy machinery including lifts, cranes, oil extracting mills, textiles are just a few of the major applications where induction motor use case is exponential & most preferable. Various signature analyses of vibrations & motor currents have also been considered in the field of research to guarantee improved reliability (Soualhi et al., 2013). Railway components are quite essential for passenger safety purposes. There are lots of faults on this rail line. As a result of failures, the train and rail track components may be blemished (Karakose & Gencoglu, 2013; Santur et al., 2016; Yaman et al., 2017a). Rail flaws have been published by the International Railway Association (UIC) with UIC 712 R code (UIC-712 R, 2002).

Every machine has its characteristics and is prone to show abnormalities in various cases due to numerous problems. It is imperative to assess these unusual behaviors of the machines & incorporate corrective measures to restore the faulty machines within the shortest possible amount of time. Due to so many fault cases not limited to the stator side and rotor side, there are some techniques developed by the scientists along with other industry specialists. Fault Diagnosis & Detection (FDD) is one of the techniques that is being used for a long time. It is the process of uncovering faults or errors in the system while attempting to identify the source of the problem.

However, Conventional Fault Diagnosis & Detection (FDD) techniques have only been used to check the pattern of the process to point out any anomaly throughout the pattern. This technique allows the engineers to monitor whether the system is behaving as per the standard operating procedure defined in general. Signal processing is always an essential part of the three sectors. These signal processing techniques can be classified into the time domain, frequency domain & time-frequency domain (Yuan et al., 2012). Unfortunately, this traditional technique might not be able to expose several hidden attributes and the faults associated with those attributes, which make it difficult for the industries to not only address but also to meet the demand of the business through operational excellence.

LITERATURE REVIEW

In (Hasan et al., 2019), this paper is for diagnosing vibration of bearing signal, texture analysis based new approach was taken. From different textures, signals of different bearing failures were understood by converting the bearing vibration signals into grayscale images. Then texture features were obtained by using local binary pattern LBP. Then bearing vibration signals are classified by applying different machine learning models. They used three datasets for different speeds, bearing fault types and bearing error sizes. The first one is a success rate of 95.9%, the second and third one is 100%. It was more successful than traditional methods. In (Liu et al., 2013; Yaman et al., 2017b), a new intelligent fault diagnosis and conditioning monitoring system for the classification of mechanical equipment of different conditions that produce distinct thermal signatures for different faulty situations have been proposed in this paper. Six types of cooling radiators are considered for this proposal. The system has the following steps: thermal image acquisition, image preprocessing, image processing, two-dimensional discrete wavelet transform (2D-DWT), feature extraction, feature selection using a genetic algorithm (GA), and finally classification by artificial neural networks (ANNs). Based on the feature selection operation, their ANN layer has 16 neurons as the input layer. The topology of ANN is 16-6-6 and has finally a satisfactory accuracy and performance. Extraction from the signal's time-frequency representations (TFRs), this paper presents a new scheme for rolling bearing fault diagnosis using texture features. Firstly, adaptive optimal kernel time-frequency representation (AOK-TFR) is applied to extract TFRs of the signal, which essentially describe the energy distribution characteristics of the signal over time and frequency domain. Since the signal-dependent, radially Gaussian kernel can exactly track the minor variations in the signal and provide an excellent time-frequency concentration in a noisy environment. Secondly, the uniform local binary pattern (LBP), is used to calculate the histograms from the TFRs to characterize rolling bearing fault information. Finally, the obtained histogram feature vectors are input into the multi-SVM classifier for pattern recognition.. In (Ty et al., 2016), for reliable fault detection and classification, this paper proposed a 2D texture feature and multiclass support vector machines of induction motors. Firstly, they convert time-domain vibration signals to 2D grey images and found repetitive patterns. By generating the DNS map, they extract these features. Even in a noisy environment, they achieved 100% accuracy. In (Hasan & Kim, 2018), In this paper, their approach was for fault diagnosis of rotating machines based on thermal image histogram. A statistical approach and machine learning techniques were adopted for machine condition diagnosis. Firstly, this method was analyzed for four conditions. A proper feature was extracted through histogram features based on statistical images for thermal image data. Due to mass dimensionality, a feature extraction algorithm was used for calculated image features data. ICA showed a better clustering performance than PCA, and the most employed combo of ICA and SVM was a new finding of this paper. In (Ruiz et al., 2018), the texture features tensor was established from a sub band time-frequency image to identify the fault state of a rolling bearing. Their method was based on linear SHTM using the sub band TFT texture sensor. Their method was free from the "curse of dimensionality" problem. Then a conventional feature vector-based classifier, their linear SHTM, achieved a higher recognition rate. In (Liu et al., 2017), faults in railway condition monitoring systems can be efficiently affected and diagnosed by utilizing data mining solutions' audio data. The system enables extracting Mel-frequency cepstrum coefficients (MFCCs) from audio data with reduced feature dimensions using attribute subset selection. It employs support vector machines (SVMs) for the early detection and classification of anomalies. Accuracy exceeding 94.1% whether used alone or in combination with other known methods. Machine Learning Algorithms help reduce dimensions over a large dataset that can be fruitful in detecting faults in industrial machinery (Miraz et al., 2020). Installing sensor in motor-drive end bearing then detecting faults using CNN is proposed by M. Souza et al (Souza et al., 2021). They use a Predictive Maintenance Model combined with CNN to detect faults in bearing with 97% accuracy. Fault Diagnosis and Detection using Deep learning and Transfer learning are yet to be explored hugely. From this point of view, we have conducted this research for finding faults from vibration signals using transfer learning models.

PROPOSED METHODOLOGY

For this research, the researchers have tried to find an efficient way to detect the faults in industrial machinery earlier and reduce the loss as much as possible. An open-source dataset is used for this purpose. The dataset is divided into three sets. These datasets are fed into different transfer learning models along with LSTM. These models on the validation set have observed the result. Later the researchers used Precision, Recall, Accuracy and F1-score for measuring performance. Fig 1 shows the proposed methodology correctly

Data Collection

The researchers have used the open-source Case Western Reserve University (CWRU) dataset (Case Western Reserve University, n.d.). The dataset consists of the vibration signal of four classes, namely Normal, Ball, Inner, Outer. Except for Normal, all of them are faulty signals. The CWRU data center has evaluated 2 HP Reliance electric induction motors for normal bearings and faults in different conditions and single-point-drive end (DE) and fan end (FE).

Figure 1. Proposed diagram for measuring performance

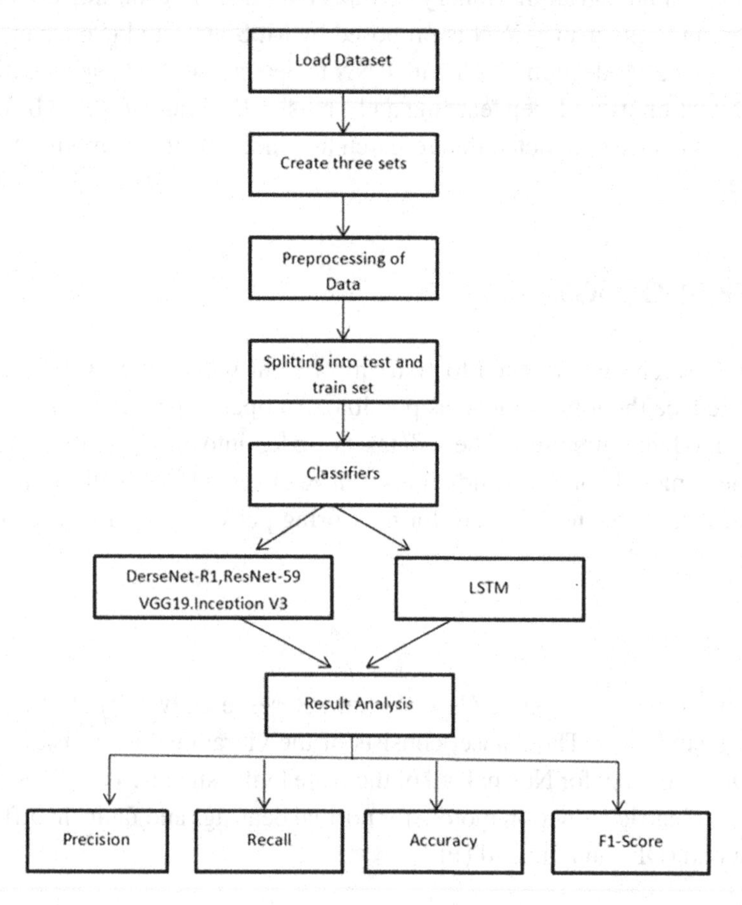

Three working environments in our dataset are available, with data collection is about 12000 samples per second. The range of motor load is 0-2. We also have a fault diameter of 0.007 inches. For creating faults on the SKF Drive End bearing (6205-2RS JEM) CWRU authority have used Electro-Discharge Machining. A varying motor speed of 1797,1772, and 1750 is available for these environments. We have plotted the histogram of all items in the data set used in this research. **Fig 1**. shows the proposed methodology of this project. The four types of sound signal is pictorially shown in **Fig 2**.

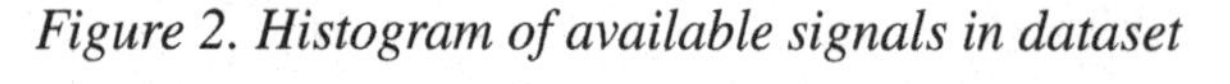

Figure 2. Histogram of available signals in dataset

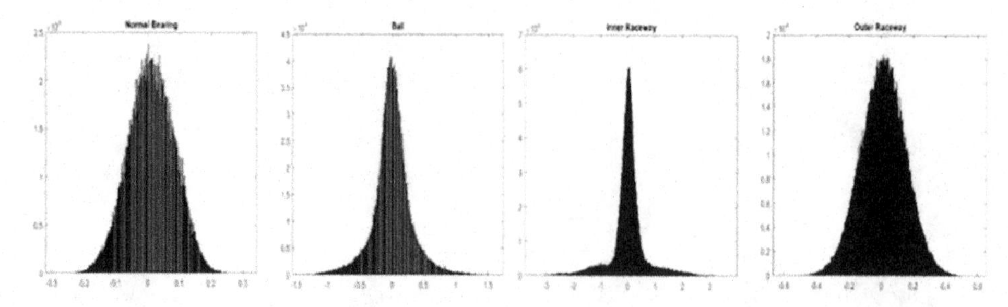

Data Preprocessing

At first, the dataset loaded with the signal data consists of three sets. All the signal data are converted into a NumPy array for faster calculation. The dataset has four types of signals. LabelBinarizer is used from Scikit Learning that accepts categorical data and returns a NumPy array. So that the signals are converted into a numerical index using this array. The researchers have used the fit_transform() method for scaling training data. Through the fit_transform() method, the researchers have also allowed the trainable models to learn about the training parameters. To increase the probability of different results, random seeds have been applied. The random seed value is set to 2. 30% of the available data in each set is used for validation purposes. In the case of applying transfer learning models, the researchers used Keras and TensorFlow library and set all the trainable layers as false. For ignoring overfitting, dropout has been applied.

Long Short-Term Memory (LSTM)

Recurrent Neural Network face difficulties carrying information from previous layer to next layer if the sequence of information is too long (Kumar et al., 2018). Vanishing gradient problem is also a major issue while propagating backwards in RNN. For controlling the flow of information LSTM play a significant role as it has some internal mechanism known as gates. The most important part is these gates which information to keep and which one to eradicate. LSTM is plays important role in case of solving short term memory problem. Gates between LSTM follow sigmoid function that keeps the value between 0 and 1. If any value gets closer to 0 then the gates usually follow that information and keep that information that are close to 1 after applying sigmoid function. LSTM usually combines information from previous hidden state with current state input. The combined information is passed to the forget layer that eliminates irrelevant information. After that the candidate layer use the combine the information that add information to a gate. The combine data is also submitted to a new input. The gate then decides which information to keep and which not to keep. After that the forget layer, input layer and candidate layer are computed. Using a vector then the output layer is generated. Pointwise multiplication of the obtained output then creates a hidden state. Figure 3 shows the flow diagram of LSTM. Detailed procedure of LSTM is shown here. LSTM plays major role in the area of computer vision

Figure 3. Flow Diagram of LSTM

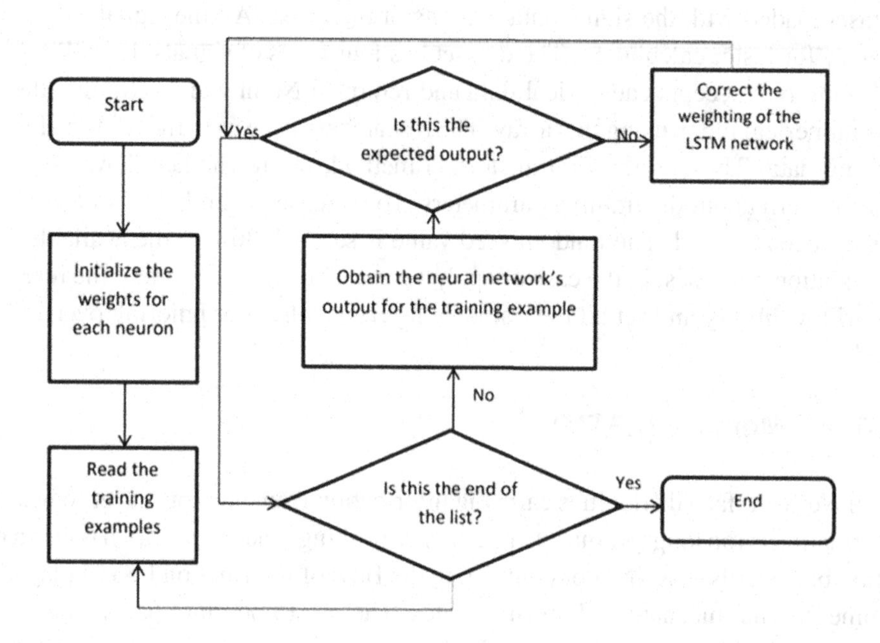

VGG19

VGG stand for Visual Geometry group, is a successor of Alex Net. VGG 19 uses weight of ImageNet. ImageNet consist of huge amount of images. Transfer learning models are generally pre-trained using these images. VGG is one of the deep CNN architecture that is hugely used in classifying images. In VGG 19 there is a total number of 19 layer (Saikia et al., 2019). The layers play significant role in terms of extracting necessary information. At first layer Convolution operation is applied where the input takes a signal or image of 224 X 224 size. The Convolution operation is applied with a Convolution matrix of size 3 X3. In the second layer the input data is again passed to a Convolution layer with a matrix size of 3 X 3. For reducing the dimensionality of feature map then Max Pooling is performed to the data obtained from second layer. After that, the data is passed into another two Convolution layers where the size of filter is 3X3. In these two Convolution layers on total 128 filters are applied with stride 3. Again, Max Pooling is performed and passed to 4 Convolution layers again of filter size 3 X 3. This time the number of filters is 256 in each layer. Max Pooling operation is again performed on the data in 11th layer. Later the images are passed into another 4 Convolution layers. This time the size of the filters is 3 X 3 with a total number of 512 filters in each layer. Again, Max Pooling operation is performed and finally passed to another four Convolution layers with a total of number of 512 filters of size 3 X 3 in each step. Finally, the input data is passed to fully connected dense layer where Softmax activation function is applied for identifying proper category of data. Figure 4 demonstrate the full flow diagram of VGG 19. The internal structure of VGG is described here precisely.

Figure 4. Proposed flow diagram of VGG 19

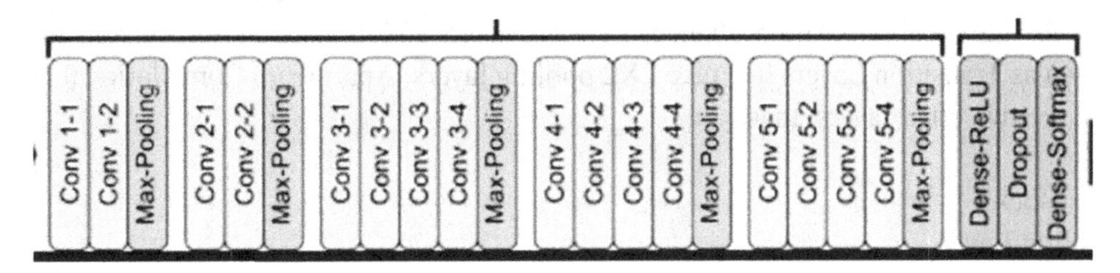

InceptionV3

InceptionV3 focuses on lesser computation, is a successor Inception family. Features such as factorization, reduction of dimensionality and regularization are added in this architecture. Factorized Convolutions tries to eliminate excessive number of parameters along with track on network efficiency. This architecture replaces bigger Convolutions with little Convolutions that help to train a model faster. During training it uses auxiliary classifiers that is in main network loss. The pooling operations applied in this architecture help to reduce the grid size and combat bottlenecks of computational cost (Hassanien et al., 2021). InceptionV3 is also trained using ImageNet and shows higher accuracy than its previous architectures. Figure 5 shows the detailed architecture of InceptionV3. There are in total 42 Convolution layer with size 3 X 3. The data is padded after convolution layers. Pooling operations are performed simultaneously for extracting the proper information by reducing the dimensionality.

Figure 5. Flow diagram of inceptionV3

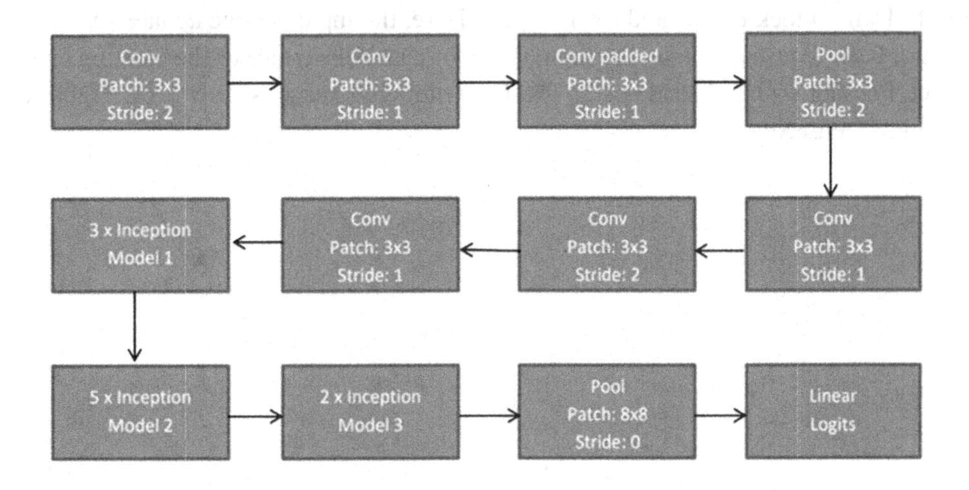

DenseNet-121

DenseNet is also known as Densely Connected Convolutional Networks (Saidur, 2010), it increases the depth of Convolutional layer. Simplifying the patter between connected layers is a unique feature of DenseNet. DenseNet takes fewer parameters than CNN because of no need to learn feature maps that are redundant. DenseNet concatenates the feature map of the layers with the incoming layers (Zhang et al., 2021). Only the number of filters changes between layers and dimension remain constant. The layers are known as Transition Layers that take 2X2 pooling layers. Apart from Convolutional and Pooling layers there are two other layers, these are

1. Dense blocks
2. Transition Layers

DenseNet focuses to start with a Convolution and pooling layer. After that the below layers are applied sequence,

1. Dense Block
2. Transition Layer
3. Dense block
4. Transition Layer
5. Dense Block
6. Transition Layer
7. Classification Layer

At the first layer there are 64 filters of size 7X7 with a stride of 2 (Zhang et al., 2021). After that Max Pooling operation is applied in the data where the pool size is 3X3 along with a stride of 2. After each Convolution block Batch Normalization is applied with ReLU activation function. Dense blocks are consisting of two Convolutions where the size of kernels is 1X1 and 3X3. The first dense block is repeated 6 times and second Dense block is repeated 12 times. Later the third Dense block is repeated 24 and fourth Dense block is repeated for 16 times. Here, the input is concatenate with the output tensor. Transition layer focuses on reducing the number of channels to half of the existing channels where 2X2 Average Pooling size and stride of 2. The Convolutional kernel size is 1X1. Figure 6 shows the flow diagram of DenseNet-121.

Figure 6. DenseNet-121 architecture diagram

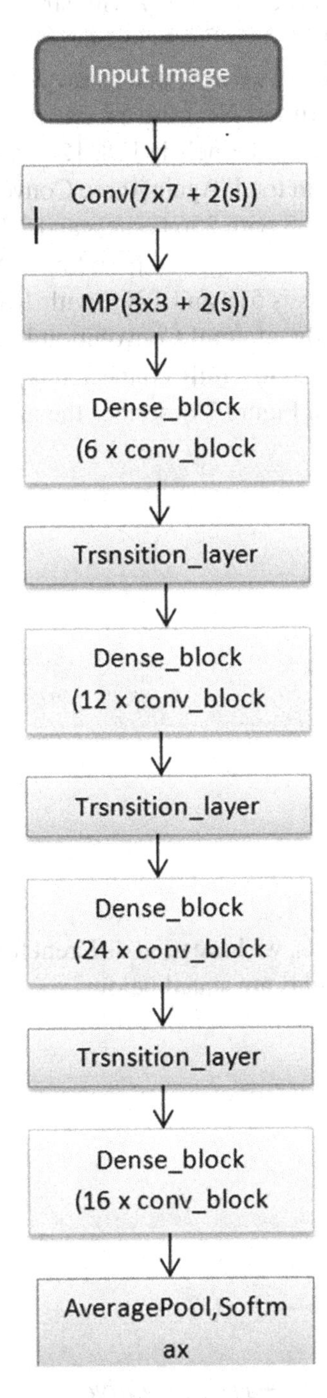

ResNet50

ResNet50 is a deep CNN architecture consist of 48 Convolution layers along with 2 pooling layers. The first pooling layer is MaxPooling layer and the second pooling layer is Average Pooling layer. ResNet50 consist of a Convolution layer at first of kernel size 7 X 7 with stride size is 2. After that Max Pooling operation is performed where the size of stride is 2. Further Convolution layer there are 64 kernels of size 1X1 following a 3 X 3 Convolution layer of 64 kernels and again a Convolution layer that consist of 256 kernels. These 3 layers are repeated for 3 times. Next three Convolution layers are applied of size 1 X 1, 3 X 3 and 1 X 1 sequence and the number of kernels is 128, 128 and 512 (Zhang & Schaeffer, 2020). The whole Convolution is applied 4 times in total. Again, three Convolution layers are used of size 1 X 1 where the number of kernels are 256, 3 X 3 with 256 kernels and 1 X 1 size of Convolution matrix of 1024 kernels. These three Convolution layers are applied for 6 times. Next three Convolution layers are of size 1 X 1, 3 X 3 with number of kernels is 512 and 1 X 1 with 1024 kernels. These three Convolution layers is applied for three times. Finally, after all Convolution layers Average Pooling operation is applied. After that the obtained data is passed to a fully connected layer where Softmax activation function is applied. There are in total 50 layers. Figure 7 describes the architecture of ResNet50. DenseNet is applicable in many sectors.

Figure 7. Flow diagram of ResNet50

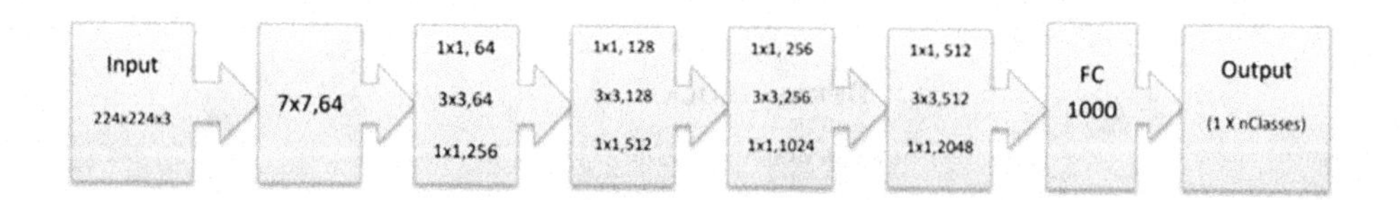

PERFORMANCE MEASURE

For measuring the efficiency of these models, we have used different formula's for calculating Precision, Recall, F1-Score and Accuracy. The formulas are stated below.

$$\frac{True\ Positive}{True\ Positive + False\ Positive} \tag{1}$$

$$\frac{True\ Positive}{True\ Positive + False\ Negative} \tag{2}$$

$$\text{Accuracy} = \frac{True\ Positive + True\ Negative}{True\ Positive + False\ Positive + True\ Negative + False\ Negative} \tag{3}$$

EXPERIMENTAL RESULT ANALYSIS

The researchers have applied LSTM at first on the dataset. The dataset consists of three training set. Table 1 shows the obtained result from LSTM. From **Table 1** it can be seen the highest accuracy shown in training set is 99.87% where in validation set the highest accuracy is 100%. Data loss is less in validation set. The lowest data loss rate is 0.0006. LSTM shows a significant result on these datasets. For testing purpose, we have used set 3

Table 1. Accuracy on training set and validation set of LSTM

Epochs	Data Loss in Training Set	Accuracy in Training Set in %	Data Loss in Test Set	Accuracy in Validation Set in %
1	1.1857	43.88	0.3021	89.60
2	0.2076	92.98	0.0128	99.89
3	0.0370	99.03	0.0013	100.0
4	0.0221	99.42	0.0008	100.0
5	0.0283	99.48	0.0007	100.0
6	0.0136	99.65	0.0010	100.0
7	0.0164	99.71	0.0007	100.0
8	0.0090	99.79	0.0010	100.0
9	0.0085	99.87	0.0008	100.0
10	0.0110	99.84	0.0006	100.0

Fig 8 plotted the obtained data accuracy in training data and validation data. Figure 8 shows the data accuracy graph of LSTM in training set and validation set.

Figure 8. Accuracy graph of LSTM on training and validation set

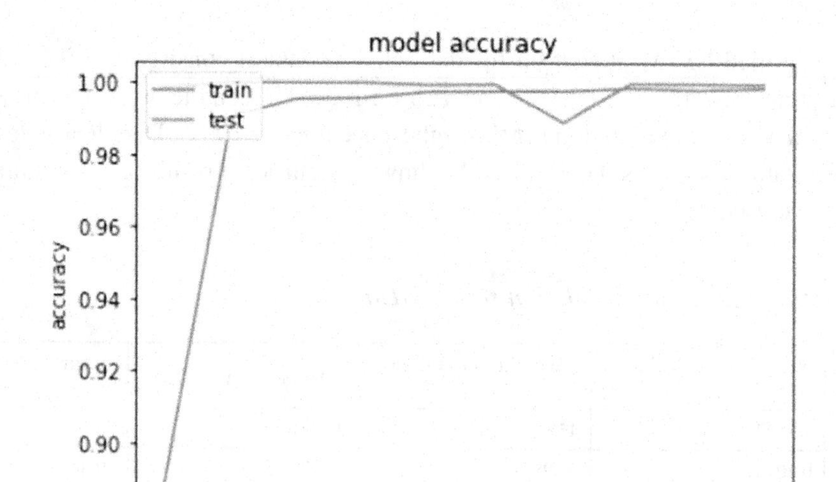

After that it has been observed the data loss graph in both training and validation set. The curves get lower with the increase of epoch number. The data loss is lesser in validation set compared to training set. The graph is shown in Figure 9.

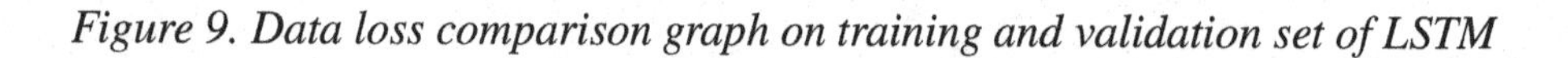

Figure 9. Data loss comparison graph on training and validation set of LSTM

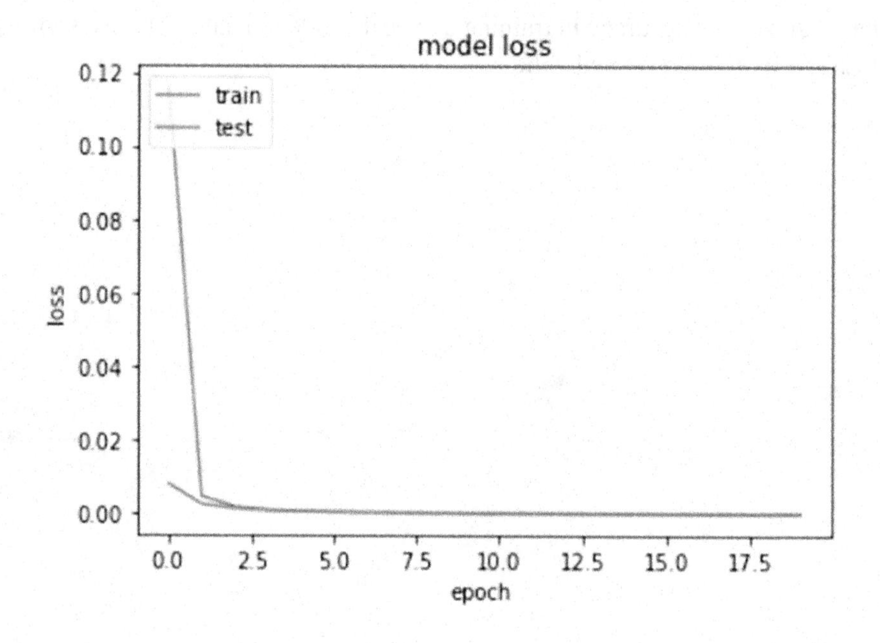

The classification report is shown in Figure 3. From Figure 10, we can see the precision is 100% in validation set. Recall is also having a higher value of 100%. Our LSTM has identified 4 classes properly.

Figure 10. Classification report of LSTM model

	precision	recall	f1-score	support
normal	1.00	1.00	1.00	5949
ball	1.00	1.00	1.00	5949
inner	1.00	1.00	1.00	5949
outer	1.00	1.00	1.00	5949
accuracy			1.00	23796
macro avg	1.00	1.00	1.00	23796

Figure 11. Heat map of LSTM model

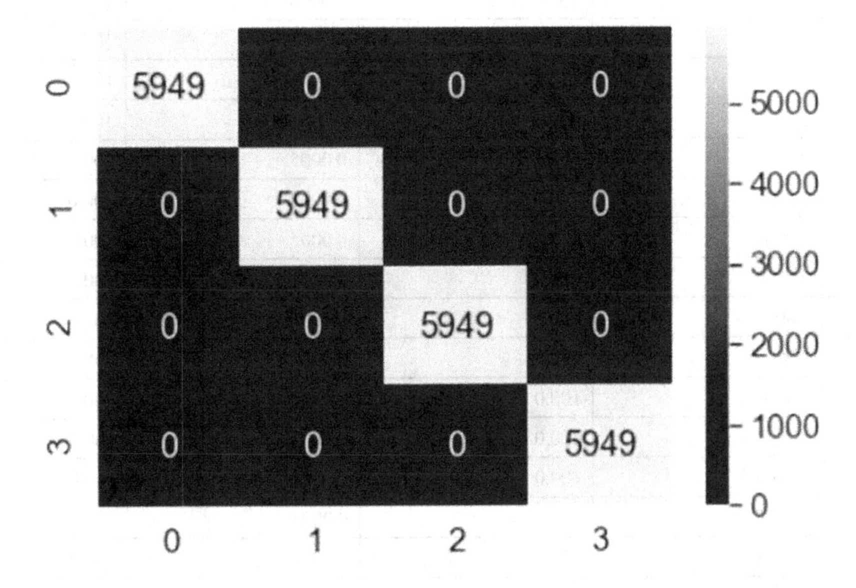

Fig 11 represents the heatmap of the validation set. After applying LSTM model, the researchers have focused on applying on transfer learning classifiers for finding accuracy on our dataset. The applied transfer learning classifiers are listed below.

1. VGG19
2. INCEPTIONV3
3. ResNet-101
4. DenseNet-121

At first, the researchers have applied transfer learning classifiers. For applying classifiers researchers have frozen out the base layer and added our own trainable layer. The trainable layer consists of 512 neurons where we applied ReLU activation function. For extracting necessary information, we have applied Global Average pooling operation. After applying pooling operation researchers have got necessary information. After that researchers have flatten the data. Final layer consists of 4 neurons for identifying four types of data. For optimizing purpose researchers have used Adam optimizer algorithm with a learning rate of 0.001. At the output layer researchers have used Softmax activation function. The batch size is 128 and observed the result unto 20 epochs. **Table 2** describes the result obtained from VGG19 after training on our dataset.

Table 2. Obtained training and validation accuracy of VGG19

Epochs	Training Set Data Loss	Training Set Accuracy in %	Validation Set Data Loss	Validation Set Accuracy In %
1	0.2925	92.36	0.0055	99.99
2	0.0041	99.98	0.0017	100.0
3	0.0012	100.0	0.0007	100.0
4	0.0010	100.0	0.0006	100.0
5	0.0009	100.0	0.0005	100.0
6	0.0007	100.0	0.0006	100.0
7	0.0010	100.0	0.0004	100.0
8	0.0008	100.0	0.0005	100.0
9	0.0009	100.0	0.0007	100.0
10	0.0007	100.0	0.0005	100.0
11	0.0006	100.0	0.0008	100.0
12	0.0005	100.0	0.0005	100.0
13	0.0006	100.0	0.0007	100.0
14	0.0007	100.0	0.0005	100.0
15	0.0006	100.0	0.0003	100.0
16	0.0008	100.0	0.0007	100.0
17	0.0005	100.0	0.0005	100.0
18	0.0008	100.0	0.0007	100.0
19	0.0007	100.0	0.0003	100.0
20	0.0007	100.0	0.0006	100.0

From Table 2 we can see the maximum accuracy in training set is 100% with a lowest data rate of 0.0005. On the other hand, maximum accuracy in validation set is 100% with a lowest data loss rate of 0.0003. Figure 12 explains the comparative accuracy graph of training set and validation set on our dataset.

In Figure 13, we can see data loss rate is much lower in validation set rather than in training set. For this research we have used set 3 as our validation set. In validation set it shows an average accuracy of

100%. The classification report is presented in Figure 14. In the classification report is totally based on the validation dataset.

Figure 12. Accuracy comparison graph of VGG19 on training and validation dataset

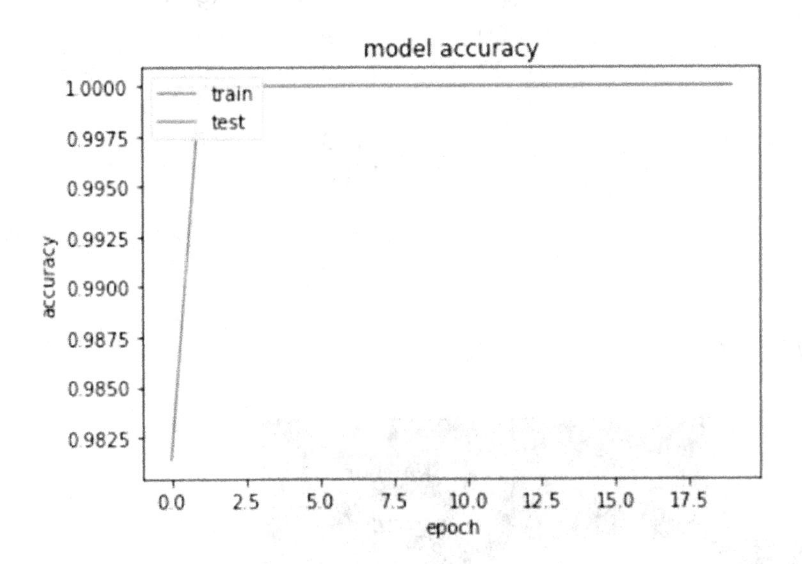

Figure 13. Data loss graph on training and validation set of VGG19

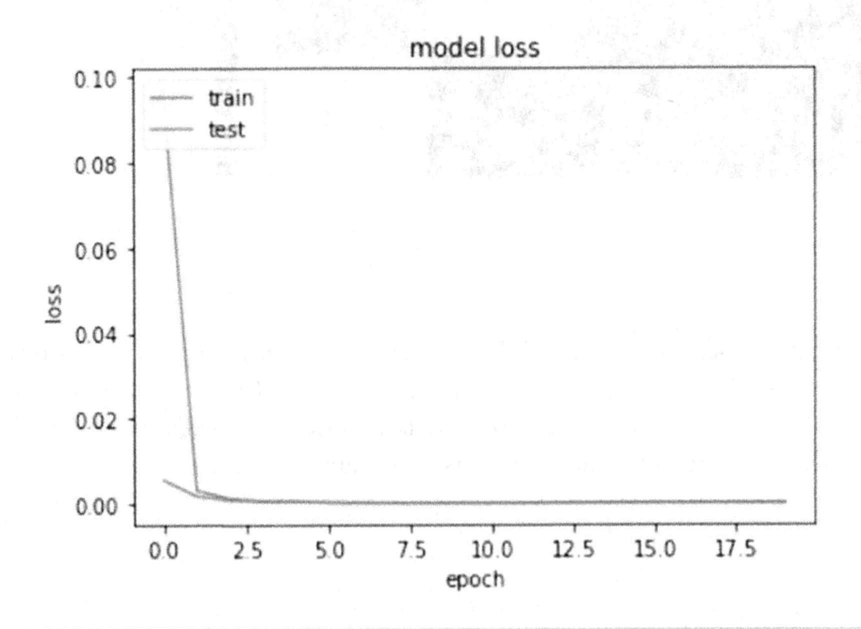

Figure 14. Classification report of VGG19

	precision	recall	f1-score	support
normal	1.00	1.00	1.00	5949
ball	1.00	1.00	1.00	5949
inner	1.00	1.00	1.00	5949
outer	1.00	1.00	1.00	5949
accuracy			1.00	23796
macro avg	1.00	1.00	1.00	23796

Figure 15. Heat map of VGG19

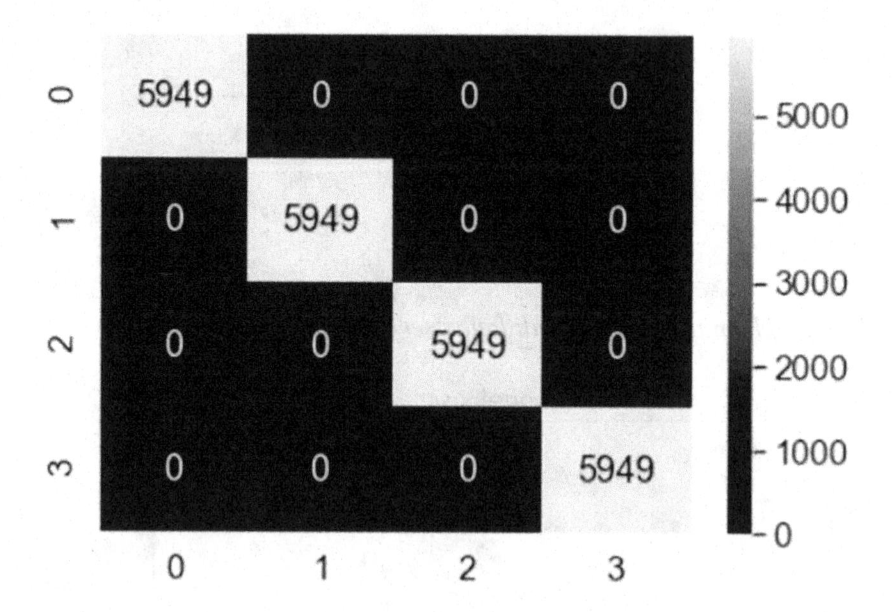

After that we have applied InceptionV3 model, Table 3 describes the training and validation accuracy after applying InceptionV3. The highest accuracy in training set is 100% with data loss rate of 0.0006. Highest accuracy in validation set is 100% with lowest data loss rate of 0.0005. Figure 16, Figure 17, Figure 18 shows the necessary graph to make comparison of training data set accuracy and validation data set accuracy.

Table 3. Accuracy obtained in training and validation set of InceptionV3

Epochs	Training Set Data Loss	Training Set Accuracy in %	Validation Set Data Loss	Validation Set Accuracy In %
1	0.3578	89.98	0.0094	100.0
2	0.0067	99.99	0.0030	100.0
3	0.0023	100.0	0.0014	100.0
4	0.0020	100.0	0.0012	100.0
5	0.0018	100.0	0.0011	100.0
6	0.0019	100.0	0.0015	100.0
7	0.0017	100.0	0.0017	100.0
8	0.0015	100.0	0.0010	100.0
9	0.0015	100.0	0.0008	100.0
10	0.0014	100.0	0.0006	100.0
11	0.0013	100.0	0.0012	100.0
12	0.0015	100.0	0.0007	100.0
13	0.0014	100.0	0.0006	100.0
14	0.0012	100.0	0.0008	100.0
15	0.0011	100.0	0.0005	100.0
16	0.0009	100.0	0.0009	100.0
17	0.0006	100.0	0.0005	100.0
18	0.0008	100.0	0.0007	100.0
19	0.0006	100.0	0.0006	100.0
20	0.0008	100.0	0.0005	100.0

Figure 16. Accuracy comparison graph of inceptioV3

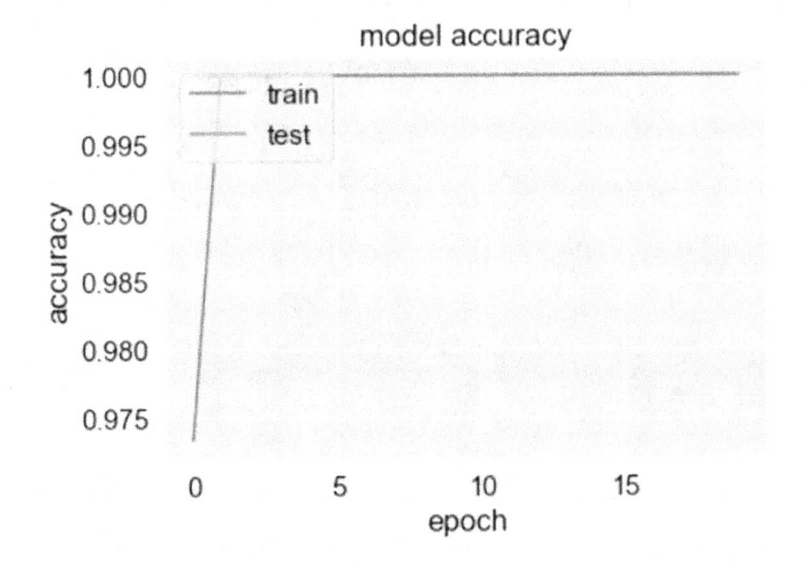

Figure 17. Data loss comparison graph of inceptionV3

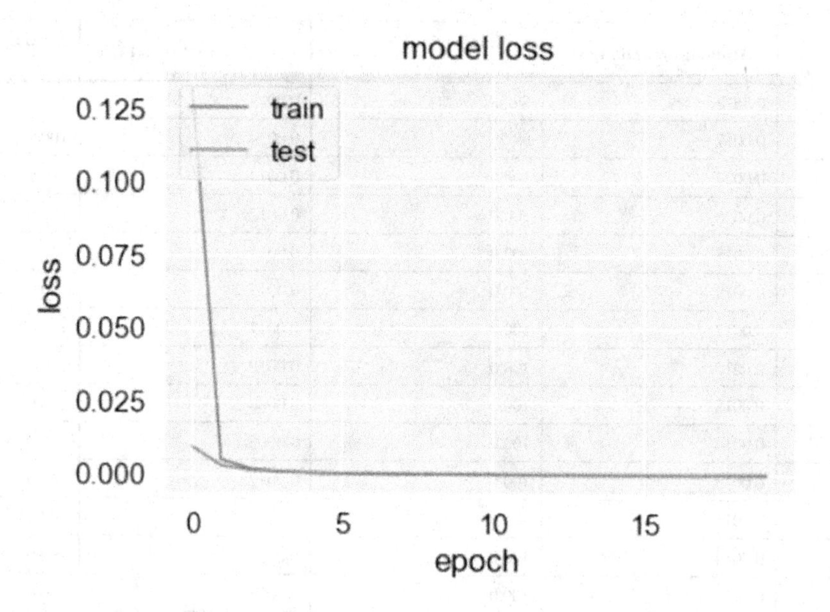

Figure 18. Classification report of inceptionV3

	precision	recall	f1-score	support
normal	1.00	1.00	1.00	5954
ball	1.00	1.00	1.00	5948
inner	1.00	1.00	1.00	5949
outer	1.00	1.00	1.00	5945
accuracy			1.00	23796

Figure 19. heat map of inceptionV3

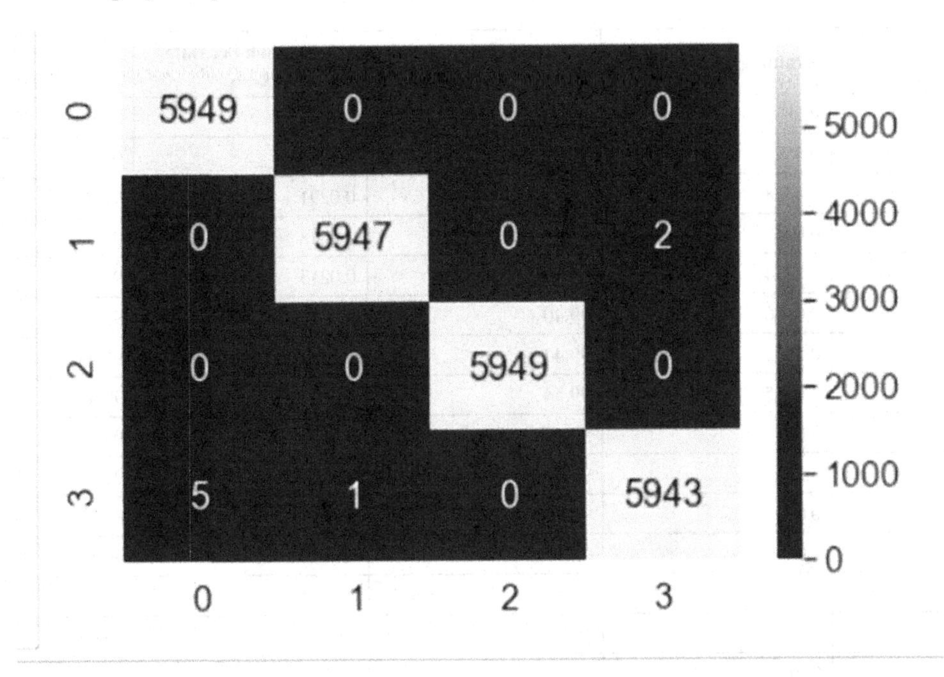

Fig 19 demonstrate that only 2 data have been classified wrongly that belongs to class 1 or inner race.

After applying InceptionV3, researchers have applied ResNet-101. **Table 4** shows the accuracy shown by ResNet-101 on training dataset and validation dataset. From Table 4 it can be seen the highest accuracy in training set is 99.85 with a lowest data loss rate of 0.0059 and in validation set maximum accuracy is 99.73 and lowest data loss rate is 0.0116.

Table 4. Accuracy obtained in training and validation set of ResNet-101

Epochs	Training Set Data Loss	Training Set Accuracy in %	Validation Set Data Loss	Validation Set Accuracy In %
1	0.5838	85.82	0.0624	98.60
2	0.0527	98.74	0.0360	99.12
3	0.0338	99.12	0.0291	99.20
4	0.0262	99.23	0.0248	99.34
5	0.0214	99.36	0.0213	99.34
6	0.0186	99.40	0.0192	99.41
7	0.0165	99.47	0.0184	99.41
8	0.0155	99.54	0.0183	99.45
9	0.0117	99.67	0.0168	99.44
10	0.0109	99.61	0.0145	99.52
11	0.0098	99.68	0.0149	99.51
12	0.0105	99.74	0.0148	99.54
13	0.0107	99.73	0.0134	99.52
14	0.0071	99.67	0.0129	99.63
15	0.0134	99.75	0.0142	99.64
16	0.0062	99.54	0.0119	99.61
17	0.0062	99.85	0.0119	99.73
18	0.0071	99.85	0.0123	99.64
19	0.0070	99.80	0.0119	99.62
20	0.0059	99.85	0.0116	99.70

Figure 20. Accuracy comparison graph of ResNet-101

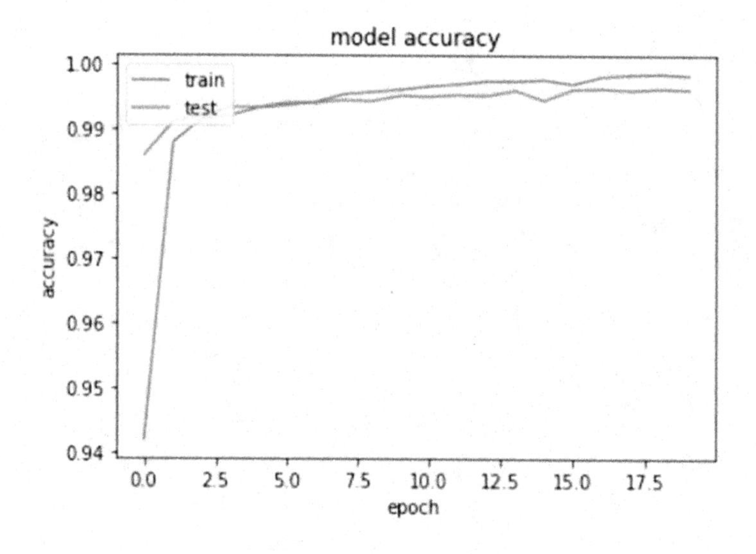

Figure 21 shows the data loss graph for each epoch for both datasets. We have used set 3 for testing purpose.

Figure 21. Data loss comparison graph for ResNet-101

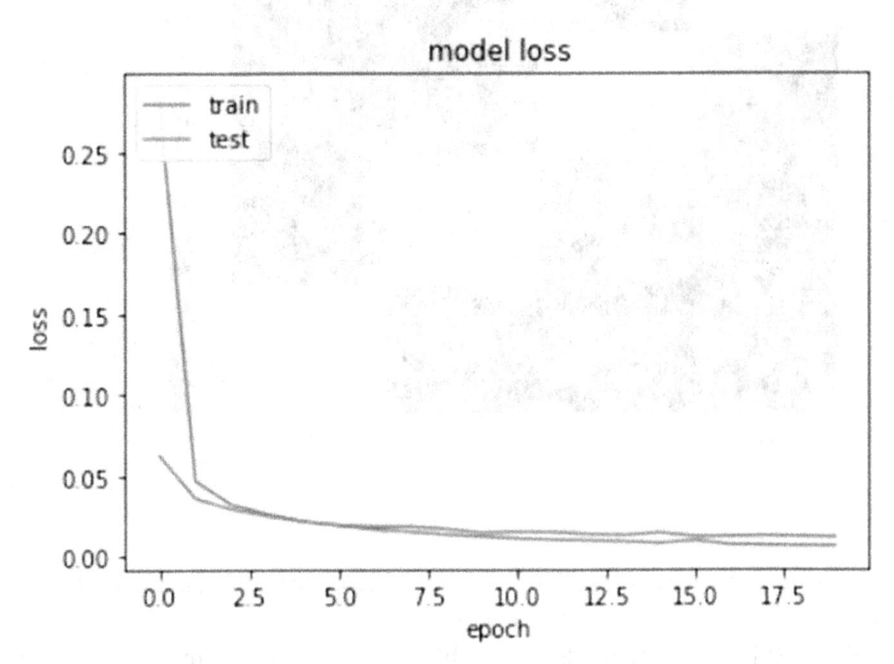

Figure 22. Classification report of ResNet-101

	precision	recall	f1-score	support
normal	0.99	1.00	0.99	5891
ball	0.98	0.92	0.95	6359
inner	0.96	0.98	0.97	5858
outer	0.94	0.99	0.96	5688
accuracy			0.97	23796
macro avg	0.97	0.97	0.97	23796

From Figure 22 we can see the precision rate is 0.99 for class normal. The classifier shows an average accuracy of 97%. ResNet-101 shows less accuracy than previous two transfer learning classifiers.

Figure 23 shows the Heatmap of ResNet-101.

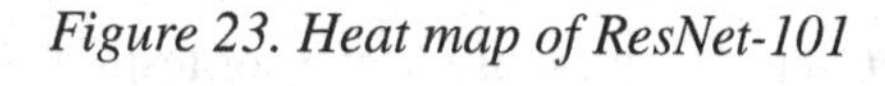

Figure 23. Heat map of ResNet-101

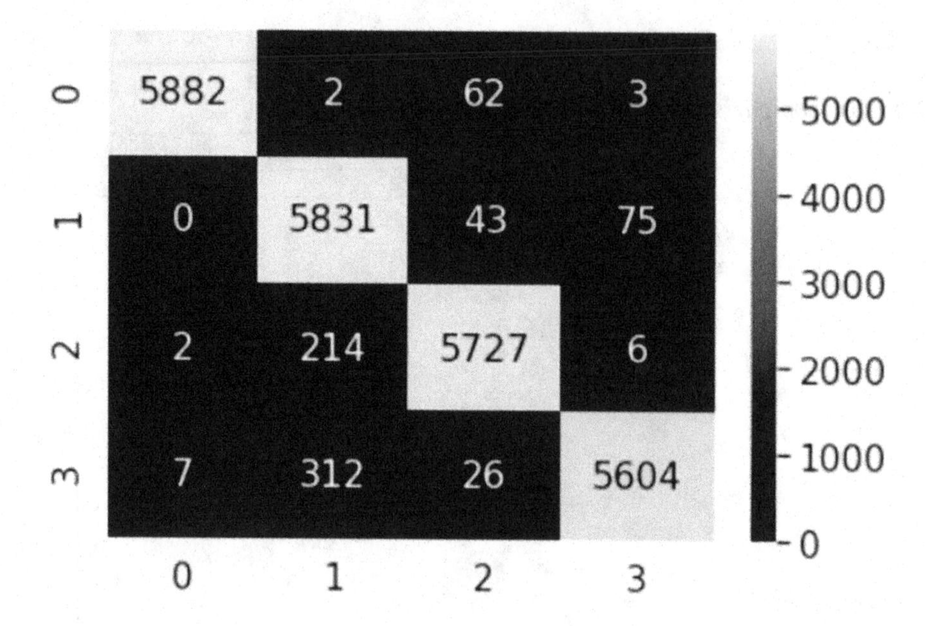

Fig 23 represents the heat map of ResNet-101 where the number of misclassified values are more than InceptionV3.

Finally, DenseNet-121 is applied on the dataset. **Table 5** describes the obtained result of DenseNet-121. The highest accuracy in training set is 99.84 and in validation set is 100%. The lowest data loss rate is 0.0010 in validation dataset. **Fig 24** shows the accuracy comparison graph between training set and validation set.

Table 5. Accuracy obtained in training and validation set of DenseNet-121

Epochs	Training Set Data Loss	Training Set Accuracy In %	Validation Set Data Loss	Validation Set Accuracy In %
1	1.3083	33.26	0.7608	65.89
2	0.5779	78.29	0.3067	89.51
3	0.1855	94.66	0.0827	99.85
4	0.0729	97.52	0.0095	100.0
5	0.0524	98.56	0.0026	100.0
6	0.0324	99.19	0.0015	100.0
7	0.0320	99.32	0.0063	100.0
8	0.0208	99.44	0.0604	100.0
9	0.0264	99.37	0.0005	100.0
10	0.0260	99.49	0.0039	100.0
11	0.0098	99.74	0.0028	100.0
12	0.0097	99.77	0.0025	100.0
13	0.0150	99.69	0.0021	100.0
14	0.0108	99.72	0.0020	100.0
15	0.0279	99.62	0.0018	100.0
16	0.0071	99.78	0.0013	100.0
17	0.0105	99.76	0.0013	100.0
18	0.0101	99.84	0.0015	100.0
19	0.0073	99.79	0.0012	100.0
20	0.0179	99.81	0.0010	100.0

Figure 24. Accuracy comparison graph of DenseNet-121

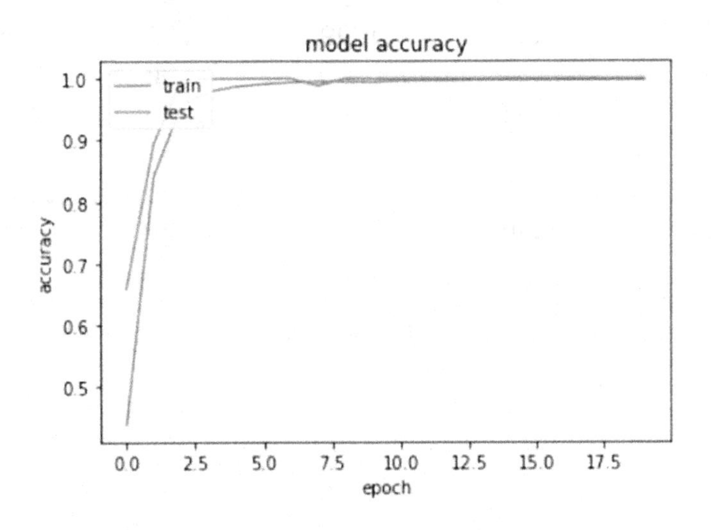

After that we have plotted the Data loss graph basis on the number of epochs. From Figure 25, we have observed data loss is higher in training set. Figure 26 we can see the classification graph where average accuracy is about 98%. In class 1 the precision level is 100% and in detecting ball fault the precision remains lowest. Figure 27 shows the Heat map DenseNet classifier.

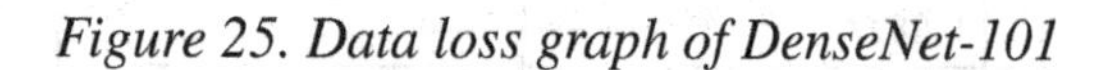

Figure 25. Data loss graph of DenseNet-101

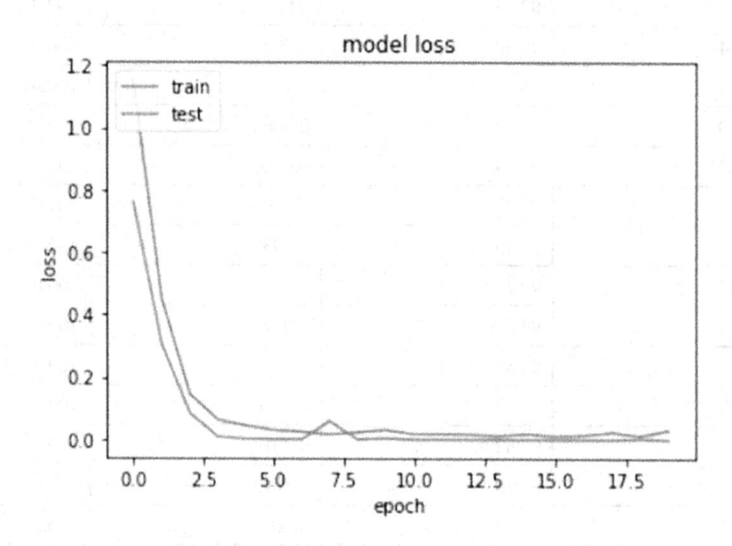

Figure 26. Classification report of DenseNet-101

	precision	recall	f1-score	support
normal	1.00	0.99	0.99	6030
ball	0.94	0.99	0.96	5609
inner	0.99	1.00	1.00	5916
outer	0.99	0.94	0.96	6241
accuracy			0.98	23796
macro avg	0.98	0.98	0.98	23796

Figure 27. Heat map of DenseNet-101

	0	1	2	3
0	5949	0	0	0
1	0	5576	0	373
2	0	33	5916	0
3	81	0	0	5868

Table 6 shows the average accuracy of each of the transfer learning classifiers and LSTM.

Table 6. Accuracy comparison table for different classifiers

Classifier Name	Average Accuracy on Set 2	Average Accuracy on Set 3
LSTM	100%	100%
VGG19	100%	100%
InceptionV3	100%	100%
ResNet-121	97.0%	98.0%
DenseNet-101	98.0%	98.0%

From Table 6 we have seen ResNet-121 shows less accuracy than any other models. For validation we have used set 2 and set3 where for training we have used set 1.

FUTURE RESEARCH DIRECTIONS

In the following research, researchers will try to apply machine learning algorithms to our dataset and compare them with our proposed transfer learning classifiers. The future target is to use more than one dataset and observe the accuracy level for different classifiers.

CONCLUSION

Researchers have used Long Short-Term Memory and four transfer learning classifiers for detecting faults in industrial machines. Researchers have differentiated the dataset into four classes. All four models are well capable of detecting defects in industrial machines. Earlier detection of faults can help in separating machines. Among transfer learning models InceptionV3 VGG19 have shown the highest average accuracy. ResNet-121 has demonstrated an accuracy of 97% on validation set 2, and 98% on validation set 3. DenseNet has shown an accuracy level of 98% on set 2 and 98% on set 3. LSTM has a magnificent efficiency in detecting faults as it offers an accuracy of 100% on both validations sets. The prime target is to develop Smart Factories through quick learning classifiers that can detect defects earlier.

REFERENCES

Amin, H., Darwish, A., & Hassanien, A. E. (2021). Classification of COVID19 X-ray Images Based on Transfer Learning InceptionV3 Deep Learning Model. In A. E. Hassanien & A. Darwish (Eds.), *Digital Transformation and Emerging Technologies for Fighting COVID-19 Pandemic: Innovative Approaches. Studies in Systems, Decision and Control* (Vol. 322). Springer. doi:10.1007/978-3-030-63307-3_7

Biswas, A., Chakraborty, S., Rifat, A. N. M. Y., Chowdhury, N. F., & Uddin, J. (2020). Comparative Analysis of Dimension Reduction Techniques Over Classification Algorithms for Speech Emotion Recognition. In M. H. Miraz, P. S. Excell, A. Ware, S. Soomro, & M. Ali (Eds.), *Emerging Technologies in Computing. iCETiC 2020. Lecture Notes of the Institute for Computer Sciences, Social Informatics and Telecommunications Engineering* (Vol. 332). Springer. doi:10.1007/978-3-030-60036-5_12

Case Western Reserve University. (n.d.). *Open Source Data Set*. Case Western Reserve University. https://csegroups.case.edu/bearingdatacenter/pages/welcome-case-western-reserve-university-bearing-data-center-website

Hasan, J., Kim, J.-M., & Manjurul Islam, M. M. (2019). Acoustic spectral imaging and transfer learning for reliable bearing fault diagnosis under variable speed conditions. *Measurement, 138,* 620-631. doi:10.1016/j.measurement.2019.02.075

Hasan, M. J., & Kim, J.-M. (2018). Bearing Fault Diagnosis under Variable Rotational Speeds Using Stockwell Transform-Based Vibration Imaging and Transfer Learning. *Applied Sciences (Basel, Switzerland), 8*(12), 2357. doi:10.3390/app8122357

Karakose, E., & Gencoglu, M. T. (2013). An analysis approach for condition monitoring and fault diagnosis in pantograph-catenary system. *IEEE International Conference on In Systems, Man, and Cybernetics (SMC),* (pp. 1963-1968). IEEE. 10.1109/SMC.2013.337

Khan, S. A., & Kim, J. M. (2016). Automated bearing fault diagnosis using 2d analysis of vibration acceleration signals under variable speed conditions. *Shock and Vibration, 2016,* 8729572. doi:10.1155/2016/8729572

Kumar, J., Goomer, R., Singh, A. K., & Recurrent, L. S. T. M. (2018). Neural Network (LSTM-RNN) Based Workload Forecasting Model For Cloud Datacenters. *Procedia Computer Science, 125*, 676-682. Science Direct. doi:10.1016/j.procs.2017.12.087

Liu, H., Xie, T., Ran, J., & Gao, S. (2017). An Efficient Algorithm for Server Thermal Fault Diagnosis Based on Infrared Image. *Journal of Physics Conference Series, 910*, 1240-1256. . doi:10.1088/1742-6596/910/1/012031

Liu, J., Liu, Y., Cheng, J., & Feng, F. (2013). Extraction of Gear Fault Feature Based on the Envelope and Time-Frequency Image of S Transformation. *Chemical Engineering Transactions*, *33*, 55–60. doi:10.1016/j.ces.2013.01.060

Ruiz, M., Mujica, L. E., Alférez, S., Acho, L., Tutivén, C., Vidal, Y., Rodellar, J., & Pozo, F. (2018). *Wind turbine fault detection and classification by means of image texture analysis* (Vol. 107). Mechanical Systems and Signal Processing. doi:10.1016/j.ymssp.2017.12.035

Saidur, R. (2010). A review on electrical motors energy use and energy savings. *Renewable & Sustainable Energy Reviews*, *14*(3), 877–898. doi:10.1016/j.rser.2009.10.018

Saikia, A. R., Bora, K., Mahanta, L. B., & Das, A. K. (2019). *Comparative assessment of CNN architectures for classification of breast FNAC images* (Vol. 57). Tissue and Cell., doi:10.1016/j.tice.2019.02.001

Santur, Y., Karaköse, M., & Akın, E. (2016). Learning Based Experimental Approach For Condition Monitoring Using Laser Cameras In Railway Tracks. *International Journal of Applied Mathematics, Electronics and Computers*, *4*(Special Issue-1), 1–5. doi:10.18100/ijamec.270656

Soualhi, A., Clerc, G., & Razik, H. (2013). Detection and diagnosis of faults in induction motor using an improved artificial ant clustering technique. *IEEE Transactions on Industrial Electronics*, *60*(9), 4053–4062. doi:10.1109/TIE.2012.2230598

Souza, R., Nascimento, E., Miranda, U., Silva, W., & Lepikson, H. (2021). Deep learning for diagnosis and classification of faults in industrial rotating machinery. *Computers & Industrial Engineering, 153*. doi:10.1016/j.cie.2020.107060

Ty, J., Chen, P. H., Te, J., Nanfei, D. W. (2016). *The Fault Feature Extraction of Rolling Bearing Based on EMD and Difference Spectrum of Singular Value*. doi:. doi:10.1155/2016/5957179

UIC-712 R. (2002). *Rail defects, International Union of Railways (UIC)*. Fransa.

Yaman, O., Karakose, M., & Akin, E. (2017a). Improved Rail Surface Detection and Condition Monitoring Approach with FPGA in Railways. *International Conference on Advanced Technology & Sciences (ICAT'17)*, (pp. 108-111). IEEE.

Yaman, O., Karakose, M., & Akin, E. (2017b). A Fault Diagnosis Approach for Rail Surface Anomalies Using FPGA in Railways. *International Journal of Applied Mathematics Electronics and Computers*, *2017*(Special Issue), 42–46. doi:10.18100/ijamec.2017SpecialIssue30469

Yuan, H., Li, F., & Wang, H. (2012). Using Evaluation and Leading Mechanism To Optimize Fault Diagnosis Based on Ant Algorithm. *Energy Proscenia*, *1*(6), 112–116.

Zhang, L., & Schaeffer, H. (2020). Forward Stability of ResNet and Its Variants. *Journal of Mathematical Imaging and Vision*, *62*(3), 328–351. doi:10.100710851-019-00922-y

Zhang, Y. D., Satapathy, S. C., Zhang, X., & Wang, S.-H. (2021). COVID-19 Diagnosis via DenseNet and Optimization of Transfer Learning Setting. *Cognitive Computation*. doi:10.100712559-020-09776-8 PMID:33488837

Chapter 4
Next-Generation Industrial Robotics:
An Overview

Khalid H. Tantawi
https://orcid.org/0000-0002-2433-6815
University of Tennessee at Chattanooga, USA

Victoria Martino
University of Tennessee at Chattanooga, USA

Dajiah Platt
University of Tennessee at Chattanooga, USA

Yasmin Musa
Motlow State Community College, USA

Omar Tantawi
Motlow State Community College, USA

Ahad Nasab
University of Tennessee at Chattanooga, USA

ABSTRACT

The Covid-19 pandemic resulted in a disruption across all industries; market data suggests that the demand for industrial robotics has steadily increased despite the pandemic, and possibly due to the impact of the pandemic. Particularly in industries that were not historically robotic markets, such as hospitals and distribution industries. In addition to that, non-traditional markets such as electronics and pharmaceuticals have become the dominant markets for industrial robots, taking over that position that has always been held by the automotive industry. The main challenge that faces increased artificial-intelligence (AI)-based technologies in next-generation collaborative robotics is the need for established preventive and corrective maintenance protocols on the AI-based technologies, as well as the need for established technician training programs on the new technologies and the wide availability of trained technicians. Another challenge that faces deploying mobile collaborative robotics in industry is the lack of safety standards for that technology.

DOI: 10.4018/978-1-7998-7852-0.ch004

Copyright © 2023, IGI Global. Copying or distributing in print or electronic forms without written permission of IGI Global is prohibited.

INTRODUCTION

In the year 2020, a breakthrough was marked in the history of industrial robotics, with the Electronics industry, for the first time, becoming the biggest market of industrial robots (International Federation of Robotics, 2021), a position that has always been firmly held by the automotive industry (Tantawi, Fidan, & Tantawy, 2019) (See **Figure 1** below). In addition to that, the industrial robotics market witnessed a historical growth in the metal and machinery, Chemicals, and the food industries.

Figure 1. Market size of industrial robotics from 2017 to 2020

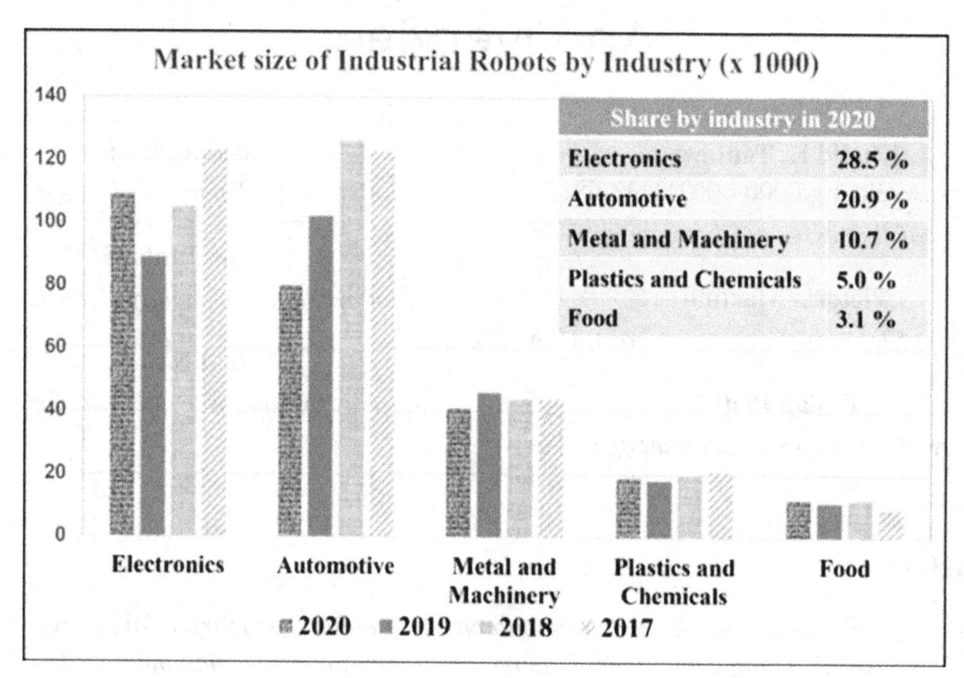

The data in the figure also show that a slight shrink in the world industrial robotics market took place in 2018 and 2019, due to a slowdown in the world economy and the ongoing trade conflict that involves the U.S., Europe, and China.

Another point that can be seen in the figure is that the automotive sector witnessed the strongest impact from the CoVid-19 pandemic. This is attributed to the switch to work-from-home in many industry sectors that continues until today, while the demand for computers and high-tech equipment at homes resulted in a sharp increase in the electronics sector.

The taking over of the largest robotics market share by the electronics sector was expected regardless of the state of the pandemic, as was evident in the year 2016, when the electronics industry exceeded the automotive industry in demand for industrial robotics in China, Japan, and Korea. Worldwide, the electronics sector's share of the robotics market rose steadily to 32% in 2017, almost equal to the automotive sector (33%). The rapid growth in the electronics and metal industries is resulting in a change in dominance of the robotics market. One important point that can be concluded from this change, is that new industry markets are emerging and overshadowing the historical automotive market. This means that there will be even greater need for skilled technicians who are trained to handle procedures and robotic manipulators commonly used in the electronics and pharmaceutical industries such as high-speed pick-and-place operations as opposed to spot welding procedures, which are more common in the automotive industry. This is in addition to the growing need for skilled workers in other smart manufacturing technologies such as Internet-of-Things (Musa, et al., 2019) (Terry, et al., 2020) (Tantawi, Fidan, Chitiyo, & Cossette, 2021).

Overall, the CoVid-19 Pandemic, resulted in a dramatic increase in the demand for industrial robotics, with skyrocketing demands for no-contact material handling in applications beyond the manufacturing industry such as in hospitals (Tantawi, Ashcroft, Cossette, Kepner, & Friedman, Investigation of the Post-Pandemic STEM Education (STEM 3.0), 2021). One of the major impacts of the pandemic was the exposure of the urgent need for more preparedness in developing self-sufficiency in the supply chain (Ayati, Saiyarsarai, & Nikfar, 2020) (Tantawi, Ashcroft, Cossette, Kepner, & Friedman, Investigation of the Post-Pandemic STEM Education (STEM 3.0), 2021) (Tantawi, 2020). Consequently, the supply chain disruptions that were seen during the pandemic have abruptly accelerated the use of next-generation intelligent robotics (Tantawi, Sokolov, & Tantawi, 2019) (Tantawi, Fidan, & Tantawy, 2019).

The primary technology that drove the deployment of next-generation of industrial robotics was artificial intelligence (AI), this industry in turn, has experienced a rapid growth in the last few years. Forbes reports that the AI industry achieved a 70% growth rate in 2018 (Coleman, 2018). The term "Intelligent Industrial Robotics" refers to industrial robotics that incorporate artificial intelligence in them, which then initiated what came to be known as Human Robot Collaboration (HRC) (Lambrecht, Kästner, Guhl, & Krüger, 2021). These technologies had undergone significant progress in the recent years, employing methods that primarily rely on algorithms based on neural networks. They can simplify troubleshooting and robot teaching processes and reduce costs significantly for manufacturers. Machine intelligence has already been established in many other technologies and applications (Kusiak, 2017) (Berggren, Nilsson, & Robinson, 2007).

Although intelligent robotics have been used in medical and military applications, they have only recently started penetrating industrial applications. The main reason for this is that industry requires established and standardized protocols for corrective and predictive maintenance, as well as established training programs and trained, widely available technicians. The introduction of intelligent robotics means manufacturers will need technicians trained in this new technology.

However, the incorporation of AI in industrial robotics, despite the slow progress, is a natural evolution in the field, however, it is still far behind when compared to service robots used in medical and military applications, in which AI is given a high priority. In the publication translated as "The Quietly Emerging AI Arms Race" (人工智能军备竞赛正在悄然兴起) (The Quietly Emerging Artificial Intelligence Arms Race, 2019) military officers from the National University of Defense Technology in China conclude that AI will be the primary defining factor for the nation that will take the lead on the world stage, and that AI will form the third military revolution (after gunpowder and nuclear power).

When analyzing the progress of deployment of the next-generation intelligent industrial robotics such as collaborative robots, the IFR 2022 report shows that there is a fast increase in demand for these robots when compared to traditional robots. For example in 2017, collaborative robots only had 2.75% of the share of the industrial robots market, the share increased to 5.73% in 2020. The IFR forecasts that intelligent robots will take the lead in the robotics industry in the near future.

Before the Covid-19 pandemic, the Intelligent Industrial Robotics market was expected to grow to $14.29 billion in 2023 from $4.94 billion in 2018 (Demaitre, 2018) (Tantawi, Sokolov, & Tantawi,

2019). The 2021 report of the International Federation of Robotics (IFR), shows that there were more than 3 million industrial robotic units in the world in 2020. ***Table 1*** lists the top 10 countries in growth of industrial robotics for the years 2016, 2019, and 2021. The data shows that the market size of China alone constituted 46% of the top 10 market size in 2016, and steadily grew to 52% in 2021.

Table 1. Top 10 markets of industrial robotics in 2016, 2019, and 2021

Year:	2016		2019		2021	
Rank	**Country**	**Units**	**Country**	**Units**	**Country**	**Units**
1	China	87,000	China	140,492	China	168,400
2	Korea	41,400	Japan	49,908	Japan	38,700
3	Japan	38,600	United States	33,339	United States	30,800
4	United States	31,400	Korea	27,873	Korea	30,500
5	Germany	20,000	Germany	20,473	Germany	22,300
6	Taiwan	7,600	Italy	11,100	Italy	8,500
7	Italy	6,500	France	6,700	Taiwan	7,400
8	Mexico	5,900	Taiwan	6,400	France	5,400
9	France	4,200	Mexico	4,600	Singapore	4,600
10	Spain	3,900	India	4,300	Spain	3,400

DEFINITIONS AND TERMINOLOGY

The definition of a robot is still very broad. Joe Engelberger, who contributed significantly in the advancement of robots, says, "I can't define a robot, but I know one when I see one" (Estolatan, Geuna, Guerzoni, & Nuccio, 2018). One way to describe robots' states that all robots share three basic components:

1. A sensory system
2. data processing capability
3. Motion

The three components are described by the formula: sense-think-act (Estolatan, Geuna, Guerzoni, & Nuccio, 2018). Therefore, under this formula, everything from milling and lathe machines to autonomous robotic vacuum cleaners, unmanned aerial vehicles (UAVs), and even smart washing machines and ATM machines are considered robots.

A more accurate definition is that used by the International Organization for Standardization (ISO) under the ISO 8373:2012 standard, which uses the following definitions for robots, industrial robots, and service robots (International Organization for Standardization, 2012):

Robot: Actuated mechanism programmable in two or more axes with a degree of autonomy, moving within its environment, to perform intended tasks. A robot includes the control system and interface of the control system. The classification of robot into industrial robot or service robot is done according to its intended application.

Industrial Robot: Automatically controlled, reprogrammable, multipurpose manipulator, programmable in three or more axes, which can be either fixed in place or mobile for use in industrial automation applications. The industrial robot includes: the manipulator, including actuators; and the controller, including teach pendant and any communication interface (hardware and software).

Service Robot: A robot that performs useful tasks for humans or equipment excluding industrial automation applications.

Therefore, under these definitions, if for example, an articulated robot is used in a manufacturing plant, then the robot is classified as an industrial robot, but if the same robot is used for serving drinks in a coffee shop, the robot then falls under the service robots category.

In this chapter, we will adopt the definition used by the International Organization for Standardization (ISO) and accepted by the International Federation of Robots. Throughout the chapter, the term "robotics" will be used to encompass robotics used for industrial applications, which the IFR and the ISO 8373 standard refer to as "Industrial Robots".

COMPARATIVE STUDY OF NEXT-AND CURRENT-GENERATION INDUSTRIAL ROBOTICS

Overview of the Two Technologies

The use of intelligent industrial robotics can cause a significant impact in four primary fields (Okuda, Haraguchi, Domae, & Shiratsuchi, 2016):

- Collision avoidance in real time,
- Performing Human-Robot Cooperative (HRC) tasks (Okuda, Haraguchi, Domae, & Shiratsuchi, 2016) (Pellegrinelli, Orlandini, Pedrocchi, Umbrico, & Tolio, 2017),
- Three-Dimensional vision
- Sensing and environment perception capability, and Force control and inspection.

Based on machine intelligence, the next generation of industrial robotics, Collaborative robots (co-bots), are more dynamic and compact in size than traditional industrial robots. They are equipped with an advanced sensory system to sense human presence around them. Reports indicate that deploying co-bots resulted in increased production in some of the major automotive manufacturers (Toyota, Ford, and Mercedes Benz) (Estolatan, Geuna, Guerzoni, & Nuccio, 2018). During the Covid-19 pandemic, cobots can support the implementation of social distancing in factories (Malik, Masood, & Kousar, 2020).

Automated mobile robots have also seen a surge in demand due to the pandemic in medical applications and in manufacturing as well. Large manufacturing companies such as the Ford Motor Company used their factories to produce medical respirators and face masks (Cardona, Cortez, Palacios, & Cerros, 2020).

There are two primary advantages of deploying the next-generation intelligent industrial robotics, in place of the current generation of robots:

The Ability of Intelligent Robotics to Self-Learn

The Intelligent robots can self-learn the optimal process and solve problems. For example, an intelligent robot can self-learn to twist itself for the best grip of a part, and the part in front of it does not have to be in an exact orientation to be handled by the robot. This is contrast to the current-generation robotics which cannot self-learn, and any slight change in the orientation of the part to be picked, can cause the robot to jam and cause a production line to stop. The improvement in the process after the machine self-learns can be as much as a "million fold" (The Robot Revolution: The New Age of Manufacturing | Moving Upstream S1-E9, 2018).

The Mobility, Free Movement, And Ability to Collaborate

The high flexibility achieved from intelligent robotics improves productivity and safety significantly. The current generation of industrial robots are isolated in cages, and only operate in highly- controlled and deterministic environments for safety (Anderson, 2016) (Liu & Tomizuka, 2017). As a result, the National Institute of Standards predicts that intelligent robotics can save manufacturers at least $40.4 billion annually (Anderson, 2016).

Improved energy efficiency is also a characteristic feature in the next generation of industrial robots. Energy efficiency in robotized industrial plants has historically been optimized by improvements in hardware, software, or a combination of both (Carabin, Wehrle, & Vidoni, 2017). Optimizing energy efficiency through hardware improvement can be achieved by proper selection of the robotic system (Glodde & Afrough, 2014), replacement of hardware components with more efficient components (Albu-Schaffer, et al., 2007), or addition of components for energy storage and recovery such as flywheels (Gale, Eielsen, & Gravdahl, 2015) (Carabin, Wehrle, & Vidoni, 2017). While energy efficiency optimization through software, is achieved by optimizing the trajectory of the robotic arm or through improving operation schedules (Paes, Dewulf, VanderElst, Kellens, & Slaets, 2014) (Carabin, Wehrle, & Vidoni, 2017).

However, the main disadvantage in industrial robotic systems was the fact that industrial robots have to be isolated in cages, and only operate in highly- controlled and deterministic environments for safety (Anderson, 2016) (Liu & Tomizuka, 2017). In the next-generation intelligent industrial robotics, the approach of improving energy consumption is radically changed, due to their unique features over current-generation robotics in the advanced sensory and perception systems (Terry, Fidan, Zhang, & Tantawi, 2019), control algorithms, as well as the enhanced data processing capability (Lai, Lin, & Wu, 2018).

TECHNICAL SPECIFICATIONS IN COLLABORATIVE ROBOTS

As in traditional articulated robots, collaborative robots have the axes labeled as follows:

Robot Axes

Axis 1-The Base Axis

As the name suggests, this axis is located at the base of the robot and allows it to move left and right. This axis is referred to as the J1 axis (Denso, ABB, and Fanuc robots) or Swing (Yaskawa Motoman), or the Joint 0 axis (Universal).

Axis 2- The Waist Axis

This axis allows the lower arm of the robot to move forward and backward, in the same way that the waist allows a person to bend forward and backward. It is called the J2 axis (ABB, Denso, and Fanuc) or the Lower axis (Yaskawa) or the Shoulder axis (Universal Robotics).

Axis 3- The Shoulder Axis

This axis powers the movements of the upper arm. It allows it to extend the reach of the robot in the radial direction from the robot center. In some cases, it allows the robot to reach points behind it. This axis is referred to as the J3 axis (ABB, Denso, and Fanuc) or the Upper axis (Yaskawa) or the Elbow axis (Universal Robotics).

Axis 4- The Wrist Yaw or Wrist Roll Axis

There are two types of this joint, the more common type is that where the axis allows the upper arm of the robot to twist, which causes the end-of-arm-tooling (EOAT) to rotate about the Z-axis of the end-effector, causing the EOAT to rotate from horizontal to vertical position and vice versa. In which case the axis performs a Wrist Roll. In the second type, the axis rotates the EOAT with the wrist such as in the robots of Universal Robotics. This axis is referred to as the J4 axis (ABB, Denso, and Fanuc) or the Rotate axis (Yaskawa) or the Wrist 1 axis (Universal Robotics).

Axis 5- The Wrist Pitch Axis

This axis powers the movement of the wrist in the up and down directions. It is referred to as the J5 axis (ABB, Denso, and Fanuc) or the Bend axis (Yaskawa).

Axis 6- The Wrist Roll (if a Yaw is present) or EOAT Roll Axis.

This axis is similar to Axis 4 in that it powers a twisting motion, but here the part that twists is the wrist itself, causing the end-of-arm tooling to twist in a circular motion. It is referred to as the J6 axis (ABB, Denso, and Fanuc) or the Twist axis (Yaskawa).

Axis 7-The Elbow Axis

Only exists on some robots such as the ABB IRB 1400 Yumi robot, this axis allows for an extra flexibility in the arm. It is also referred to as the J7 axis (ABB, Denso, and Fanuc).

The seven axes are illustrated in **Figure 2** below.

Figure 2. The 7 axes of the ABB YUMI 1400 and the six axes of the Denso Cobotta collaborative robots

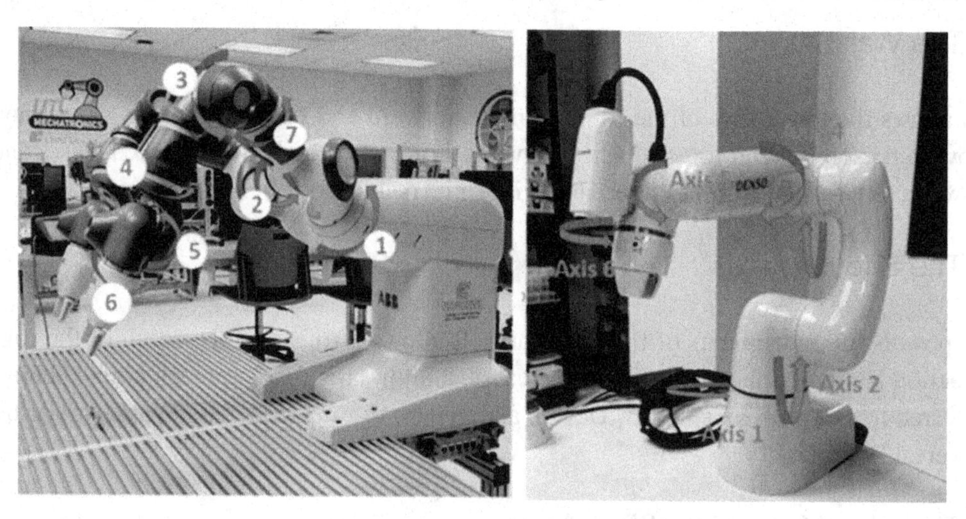

The technical specifications of eight collaborative robots that are commonly used in industry are shown in **Table 2.**

Table 2. Technical specifications of some collaborative robots

	Denso Cobotta	Universal UR3	KUKA iiwa 7	Fanuc CR-35iA	ABB Yumi (IRB 14000)	ABB SWIFTI CRB 1100	ABB GoFa	FANUCR-7iA
Weight:	4 kg	15 kg	22.3 kg	990 kg	38 kg	21 kg	27kg	55kg
Payload:	0.5 kg	3 kg	7 kg	35 kg	0.5 kg	4 kg	5 kg	7 kg
Repeatability:	± 0.05 mm	± 0.1 mm	± 0.1 mm	± 0.03 mm	± 0.02 mm	± 0.01 mm	± 0.05 mm	± 0.1 mm
Maximum reach (mm):	342.5	500	800	1813	559	580	950	911
Number of axes:	6	6	7	6	14 (7 per arm)	6	6	6
Maximum wrist torque (N.m):	J4: 0.56 J5: 0.47 J6: 0.24	J4:13 J5:13 J6:13	J4:110 J5:40 J6: 40	J4: 110 J5: 110 J6: 60	J4: J5: 0.64 J6: 0.23	J4: J5: J6:	J4: J5: J6:	J4:16.6 J5:16.6 J6:9.4
Maximum TCP speed:	1,500 mm/s	1,000 mm/s	1,300 mm/s	750 mm/s	1,500 mm/s	5,000 mm/s	2,200 mm/s	

To compare the technical specifications of collaborative robots to traditional robots, the technical data of four common traditional robots are shown in **Table 3**. Pictures of the four traditional robots are shown in **Figure 3**.

Table 3. Technical specifications of traditional robots

	ABB IRB 140	FANUC LR Mate 200iD 4s	KUKA KR5sixx R650	MOTOMAN HP3JC
Weight(kg):		20	28	27
Payload:	6kg	4 kg	5 kg	3 kg
Repeatability:	±0.03 mm	± 0.01 mm	±0.02 mm	± 0.03 mm
Maximum Reach:	800 mm	550 mm	653 mm	532 mm
Number of Axes:	6	6	6	6
Maximum Wrist Torque in units of N.m	J4: 8.58 J5: 8.58 J6: 4.91	J4: 8.86 J5: 8.86 J6:4.9	J4: 1100 J5:1100 J6: 1100	R: 7.25 B: 7.25 T:5.21
Maximum Moment of Inertia in units of kg.m^2	J4: J5: J6:	J4: 0.2 J5: 0.2 J6: 0.2	J4: 0.045 J5: 0.045 J6: 0.045	R(J4): 0.3 B(J5): 0.3 T(J6): 0.1
Maximum TCP speed:	2.5 m/s	11 m/s	1.2 m/s	
Power (kW):	0.4 kW	0.5 kW	0.6 kW	0.8 kW

Figure 3. From left: The ABB IRB140 robot, Fanuc LR Mate 200iD, Kuka KR5sixx R650, MOTOMAN HP3JC

When comparing the data for the collaborative robots in **Table 2** to those of the traditional robots in **Table 3**. It can be inferred that the primary differences lie in three aspects:

1. The Tool Center Point Speed is significantly lower in collaborative robots than in traditional robots. On average the maximum speed is five to six times lower than in traditional robots.
2. The payload capacity is considerably lower in collaborative robots than in traditional robots.

3. The accuracy (or more accurately, the position repeatability) is relatively lower in collaborative robots than in traditional robots.

The first two aspects (speed and payload capacity) are a direct result of the safety concerns due to the nature of operations in which the collaborative robots are expected to operate.

The primary drawback in traditional industrial robots was the necessity to keep them caged in highly-controlled and highly-predicted environments for operator safety. Therefore, searching for ways to make industrial robots safe and therefore, remove the workspace constraints, is the subject of active research. As reducing the workspace constraints on industrial robotics can significantly improve energy efficiency and productivity. Despite the advances in the sensory and perception systems in robots, operator safety remains the main concern.

The most recently produced collaborative robots, ABB SWIFTI and GoFa robots, set new expectations for collaborative robots by safely operating at higher speeds and having higher payload capacities. The SWIFTI, a 6-axis cobot, can carry a payload of up to 4 kg and has a tool center point (TCP) speed of 5 m/s (ABB). The GoFa, similar to the SWIFTI, has 6-axes, but has a TCP of 2.2 m/s, a 5 kg payload maximum, and a reach of 950 mm (ABB). The development of collaborative robots has allowed for users to work with robots as opposed to solely operating them, as research progresses, larger industrial collaborative robots can safely work with operators (Gopinath, Kerstin, Derelov, Gustafsson, & Axelsson, 2021).

RIA AND ISO STANDARDS RELATED TO COLLABORATIVE INDUSTRIAL ROBOTICS

The Robotic Industries Association (RIA) defines the hazard spaces that should be considered when operating a robot as:

Maximum Space

The amount of space which the manufacturer deems a risk of being hit by the moving pieces of robots as well as the space that may be in danger of getting hit by the end-effector (gripper, soldering gun, screwdriver, etc.) and/or any other counterparts that are involved with the task that the robot is set to complete.

Restricted Space

A portion of Maximum space that uses limiting devices that restrict the bounds of the robot movement and are not to be exceeded.

Operating Space

The actual space that is used by a robot when performing a task.

Safeguarded Space

The safe space within the bounds of a safeguard perimeter.

Collaborative Workspace

The space in within the safeguard that has no impending danger to a human during operation.

The ISO/TS 15066 standard was created to sophisticate the requirements in ISO 10218-1 and 10218-2. In regard to collaborative robots and monitoring speed and separation to ensure safety, ISO/TS 15066 states that the minimum protective distance, S, at time t_0 is modeled by the following equation (Marvel & Norcross, 2017):

$$S(t_0) \geq \left(\int_{\tau=t_0}^{\tau=t_0+T_R+T_S} v_H(\tau)\,d\tau\right) + \left(\int_{\tau=t_0}^{\tau=t_0+T_R} v_R(\tau)\,d\tau\right) + \left(\int_{\tau=t_0+t_R}^{\tau=t_0+T_R+T_S} v_S(\tau)\,d\tau\right) + (C + Z_S + Z_R)$$

Where:

v_H is the speed at which the operator is traveling toward the robot,

v_R is the speed at which the robot is traveling towards the operator,

v_S is the speed of the robot over the course of stopping.

T_R is the response time, or the time it takes for the robot to react to the presence of a operator,

T_S is the time the robot takes to come to a controlled stop.

C, Z_R, and Z_S represent the intrusion distance safety margin based on expected human reach, robot position uncertainty, and sensor uncertainty, respectively.

Other Safety Standards That Relate to Industrial Robotics

ANSI/RIA R15.06-2012

The technical standards that relate to collaborative industrial robotics are detailed in the Technical Standard number 15066:2016 of the International Standards Organization (ISO) and in the Technical Report 15.606-2016 of the U.S.-based Robotic Industries Association (RIA). ISO standard 10218-2011 and the RIA standard ANSI/R15.06-2012 detail the technical standards for industrial robotics. Both standards are undergoing an update that is projected to be completed by 2021 and 2022.

ANSI/UL 1740: 1998

This set of standards cover the robots and robotic equipment that require less than 600 V which are installed with consideration of the National Electrical Code: ANSI/NFPA 70. The end use of robots varies by the designated task it is needed to accomplish which leads to a variety of guidelines for installation which can be evaluated in the coordinating sections of ANSI/RI R25.06: Standard for Industrial Robots and Robots Systems-Safety Requirements. Summarized also in ANSI/RIA R15.06.

CAN/CSA Z434-03

This set of standards applies to the methods needed to follow to ensure the safety of humans working close with robots and robotic systems. These standards cover: safeguarding, maintenance, repair, manufacturing, testing, and training requirements for the human operation of manufacturing robots.

ISO 10218-1: 2011

These standards identify the measures to take to safely design industrial robots and robotic systems. These standards also identify the hazards identified in industrial robot design, the risks involved, and how to prevent such risks.

ISO 10218-2: 2011

This set of standards cover the hazards of the installation of industrial robots and robotic systems and the precautions that need to be taken in the design process.

NFPA 79: 2012

This set of standards covers the electrical wiring protocols associated with the National Electrical Code (NEC) which regulate the correct procedures for building and manufacturing industrial robots. This covers the required protection for the machine operators, equipment needed,

OSHA Technical Manual- Section IV "Safety Hazards"

The section IV of the safety manual covers: the hazard spaces of operating a robot or robotic system, the type of environment it should be operated in, the possible hazards involved, the possible errors and failures when operating robots, the investigation protocols, the proper safeguarding guidelines, and the requirements of design for manufacturing robots to prevent the above.

Section A includes studies of past errors that have occurred in manufacturing facilities located in Japan and Sweden. These studies indicate that many manufacturing robot related accidents are linked to abnormal operating conditions such as when recalibration, programming, or maintenance is taking place.

In addition to them, RIA TR15.806-2018 details the testing procedure for the power force limiting systems found on most collaborative robots. There is currently no safety standard for mobile collaborative robots, the only work available is a "Guidance for Manufacturers" that was recently published by the RIA under R15.08

Table 4. Safety standards related to industrial robotics

Safety Standard	Description	Year Published
ISO 15066 ANSI-RIA TR 15.606	Collaborative industrial robotics	2016
RIA TR15.806	Testing procedure for PFL systems	2018
ISO 10218-2011 ANSI/R15.06-2012	Industrial robotics	2011 and 2012
ANSI/ITSDF B56.5:2019	Automated Guided Vehicles (AGVs)	2019
EN 1525 EN 1526 ISO 3691-4	Driverless industrial trucks	
ISO 13482:2014	Service robots	2014
RIA R15.08	Mobile Industrial Robots	2020
ANSI/UL 1740	End-Use of Robots	1998
CAN/CSA Z434-03	Methods to ensure safety of humans working with manufacturing robots.	2003
ISO 10218-1	Reducing the risks of the Hazards Associated with robots.	2011
ISO-10218-2		2011
NFPA 79	Electrical Standards Industrial Robots	2012
OSHA Technical Manual Section IV	Safety Hazards of specific mechanical processes*	2020

Standard ANSI/ITSDF B56.5:2019 addresses safety for automated guided vehicles (AGVs) and automated maned vehicles, that typically follow prescribed paths, and therefore the standard does not apply to industrial robots and mobile collaborative robots. The European standards EN 1525 and EN 1526, and the ISO standard 3691-4 address driverless industrial trucks. Standard ISO 13482:2014 address the safety requirements of mobile service robots, and explicitly states that industrial robotics are not covered by the standard (Gallagher, 2019).

The standard update ANSI/RIA R15.08-1-2020 was only completed in 2020, and it formally defined the term Autonomous Mobile Robot (AMR), and referred to it as Industrial Mobile Robot (IMR). In the standard update, the safety requirements for IMRs were introduced. Prior to the update R15.08-1-2020, there was only a "Guidance for Manufacturers" by the RIA.

CONCLUSION

In this chapter, an analysis of the next-generation industrial robotics was presented in comparison to traditional industrial robotics. Market data suggest that the demand for industrial robotics in general, is steadily increasing in markets that were not historical markets for robotics, particularly after the CoVid-19 pandemic, such as hospitals and distribution industries.

In addition to that, non-traditional markets such as electronics and pharmaceuticals have become the dominant markets for industrial robots, taking over that position from the automotive industry. This change in market shares can indicate that the need for skilled workforce will undergo a shift in needed skills.

For example, there will be more demand for workforce skilled in wave-soldering and high-throughput pick-and-place operations over the demand for spot-welding skills.

The main challenge that faces increased artificial-intelligence (AI)-based technologies in next-generation collaborative robotics is the need for established preventive and corrective maintenance protocols on the AI-based technologies, as well as the need for established technician training programs on the new technologies and the wide availability of trained technicians.

Another challenge that faces deploying mobile collaborative robotics in industry is the lack of safety standards for that technology.

ACKNOWLEDGMENT

This research was supported by the National Science Foundation [grant number 2000685].

REFERENCES

ABB. (n.d.a). *GoFa CRB 15000*. Retrieved from https://assets.ctfassets.net/gt89rl895hgs/1MBowsjHDvAEykEwKKBWwl/e2d1a0446a0b0ab046c5b8a510730121/GoFa_CRB15000-datasheet.pdf

ABB. (n.d.b). *SWIFTI CRB 1100*. Retrieved from https://assets.ctfassets.net/gt89rl895hgs/7lxW2lwo38EpADo8WYr0ja/0f36e51c736ccd9800b2efa59bcb17a3/SWIFTI_CRB1100-datasheet.pdf

Anderson, G. (2016). *The Economic Impact of Technology Infrastructure for Advanced Robotics*. National Institute of Standards and Technology. doi:10.6028/NIST.EAB.2

Ayati, N., Saiyarsarai, P., & Nikfar, S. (2020). Short and long term impacts of COVID-19 on the pharmaceutical sector. *Daru: Journal of Faculty of Pharmacy, Tehran University of Medical Sciences, 28*(2), 799–805. Advance online publication. doi:10.100740199-020-00358-5 PMID:32617864

Berggren, M., Nilsson, D., & Robinson, N. D. (2007). Organic materials for printed electronics. *Nature Materials, 6*(1), 3–5. doi:10.1038/nmat1817 PMID:17199114

Cardona, M., Cortez, F., Palacios, A., & Cerros, K. (2020). Mobile Robots Application Against Covid-19 Pandemic. 2020 IEEE ANDESCON.

Coleman, L. (2018). *Inside Trends And Forecast For The $3.9T AI Industry*. Forbes. (2020). *COVID-19 Impact on the Industrial Robotics Market by Type (Articulated, SCARA, Parallel, Cartesian Robots), Industry (Automotive; Electrical and Electronics; Food & Beverages; Pharmaceuticals and Cosmetics), and Region – Global Forecast to 2025*. Research and Markets.

Demaitre, E. (2018). *RBR50 2018 Names the Leading Robotics Companies of the Year*. Robotics Business Review.

Denso-Wave. (2017). *Denso Cobotta user manual.* Denso. Retrieved from http://eidtech.dyndns-at-work.com/support/Cobotta_Manual/007260.html

Estolatan, E., Geuna, A., Guerzoni, M., & Nuccio, M. (2018). *Mapping the evolution of the robotics industry: a cross country comparison.* University of Toronto. Innovation Policy White Paper Series.

FANUC America Corporation. (n.d.). *Basic Robot Operations*. FANUC America Corporation.

Gopinath, V., Kerstin, J., Derelov, M., Gustafsson, A., & Axelsson, S. (2021, February). Safe Collaborative Assembly on a Continuously Moving Line with Large Industrial Robots. *Robotics and Computer-integrated Manufacturing*, *67*, 67. doi:10.1016/j.rcim.2020.102048

IFR Press Conference. (2018). Robots double worldwide by 2020. International Federation of Robotics.

International Federation of Robotics. (2020). *IFR Executive Summary Report for 2019*. IFR.

International Federation of Robotics. (2021). *Executive Summary World Robotics 2021 Industrial Robots*. IFR.

International Organization for Standardization. (2012). *ISO 8373:2012 Robots and robotic devices — Vocabulary*. ISO.

Kusiak, A. (2017). Smart manufacturing must embrace big data. *Nature*, *544*(7648), 23–25. doi:10.1038/544023a PMID:28383012

Lai, R., Lin, W., & Wu, Y. (2018). Review of Research on the Key Technologies, Application Fields and Development Trends of Intelligent Robots. *International Conference on Intelligent Robotics and Applications*, *10985*, 449-458. 10.1007/978-3-319-97589-4_38

Lambrecht, J., Kästner, L., Guhl, J., & Krüger, J. (2021). Towards commissioning, resilience and added value of Augmented Reality in robotics: Overcoming technical obstacles to industrial applicability. *Robotics and Computer-integrated Manufacturing*, *71*, 71. doi:10.1016/j.rcim.2021.102178

Liu, C., & Tomizuka, M. (2017). *Towards Intelligent Industrial Co-robots- Democratization of Robots in Factories*. Berkeley Artificial Intelligence Research.

Malik, A. A., Masood, T., & Kousar, R. (2020). Repurposing factories with robotics in the face of COVID-19. *Science Robotics*, *5*(43), eabc2782. doi:10.1126cirobotics.abc2782 PMID:33022618

Marvel, J. A., & Norcross, R. (2017, April). Implementing speed and separation monitoring in collaborative robot workcells. *Robotics and Computer-integrated Manufacturing*, *44*, 144–155. doi:10.1016/j.rcim.2016.08.001 PMID:27885312

Musa, Y., Tantawi, O., Bush, V., Johson, B., Dixon, N., Kirk, W., & Tantawi, K. (2019). Low-Cost Remote Supervisory Control System for an Industrial Process using Profibus and Profinet. 2019 IEEE SoutheastCon, 1-4.

Okuda, H., Haraguchi, R., Domae, Y., & Shiratsuchi, K. (2016). Novel Intelligent Technologies for Industrial Robot in Manufacturing - Architectures and Applications. *Proceedings of ISR 2016: 47st International Symposium on Robotics*.

Pellegrinelli, S., Orlandini, A., Pedrocchi, N., Umbrico, A., & Tolio, T. (2017). Motion planning and scheduling for human and industrial-robot collaboration. *CIRP Annals*, *66*(1), 1–4. doi:10.1016/j.cirp.2017.04.095

Robotics, I. F. (2017). *How robots conquer industry worldwide.* Author.

Tantawi, K. (2020). Literature Review: Rethinking BioMEMS in the aftermath of CoVid-19. *Biomedical Journal of Scientific & Technical Research*, *31*(1), 23944–23946. doi:10.26717/BJSTR.2020.31.005053

Tantawi, K., Ashcroft, J., Cossette, M., Kepner, G., & Friedman, J. (2022). Investigation of the Post-Pandemic STEM Education (STEM 3.0). *Journal of Advanced Technological Education*, *1*(1).

Tantawi, K., Fidan, I., Chitiyo, G., & Cossette, M. (2021). Offering Hands-on Manufacturing Workshops Through Distance Learning. *ASEE Annual Conference*.

Tantawi, K., Fidan, I., & Tantawy, A. (2019). Status of Smart Manufacturing in the United States. 2019 IEEE SoutheastCon.

Tantawi, K., Sokolov, A., & Tantawi, O. (2019). Advances in Industrial Robotics: From Industry 3.0 Automation to Industry 4.0 Collaboration. *4th Technology Innovation Management and Engineering Science International Conference (TIMES-iCON)*. 10.1109/TIMES-iCON47539.2019.9024658

Tantawi, K., Sokolov, A., & Tantawi, O. (2019). Advances in Industrial Robotics: From Industry 3.0 Automation to Industry 4.0 Collaboration. *4th Technology Innovation Management and Engineering Science International Conference (TIMES-iCON)*. 10.1109/TIMES-iCON47539.2019.9024658

Terry, S., Fidan, I., Zhang, Y., & Tantawi, K. (2019). *Smart Manufacturing for Energy Conservation and Savings*. NSF-ATE Conference.

Terry, S., Lu, H., Fidan, I., Zhang, Y., Tantawi, K., Guo, T., & Asiabanpour, B. (2020). The Influence of Smart Manufacturing Towards Energy Conservation: A Review. *Technologies*, *8*(2), 31. doi:10.3390/technologies8020031

The Quietly Emerging Artificial Intelligence Arms Race. (2019). *China Youth Daily*. Retrieved from https://m.chinanews.com/wap/detail/zw/gn/2019/10-17/8981224.shtml

The Robot Revolution: The New Age of Manufacturing | Moving Upstream S1-E9. (2018). *Wall Street Journal*.

Chapter 5
A Review of Big Data Analytics for the Internet of Things Applications in Supply Chain Management

Kamalendu Pal
https://orcid.org/0000-0001-7158-6481
City, University of London, UK

ABSTRACT

The internet of things (IoT) presents opportunities that enable communication between virtual and physical objects. It produces new digitized services that improve supply chain performance. Moreover, artificial intelligence (AI) techniques resolve unpredictable, dynamic, and complex global product development and supply chain-related problems. In this operating environment, heterogeneous enterprise applications, either manufacturing or supply chain management, either inside a single enterprise or among network enterprises, require sharing information. Thus, data management and its analytical interpretation have become a significant drivers for management and product development in networked enterprises. This chapter describes an information systems framework for the global product development purpose, and it also highlights how businesses can use business intelligence from gathered data from IoT applications. Finally, the chapter describes important categories of Big Data analytics applications for the supply chain operations reference (SCOR) model, and it also presents a data processing framework for supply chain management (SCM).

INTRODUCTION

Commercial trading between different countries and across various continents has started since ancient times of human civilization. In the early days of civilization, cross-border trade happened mainly for goods such as rice gain, wheat, spices, textile, metals, petroleum products, and other essential commodities. Through the Silk Road, the Spice Route, and various other interconnected trade transportation

DOI: 10.4018/978-1-7998-7852-0.ch005

Copyright © 2023, IGI Global. Copying or distributing in print or electronic forms without written permission of IGI Global is prohibited.

networks, human civilizations have initiated trade patterns that have blossomed with industrialization and globalization (Bardhan, 2003) (Stiglitz, 2017).

While the central precept of global trade is simple and unchanged to move goods and money from point A in one country to point B in another country, however in reality, it is much more complex to deploy this business practice. Its activities include financing trade, tracking and tracing goods within supply chains, and verifying the quality of those goods through provenance and product pedigree. As global trade has grown in scale, operational business processes have become more numerous and complex. In addition, these business practices strain to accommodate the demands of increased trade volume among more participants – business financiers, partners as importers and exporters servicing different market segments, freight forwarders, customs and port authorities, regulatory governing bodies, and insurance providers. In this way, business communities realize and appreciate the concept of supply chain operations, their respective values, and their management from ancient times. Consequently, supply chain management and automation of its business operations demanded more priority in corporate strategy formation and execution.

With the advent of technological innovation, intelligent applications such as smart factories, self-decision-making machinery in manufacturing plants, and intelligent supply chain equipped with *various sensors* and radio frequency identification (RFID) tags are making regular operations smarter. These smart applications require a vast amount of data and transmit data that generate a high traffic volume, making businesses' ability to modern data analytics essential. On the other hand, *artificial intelligence* (AI) based algorithms can be utilized to resolve unpredictable, dynamic, and very complex problems to yield suitable business decisions. Recently, researchers have shown immense interest in enhancing the business's overall performance with modern information system architecture (ISA) and software system technology (SST).

The SST, particularly *data analytics*, is now significantly impacting the manufacturing industry. However, manufacturing professionals have been slow to exploit the full potential of SST. Instead of using SST to maximize productivity and revenue-generation ability, SSTs have been used mainly for *enterprise resource planning* (e.g., accounting, inventory management, human resource management) purposes within the manufacturing industry. As a result, the manufacturing industry has yet to exploit SST as an effective tool.

In addition, the advantage of globalization has simulated different initiatives in global product manufacturing and marketing business activities. For example, in the 1980s, the "*quick response*" strategy was developed to maintain a competitive advantage (Porter, 1985) for the domestic manufacturing of products. Technological innovations have made fast electronic communication a global phenomenon (Pal, 2022), and the rapid acquisition of technical skills in various countries has meant that many professional tasks could be outsourced (quality control, raw materials purchasing, sample making). Researchers (Gereffi, 1999) (Pal & Yasar, 2020) identified some of the trends in the manufacturing business. Also, the globalization trends have continued, and the radical social reform idea of making more from fewer resources (known as *Gandhian Engineering*) (Prahalad & Mashelkar, 2010) has become the business rule in today's global market. Also, operational planning – and appropriate information system (IS) – drives the whole business, where customers play a pivotal role.

With technological advances, manufacturing companies regularly employ data mining techniques to explore the contents of data warehouses looking for trends, relationships, and outcomes to enhance their overall operations and discover new patterns that allow companies to serve their customers better. This way, manufacturing organizations rely on business processes related to data to formulate strategy and succeed under value-based reimbursement models. The new paradigm requires data-driven insights that can help operational managers reduce unnecessary variation in business and make more informed service-line decisions across the enterprise. In this way, intelligent data processing plays a key role. This chapter presents some of these issues identifying in particular: (i) the concept of big data, (ii) data gathering, (iii) data processing, and (iv) the broader research dilemmas. Hence, the central theme of this chapter is to expose the reader to some of the more interesting insights into how data and information systems (IS) help run manufacturing supply chain management.

The evolution in computer processing power and storage capacity has enabled organizations to develop data-rich IS for daily operations, and therefore, there has been tremendous growth in data stored. In addition, business data collection itself has progressed from the transcription of paper-based records via manual data-entry processes to the use of smart cards, mobile phones (Location Data, GPS), Internet of Things (IoT) (e.g., radio frequency identification (RFID) tags, sensors), webcasting and Internet users' mouse clicks. This data generation has generated a need for new techniques and technologies to transform these data into appealing and valuable information and knowledge.

Today, big data is generated by web applications, social media, intelligent machines, sensors, mobile phones, and other intelligent hand-held devices, bacterized on the velocity, volume, and variety it produces along the supply chain. Such decision-support software applications employ pure mathematical and artificial intelligence techniques and sometimes use both methods to perform analytical operations that uncover relationships and patterns within the manufacturing supply chain generated Big Data.

Business processes along the supply chain must balance to provide customer service at no additional cost or workload. It also requires trade-offs throughout the supply chain. Therefore, it is necessary to consider a single interconnected chain rather than narrow functional business processes when considering practical mechanisms, which help find acceptable solutions at the time of need.

Real-time supply chain decision-making and coordination are essential in the international marketplace, shortening product life cycles and fast-changing trends. Technological evolution and the latest information-sharing techniques make real-time decision-making and coordination easier than in the past. In addition, the importance of integrating and coordinating supply chain business partners have been appreciated in many manufacturing industries (Pal, 2016).

Manufacturing supply chain managers are seeking ways to manage Big Data sources effectively. There are many examples of manufacturing business operations using Big Data solutions that highlight the wealth of business process enhancement scopes available through the clever use of data:

- Big Data-based applications, which help integrate strategic business planning, are recently assisting manufacturing businesses to coordinate more susceptible supply chains as they better apprehend operating market tendencies and customer desires. It forms the triangulation of a range of marketing and operating business environment data (e.g., social media discussion forums, demographic information, and other static and dynamic data from diverse sources), giving the ability to forecast and proactively formulate strategies for manufacturing supply chain businesses.
- Software-defined machines, data-driven predictive analyses, the Internet of Things (IoT), and soft-computing-based machine learning mechanisms are ushering in a new industrial revolution. These new computing power breeds are used in predictive asset maintenance to avoid unplanned downtimes in the manufacturing shop floor.

- IoT can provide real-time telemetry data to reveal the details of production processes. In addition, machine learning algorithms are used to analyze the data to reveal the details of production processes that can correctly forecast near future machine failures and appropriate actions.
- Big Data-based solutions are helping avoid delivery delays and create pollution-reduced environments by analyzing Global Positioning Systems (GPS) data with the help of traffic and meteorological data, which actively plan and find cost-effective delivery routes.
- Extensive Data-based software systems are helping manufacturing supply chains for self-critiquing (rather than a reactive) response to supply chain risks (e.g., supply chain failures due to natural or synthetic problems).

These examples give just a few insights into the numerous advantages derived from the analysis of Big Data sources to enhance supply chain reactiveness. This way, appropriate data manipulation techniques within special-purpose software, generally known as business analytics, are used for timely decision-making.

Supply chain managers are now increasingly seeking to '*win with data*'. They rely on data to monitor expenditure, look for trends in corporate operational cost and related performance, and support process control, inventory monitoring, production optimization, and process improvement efforts. In addition, many businesses are collecting vast amounts of data, with many using it to capitalize on data analytics to gain a competitive advantage (Davenport, 2006). This way, appropriate-data capture, data cleaning, different data analysis techniques, and Big Data are all considered part of an emerging competitive area that will transform how manufacturing supply chains are managed and designed.

The rest of the chapter is structured as follows: Section 2 presents an overview of the IoT applications in supply chain operations. Section 3 describes the extended supply chain network concept and includes an overview of the value chain network of participating partners in a manufacturing business. Additionally, this section explains the Supply Chain Council's SCOR (Supply Chain Operations Reference) model and its relevance for supply chain performance measurement. Section 4 discusses the era of Big Data analytics, sources of Big Data, various kinds of analytics and their techniques applicable to corporate decision-making. Section 5 identifies data-based supply chain decision-making-related research. Finally, section 6 illustrates a Big Data-based framework for SCM; and provides a few examples of diverse manufacturing supply chain analytics types. Finally, section 7 describes the future research plan. Finally, Section 6 presents some concluding remarks.

IoT APPLICATIONS AND EXTENDED SUPPLY CHAIN NETWORK

The vision behind the IoT is to create real-time connection and data communication with people or any business process along supply chain operation, anytime, anywhere, using any network. It facilitates overcoming the limitations of legacy systems and ways of data communication and processing. In this way, IoT systems embedded with industrial operations often consist of software, electronic components, actuators, sensors, detectors, and wireless connectivity that enable them to collect data from these objects can be defined as the "IoT". An IoT system has the following essential characteristics:

- IoT is a ubiquitous technology advancement that enables supply chain business-related objects to be connected to the Internet via wired or wireless networks to communicate.

- Several wireless sensor networks are available for IoT devices, including near-field communication (NFC), Zigbee, radio frequency identification (RFID), Bluetooth, and Wi-Fi.
- The sensors can be connected to various technologies, including long-term evolution (LTE), general packet radio service (GPRS), and global system for mobile communication (GSMC).
- The efficiency of an IoT system is largely determined by three main components, each of which is vital to its day-to-day operation: (i) perception layer, (ii) network and middleware (Edge, Fog, and Cloud) layer, and (iii) application layer.

Nevertheless, cloud computing offers virtually unlimited storage and system capabilities to address many IoT-related challenges. Consequently, the phrase "cloud of things" (CoT) is used to allude to the fusion of IoT and cloud computing. The CoT is a paradigm for increasing productivity and improving system performance that is widely used by most industries and manufacturers. Several researchers discussed in their research (Pal, 2023) the use of the cloud to analyze vast amounts of data (i.e., Big Data) when data storage and processing are required.

With the advent of the IoT, vast amounts of data are generated in real-time, which poses a significant concern for traditional cloud computing network topologies (Decker et al., 2008). A traditional cloud infrastructure condenses all processing, storage, and networking into a limited set of data centres, and the distance between remote devices and remote data centres is relatively wide (Wang et al., 2016). Edge computing could address this challenge since it provides access to computing resources closer to IoT edge devices and may lead to a new ecosystem for IoT innovation (Veza et al., 2015). In this way, architectural issues on IoT systems in automating supply chain operations play a significant role, and there are different types of layered architecture for supply chain industrial applications (Pal, 2022).

IoT Systems Architecture

The layered architecture is designed to meet the requirements of various industries (e.g., manufacturing, retail), enterprises, societies, institutions, and governments. The functionalities of the various layer are (Pal, 2023): (i) edge layer: This is the hardware layer and consists of sensor networks, embedded systems, RFID tags and readers or various types of sensors in different forms. Many of these hardware elements provide identification and information storage, information collection, information processing, communication, control, and actuation, (ii) access gateway layer: It takes care of message routing, publishing, and subscribing and performs cross-platform communication if required, (iii) the middleware layer interfaces the access gateway layer and the application layer. It is responsible for functions and takes care of issues like data filtering, data aggregation, semantic analysis, access control, and information discovery, such as EPC (Electronic Product Code) information service and ONS (Object Naming Service), and (iv) application layer that is responsible for delivering various applications to different users in IoT. Figure 1. represents a simplified IoT technology-based architecture for SCM.

About the technologies of IoT, (Zawadzki & Zywicki, 2016) presents the technology areas enabling the IoT: (i) identification technology: The purpose of identification is to map a unique identifier or UID (globally unique or unique within a particular scope) to an entity to make it retrievable and identifiable without ambiguity, (ii) IoT architecture challenges: Scalability, modularity, extensibility, and interoperability among heterogenous things and their environments are the essential design requirements for IoT, (iii) communication technology, (iv) network technology: The IoT deployment requires the development of suitable network technology for implementing the vision of IoT to reach out to objects in the physical world and to bring them into the Internet, (v) software and algorithms, and (vi) hardware technology which includes intelligent devices with enhanced inter-device communication.

The other essential technologies necessary for the IoT-based sensing environment are: (i) data and signal processing technology, (ii) discovering and search engine technology, (iii) relationship network management technology in managing networks that contain a vast number of heterogeneous things, and (iii) power and energy storage technology, (iv) security and privacy technologies to maintain two significant issues in IoT system, and (iv) standardization which should be designed to support a wide range of applications and address standard requirements from a wide range of industrial SCM systems.

Figure 1. A layered IoT architecture for supply chain operation

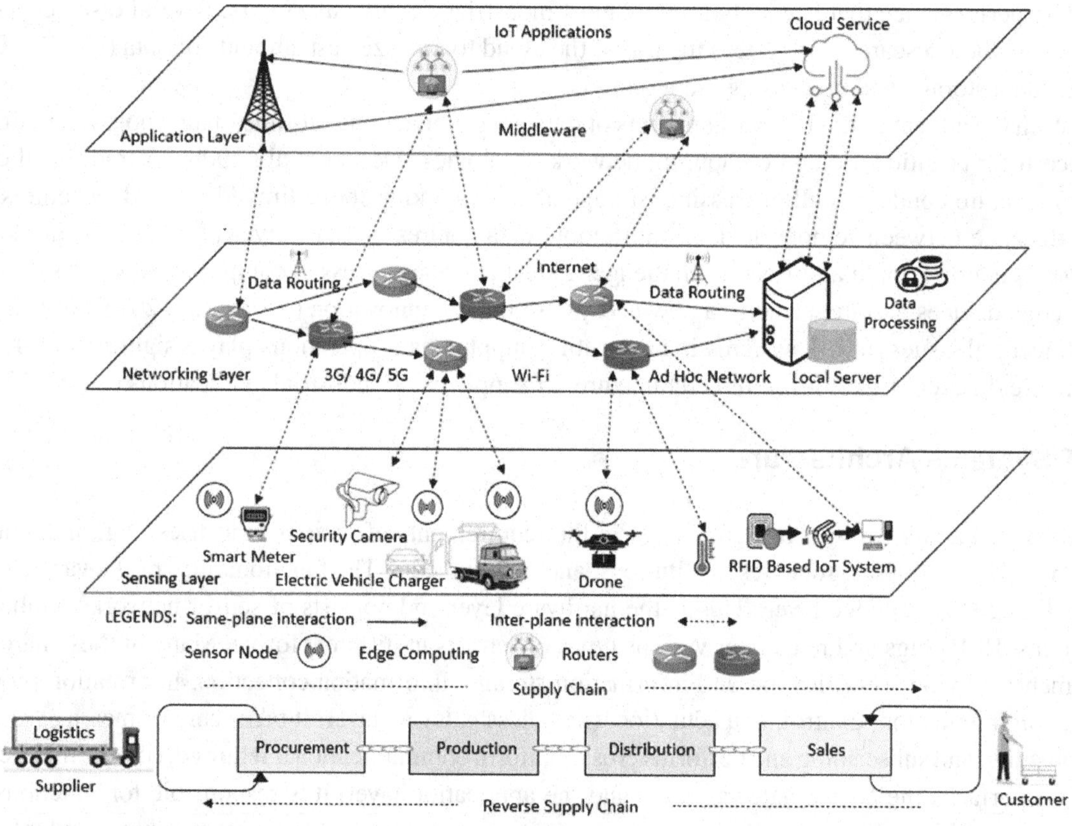

REVIEW OF RECENT LITERATURE

In this way, information technology enables effective supply chain management; and information sharing capability. This section presents a brief review of IoT technology and its uses in SCM's different application areas: (i) procurement process, (ii) make process, (iii) delivery process, (iv) return process, and (v) industry-specific deployment.

Table 1. Applications of IoT technology for supply chain operations

Procurement process: The request for materials and services by companies is the sourcing process. Planning source activities strategically across the supply chain is a sign of success. Yuvaraj and Sangeetha (Fang & Ma, 2020) highlighted how to monitor the goods anytime and anywhere by integrating RFID tags with GPS technology to track indoor and outdoor products in a supply chain environment. The impact of IoT on supplier selection was studied by Yu and colleagues (Bowman et al., 2009). Also, several advantages of IoT regarding the sourcing process have been identified. For example, analyzing the impact of the cost of sensors and notifications on the purchase cost of a unit, researchers (Fang et al., 2013) developed a simple linear cost model.
Make process: The operational areas that IoT applications can enhance and relevant to the supply chain make process involve: factory visibility as in (Li et al., 2014), management of innovative production networks as in (Paksoy et al., 2016), intelligent design and production control as in (Xing et al., 2012), systematic design of the virtual factory as in (Fang et al., 2016), smart factory in the petrochemical industry (Yan & Huang, 2009), opportunities for sustainable manufacturing in industry 4.0 (Shin et al., 2011).
Delivery process: One of the most significant logistics tasks is the delivery function. Logistics includes planning, storing and controlling goods and services flows (Ballou et al., 2007). In the supply chain, the delivery process is concerned with the warehouse, inventory and order management, and transportation. The main impact of IoT on the supply chain delivery process includes: (i) Warehousing function: the IoT enables timesaving of joint ordering via smart RFID tags (Angeles, 2005). IoT also achieves collaborative warehousing via using smart things and multi-agent systems. It also increases the safety and security of the supply chain (Liukkonen & Tsai, 2016). (ii) In order and inventory management: the IoT enables sharing of information and inventory accuracy (Bowman et al., 2009) using RFID tags. [iii] In transportation function: the IoT achieve accurate and timely delivery using sensors and networks (Fang et al., 2013). It also saves scanning and recording time using smartphones (Li et al., 2014).
Return process: A closed-loop supply chain model to meet the demand of sales collection centers using new and remanufactured products presented by a group of researchers (Paksoy et al., 2016). The e-reverse logistic framework was designed by Xing et al. (Xing et al., 2011). An integrated three-stage model for optimizing procurement, pricing, product recovery and strategy of return acquisition was proposed by Fang et al. (Fang et al., 2016).
Industry-specific deployment: IoT technology is also deployed in various supply chain applications. For example, a group of researchers used IoT for a pharmaceutical supply chain (Yan & Huang, 2009), and it has also been used for the construction industry (Shin et al., 2011), petrochemical industry (Li, 2016), retail industry and food supply chains (Verdouw & Wolfert, 2016).

The IoT technology introduces a new concept that aims to enhance the forms of communication that induvial supply chain business partners need today. The Internet is a multi-media networking tool that anyone can access using electronic devices. The main form of communication is human-to-human for a particular purpose and objective. However, IoT technology attempts to not only have humans communicate through the Internet but also have objects or devices.

Business opportunities with better incentives that drive manufacturing workload to geographically distributed locations are enabled by a few factors - the growing modernization of international business, advancements in manufacturing product manufacturing technology with related information systems capabilities, and improvement in multimodal transportation and services. In this way, global supply chains are formed where cost reduction strategies result in ultimate products and services being produced with intermediate input originating from several countries, and this practice is now widespread in many business sectors. In addition, it extends overall business activities to more globally participating alliance partners. Moreover, a manufacturing business can create a business environment in many collaborative supply chains, ultimately creating a new complex makeup of related business activities.

These new intricate arrangements are based on three main motivating influences: (1) the geographical location and nature of connections between tasks in the chain; (2) the allocation of power among dominating business alliances and other associates in the chain; and (3) the role of government agencies and policies informing business alliances and business location. The geographical location is defined by splitting production activities and their delocalization.

Management teams are puzzling to substantiate their investment in traditional legacy systems for SCM purposes. The essential purpose is that systems mainly cater to transaction-based corporate activity. However, they lack the intellectual and analytical ability needed for an integrated view of the global supply chain management system. This way, it is where Business Intelligence (BI) tools like data warehousing, ETL (Extraction, Transformation, and Loading), data mining, and OLAP (On-Line Analytical Processing) can cater capability to analyze operational effectiveness and performance management across the global supply chain.

Supply Chain Performance Measurement

Supply Chain performance measurement is crucial for a manufacturing enterprise to survive in the competitive world of business. Some well-known manufacturing enterprises use *business intelligence* (BI) software applications and in-house business practices to measure supply chain performance. It also measures the performance of the enterprise's internal business processes in consideration, but it also takes into consideration of extended enterprise activities.

In order to tackle some of these drawbacks in traditional financial accounting-based practices for measuring supply chain performance, a variety of measurement mechanisms have been introduced. They include Balance Scorecard, Supply Chain Council's SCOR (Supply Chain Operations Reference) model, Logistics Scoreboard, Activity-Based Costing (ABC), and Economic Value Analysis (EVA). In this chapter, the SCOR model has been used for manufacturing supply chain decision-making. A brief introduction of the SCOR model is described below.

SCOR Model

The Supply Chain Operations Reference (SCOR) model prescribes a general business process management approach to the supply chain community. This reference model was initially introduced by the Supply Chain Council (SCC) in 1996. The SCOR model deals with the supply chain management activities from an operational purpose. It includes corporate customer engagement information, day-to-day business transactions, and relevant market reactions. This model encapsulates business process activity management, benchmarking with market-leading enterprises, and best practices into a single framework. Many well-known companies have used the SCOR model (e.g., Intel, General Electronic, Airbus, DuPont, and IBM) (Supply Chain Council, 2010). Intel Corporation used its first SCOR for its resellers' product department in 1999. Then it used the SCOR model for its system manufacturing department. Implementing the SCOR model to handle business process management includes improved operational cycle time, fewer inventories in storage, enhanced visibility of the supply chain, and timely access to customer engagement information. General Electric (GE) has also used the SCOR model for its transportation management department and has reported operational improvement in its transportation service after using the SCOR model (Poluha, 2007).

The SCOR is a standardized multi-level framework with Key Performance Indicators (KPIs) attached to its Level. According to SCOR, five business entities are involved in supply chain performance management: the enterprise itself, the supplier of the enterprise, suppliers to suppliers, the customer of the enterprise, and customers of the customers. These five categories of business entities have their own SCOR. Individual SCOR is composed of the five-integrated corporate operational behaviours: Plan (P), Source (S), Make (M), Deliver (D), and Return (R) – from the suppliers' supplier to customers' customers, and all aligned with leading enterprise's operational strategy, material, work, and information flow. A diagrammatical representation of SCOR model is shown in Figure 2.

Figure 2. An organizational overview of the SCOR model

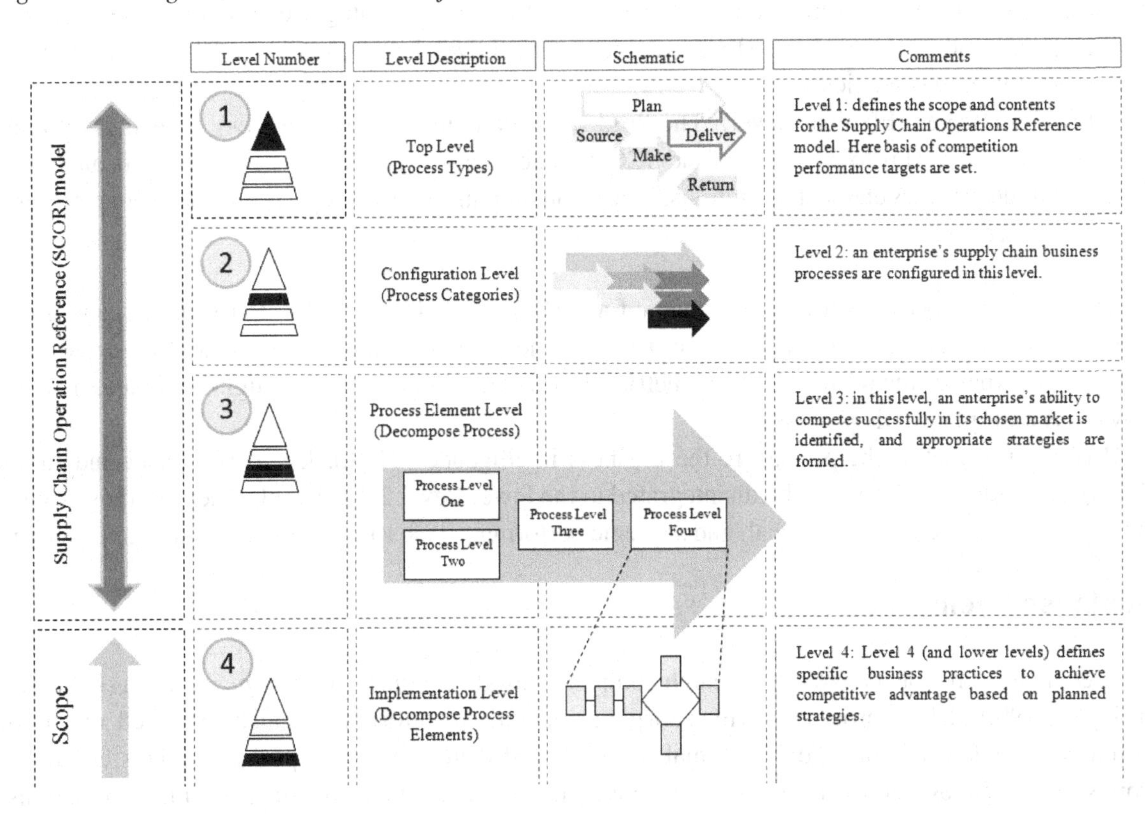

There are four levels in the SCOR framework: Top Level, Configuration Level, Process Element Level, and Implementation Level, as shown in Figure 2. It is considered a best practice framework for evaluating performance in supply chain management. The framework uses many performance measurement metrics and their relevant attributes. This way, the SCOR framework creates business process improvement using benchmarking techniques. The top three levels of the SCOR model are as follows:

1. Level-1(Process Types): This Level defines the scope and contents of the Supply Chain Operations Reference (SCOR) model. Here the basis of competition performance targets is set. To express the detailed requirements, it uses the following criteria:
    - **P**lan (P): a process that balances requests of demand-side and supply-side and has a smooth sub-process for sourcing, making, and delivery activities under the best business plan.
    - **S**ource (S): a process that procures raw materials and relevant services to fulfil strategic or practical needs.
    - **M**ake (M): transforming raw materials to finish or semi-finished goods to fulfil strategic requirements.
    - **D**eliver (D): a process that transports finished goods or services to fulfil plans or strategic requirements.
    - **R**eturn (R): a collection of returned products for several reasons.
2. Level-2 (Process Categories): The second SCOR level defines the configuration of planning and execution processes in material flow, using standard process categories such as make-to-order, engineer-to-order, or make-to-stock categories. This Level derives from Level 1, which settles on more concrete strategies.
3. Level-3 (Process Element Level): The third Level contains the main process elements and their process flows. This lower Level of the SCOR model is most significant to business data analytics. It includes process element definitions, inputs and outputs, relationships, performance metrics, and best practices.

This way, all process metrics use aspects of a performance characteristic. For example, any given process's performance characteristics are either customer-facing (reliability, responsiveness, and flexibility) or internal-facing (cost and assets) metrics. Level 1 metrics are primary, high-level measures that may cross multiple SCOR processes.

Performance metrics help measure the business intelligence (BI) for key performance indicators (KPIs). Each Level 1 metric can be disintegrated to the lower Level 2 and Level 3 metrics, thus providing standardized operational, tactical, and strategic performance measurements across the supply chain.

Business Intelligence

Gartner, an information technology research company, introduced the term "*Business Intelligence*" (BI) during the 1990s. BI is a process by which corporate operational information can be gathered, cleansed, stored, and made available to decision-makers who need it in its most helpful form. This can range from simple business reports through trend analysis to complicated dashboards and future predictions. However, data in raw form is of limited use, and manufacturing enterprises are increasingly opting to use dedicated BI software to appreciate their data's full capability. BI commercial vendors specialize in an automated Big Data analytic platform that allows an enterprise to easily collect all corporate data, manipulate them, and display them as actionable information or information that can be acted upon in making informed decisions.

Moreover, enterprises of all sizes can connect to and analyze data (even Big Data) to produce better quality business decision-making environments in recent years. Interactive visualizations and dashboards enhance analyzing and interpretation of data to make prudent business decisions in real-time. In addition, BI enables manufacturers to enhance how they do day-to-day business. Manufacturing enterprises are empowered to offer products and services at the lowest possible cost and with the highest degree of productivity and efficiency possible – while returning the highest revenues and profits.

BI platform can collect valuable data from many sources and assemble it into a central Data Warehouse (DW). It can be validated, cleaned, summarized, standardized, and even enhanced by copying the data. Moreover, BI capabilities are spreading to virtually all parts of the manufacturing enterprise as businesses strive to put critical data into the hand of business operational decision-makers who need it to do their jobs. The Big Data user community wants the following from the real-world BI systems:

- The capability to execute ad hoc queries
- Access and use multiple databases
- Scalability, affordability and reliability of data and the related operations
- Ease of integration with heterogeneous internal and external data sources

In this way, BI refers to manufacturing enterprises' methods and techniques for tactical and strategic decision-making. For example, it leverages technologies focusing on counts, statistics, and business objectives to improve business performance.

The goal of a DW is to hold the data needed by business managers to make decisions. The next step is to document the data warehouse. *Metadata* describes the source data, identifies the transformation and integration steps, and defines how the data warehouse is organized. This step is crucial to help decision-makers understand what data elements are available. Once the data requirements and sources have been identified, the data must be transformed and integrated so that decision-makers can search and analyze it efficiently. The main activities of this process flow are to extract, clean, transform, and load this data into the data warehouse. This process flow is called Extract, Transform, and Load (ETL).

Thus, the data warehouse process contains the following steps: (i) data extraction from multiple sources, databases, and files; (ii) data transformed and cleaned before loaded; and finally, (iii) data is loaded into the data warehouse. So, the first step in the data warehouse is to extract data from different business data sources; and transform it using built-in transformation processes. In addition, the data flow loading the warehouse dimensions is first passed to a short editor that removes any duplicated rows of data before being passed on to other processing units.

By itself, stored data does not create any business value, which is true of traditional databases, data warehouses, and the new Big Data technologies (e.g., Hadoop) for storage. However, once the data is appropriately stored, it can be analyzed, creating tremendous value. In addition, various analysis technologies and approaches have emerged that are especially applicable to Big Data sources within SCM.

THE ERA OF BIG DATA ANALYTICS

The main objective of processing Big Data is to generate differentiated value which can be trusted. This is done by applying *advanced analytics* against the complete data collection regardless of scale. To understand the critical issues of Big Data, one needs to understand the *data sources* of business processes within manufacturing enterprises.

New breeds of manufacturing business organizational process automation software, such as Enterprise Resource Planning (ERP), Supply Chain Management (SCM), Supplier Relationship Management (SRM), Global Positioning Systems (GPS), Social Media (SM) analysis systems, adoption of digital sensors and radio frequency identification (RFID) tags leading to the 'Internet of Things (IoT), and digitization of voice and multimedia are capable of collecting substantial data sets and unlock valuable information from it.

Before Big Data can be used for business purposes, it must be stored, managed, and analyzed appropriately for decision-making purposes. Distributed file systems, flexible storage systems, and massively parallel processing databases enable much larger volumes at low-cost infrastructure (e.g., Data Center, Private Cloud, Public Cloud, and Hybrid Cloud). In addition, processing a high volume of complex data in a distributed manner needs different techniques (e.g., time series analysis, knowledge-based classification, scenario-based analysis, and other statistical analysis methods). A diagrammatic representation of Big Data Infrastructure and its processing capabilities is shown in Figure 3.

Figure 3. Typical types of Big Data application systems

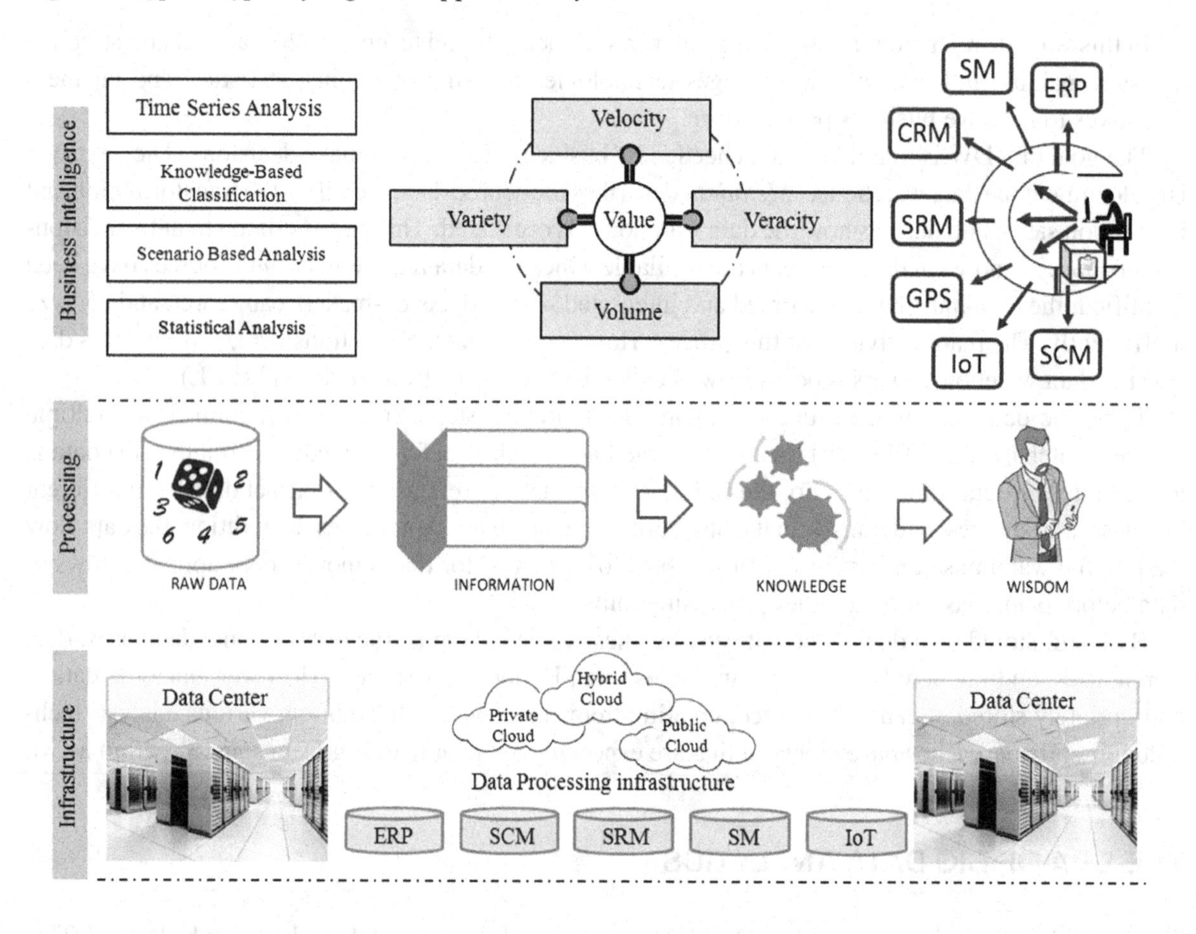

Data streams can come from internal and external sources within the global manufacturing chain. In addition, data can come in various formats, such as transaction and log data from multiple applications, structured data as database tables, semi-structured data such as XML, and unstructured data such as text, images, video streams, and audio statements. In order to process these data, other information systems are now making a revolution in sensing technology, computing and communication power that brings together a variety of resources. These resources range from networked embedded computers and mobile devices to multimodal data sources.

New breeds of manufacturing business automation software, such as Enterprise Resource Planning (ERP), Supply Chain Management (SCM), and Supplier Relationship Management (SRM), can collect substantial data sets and create valuable information for the company. Additionally, the extended enterprise is enabled by web technologies, e-business for customers, supplier portals, and information technology innovations targeted at supply chains and customer relationships.

As Big Data is one of the most 'exciting' terms in today's business computing, there is no consensus on defining it. Instead, the term is often synonymously with related concepts such as Business Intelligence (BI) and data mining. This also forms the basis for the most used definition of Big Data, the five Vs: Volume, Velocity, Variety, Value, and Veracity, as shown in Figure 3.

Volume

Data volume signifies the amount of data available to a manufacturing business, which does not necessarily have to own all of it if it can access it. Data includes text and videos, music, click streams from web queries, social media (e.g., Facebook, Twitter) update, and large image files.

Velocity

Data velocity assesses the speed of data creation, streaming, and aggregation. Electronic business (e-business or e-commerce) has rapidly increased the speed and richness of data used for business transactions (e.g., website clicks).

Variety

Data comes from different data sources. Data streams can come from internal and external sources within the global manufacturing supply chain. Data variety is a measuring mechanism of the richness of the data representation – text, images, video, and audio.

Value

A key challenge in transforming the manufacturing supply chain from infrastructure, data collection and remote sensing information system to a provision that presents actionable information and guides humans in decision making (e.g., relevant data, knowledge on data, and inferencing mechanisms) of the relevant information required for data analytics design and prediction purpose. For example, a hybrid of statistical methods and declarative knowledge is beneficial for leveraging sensor data streams for personalized production line machine care to reduce redoing rates among scheduled jobs and improve plant efficiency.

Veracity

Veracity's key issues relate to the messiness of data generated in dynamic supply chain environments. For example, just consider Twitter posts with hashtags, abbreviations, typos, and different colloquial speeches.

The advancement of understanding the behaviour of information systems and the processes and services they support has become a critical issue in the supply chain environment. This is demonstrated by the proliferation of software-based tools to analyze the process executions, system interactions, system dependencies and recent research in big data analytics.

Big Data Analytics

It is worth recognizing that the term '*analytics*' is not used consistently. It is used in at least three related ways (Watson, 2013a). First, examining its origins is an initiation point for understanding and appreciating data analytics. The Decision support systems (DSS) in the 1970s were the first computerized systems to support decision-making (Power, 2007). DSS appeared to be used to characterize special type software system applications and an academic research area. Over time, various decision support applications such as executive information systems (EISs), online analytical processing (OLAP), and dashboards/scorecards became very popular. Then in the 1990s, Howard Dresner, a business analyst at Gartner, customized the term *business intelligence* (BI). A simple definition is that "BI is a wider group of software applications, technologies, infrastructure, and processes for gathering, storing, accessing, and analyzing data to help business users make operational decisions" (Watson, 2009).

Various Kinds of Analytics

Big Data-based analytics is a rapidly developing field which already attracted attention from academics and practitioners. It is important to differentiate three diverse types of analytics because the differences have significance for the software system technologies and technical platform architectures used for Big Data analytics.

Descriptive Analytics

Descriptive analytics is backwards-looking and exposes what has happened. Descriptive data analytics, such as business reporting, dashboards, and data visualization, have been comprehensively used for some time and are the main applications of typical BI.

Predictive Analytics

Predictive analytics suggest what will occur in future. The techniques and algorithms for predictive analytics, such as statistical regression analysis, machine learning, and other soft-computing methods, have been used for some time. However, commercial software systems like SAS Enterprise Miner have made them easier to use and apprehend.

Prescriptive analytics can find optimal solutions, often for scarce corporate resources. For example, mathematical techniques for revenue management are increasingly general for manufacturing enterprises with "*perishable*" inventories (e.g., rental cars, hotel rooms, airline seats). In this way, modern data analytic techniques help enterprises to discover deeper insights from collected data.

Techniques for Analytics

Most techniques used to analyze Big Data analytics are based on applied mathematical computation and machine learning paradigms. However, Big Data analytics techniques are widespread and context-specific. These techniques can be classified and discussed according to the data under investigation. Depending on the type of data, four central Big Data analytics can be defined:

Text Analytics

This technique is used to find-out relevant information from textual data. Text analysis uses statistical computation, natural language processing techniques, and machine learning methods. Some essential text analytics techniques are information extraction, text summarization, question answering, and sentimental analysis.

Audio Analytics

Techniques deployed to extract information from unstructured audio data. Customer service telephone conversations with customers are the primary source of audio analytics. Some of the main techniques used for audio analytics include large vocabulary, continuous speech recognition and phonetics-based systems.

Video Analytics

Techniques used to monitor and analyze video streams. Video analytics has been mainly used as a replacement for labor-intensive observation. In addition, many business applications of video analytics are being used in manufacturing to collect data on customer demographics and preferences.

Social Media Analytics

Techniques used to analyze structured and non-structured data from social media channels (e.g., Twitter, Facebook). Some exciting techniques used to analyze social media data are community detection, social influence analysis and link prediction.

Many successful manufacturing enterprises have invested large sums of money in business intelligence and data warehousing tools and technologies. These manufacturers believe that up-to-date, accurate and integrated information about their supply chain, products and customers is critical for their survival.

SUPPLY CHAIN MANAGEMENT BUSINESS INTELLIGENCE RESEARCH

In SCM, there is a growing interest in *business analytics* (BA), also known as supply chain analytics (SCA), in academic literature. These analytics use data and quantitative techniques to analyze business performance in different functional areas of supply chain management (Handfield, 2006) (Davis-Sramek, Germain, & Iyer, 2010) (Davenport & O'dwyer, 2011) (O'dwyer & Renner, 2011). This section presents a brief review of Big Data-Based SCM.

In recent research, J R Stock (Stock, 2013) proposes a Big Data-driven SCM. He suggests that Big Data-Based analytics for SCM will allow decision-makers to make decisions faster with the help of circumstance-specific actions. An industry survey conducted by Mitsubishi Heavy Industries and consulting company Deloitte (Deloitte & MHI, 2014); and questioned supply chain executives about innovations that drive supply chains. The main objective was to get the views of business executives on emerging technology trends that could dramatically impact future supply chains. A critical issue that emerged from the survey was *supply chain analytics*. The survey also identified areas for analytics-based SCM. The Council for Supply Chain Management Consultants published a report (Richey et al., 2014) on Big Data in SCM based on interviews with supply chain managers.

Many commercial Big Data applications related to SCM are attracting the attention of academics and practitioners (Watson et al., 2014) (Davenport, 2006) (Davenport & Harris, 2007) (McAfee & Brynjolfsson, 2012) (Deloitte & MHI, 2016). For example, a few Big Data applications in corporate marketing management show tremendous opportunity in Big Data-based business analytics (Szilard et al., 2013). The number of publications on supply chain network design using Big Data and business analytics is also growing steady (Baesens, 2014) (Dietrich et al., 2014) (Sathi, 2012) (Siegel, 2013) (Watson et al., 2012).

There are many successful cases of SCA implementation by well-known manufacturing enterprises. For example, Proctor & Gamble and Walmart are reported to have enhanced operational performance using data and analytical decision-support software for their day-to-day functions (Davenport & Harris, 2007) (Davenport & O'dwyer, 2011) (O'Dwyer & Renner, 2011) (SAS, 2012). One of the well-known high-street manufacturers in the United Kingdom (UK), TESCO has experienced substantial cost savings using Big Data analytics (Clark, 2013).

Big Data analytics in SCM is not necessarily a new idea (Davenport & O'dwyer, 2011) since different quantitative techniques and modelling mechanisms have long been used in manufacturing companies (Turban & Sephora, 1986) (Shapiro, 2000) (Kuaka, 2006) (Turkmen et al., 2010). However, there has been considerable interest in Big Data analytics in recent years, and it has ushered in new technical challenges for research communities. These challenges include technical issues regarding data volume and its storage facilities, data Provence (e.g., syntax, semantics, pragmatic), data quality and availability; and, importantly, data uncertainties (Hanfield & Nichols, 2004) (Liberatore & Luo, 2010) (Hunter et al., 2011) (Lavallel et al., 2011) (Manyika et al., 2011).

The first significant challenge for manufacturing businesses, innovative information technology and related electronic devices used in SCM is generating and capturing massive amounts of data. This data has been used to change companies' performance (Kohli & Grover, 2008). For example, a well-known consumer goods company, Li & Fung, reported the flow of over 100 gigabytes of data through the company's supply chain network on a given day in 2009 (Economist, 2010). The opportunity to gain a competitive advantage may thus arise from how companies manage data (Vosburg & Kumar, 2011) (Forslund & Jonsson, 2007) (Oliva & Watson, 2011). The second challenge for business is the increasing uncertainty in both the demand (e.g., consumer market) and the supply side of manufacturing supply chains. One solution proposed by academics (Oliva & Watson, 2011) (Demirkan & Delen, 2012) is proper supply chain planning to tackle demand and supply uncertainty.

It is evident from the research literature that SCM's critically important organizational function will need to evolve and adapt to Big Data analytics. For example, in a recent industry report (Deloitte & MHI, 2016), Mitsubishi Heavy Industries highlighted the potential of supply chains to deliver massive economic and environmental rewards for society. However, the report suggests that technological innovation will play a vital role in fulfilling this potential. For example, big data analytics can provide step-change improvements in supply chain visibility, cost savings, and customer service. The key is generating insightful data analysis and sharing it with business partners within manufacturing supply chains to act on it.

In response to these challenges, Big Data Analytics has been proposed as a promising approach to managing data better, utilizing IT resources and preparing for effective supply chain planning (Handfield, 2006) (Davenport, Harris & Morison, 2010) (Davis-Sramek, Germain & Iyer, 2010) (Viswanathan & Sadlovska, 2010). This new generation of analytics tools can develop a manufacturing enterprise's IT infrastructure and data management capabilities to improve strategic planning and enhance operational performance (Kohl & Grover, 2008) (Shapiro, 2010) (Mithras, Ramasubba & Sambamurthy, 2011). Additionally, it has been proposed that manufacturing enterprises can use SCA from acquired data using electromechanical devices (e.g., RFID tags, Sensors) and storing repositories of packaged software (e.g., CRM, ERP, SCM) to enhance manufacturing chain planning through IT-enabled planning and scheduling systems (Davenport & O'Dea, 2011) (Dwyer & Renner, 2011).

In order to solve a problem, the Big Data-based application uses intelligent reasoning in automated software that helps users apply analytical and scientific methods to business decision-making. The software applications that focus on the manufacturing chain domain are referred to as manufacturing decision support systems, providing Big Data-based tools to support a user's global supply chain reasoning process to develop a solution to the problem. This type of reasoning can be considered an intellectual process by which manufacturing operation managers use distinct artificial intelligence-based inference mechanisms to solve global operations problems.

Figure 4. Big Data potential for SCM according to the SCOR model

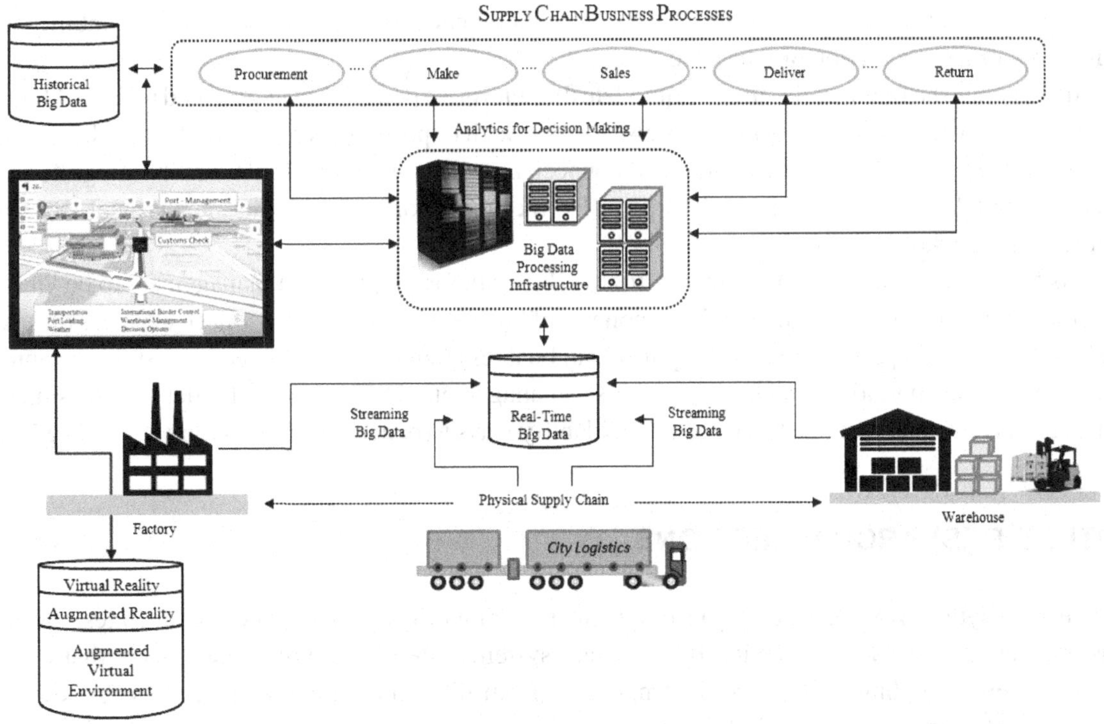

PROPOSED BIG DATA FRAMEWORK FOR SCM

As the global supply chain's information system's architecture expands over the Internet to envelop the entire supply chain, so does BI. The Internet broadens the information sources of data storage. It expands beyond what is contained within the organization's internal systems across the Internet to include business partners, suppliers, and customer relations.

In order to collaborate efficiently, business organizations must coordinate their regular activities. The essential elements of a collaborative environment are operational data store, data warehouse (e.g., data and metadata information), data mart (data warehouse, which concentrates on a particular subject area within the business operations), ETL tools, OLAP engine, analytical tools (reporting, data mining, and so on) and web portals. Combinations of these elements make different scenarios that depend on an organization's organizational administration, informational architecture, and supply chain. A data warehouse system must:

- Make an organization's information easily accessible,
- Present the organization's information consistently,
- Be adaptive and resilient to change,
- Serve as the basis of improved decision-making.

Figure 4 shows how the SCOR reference model can be utilized in Big Data for supply chain management. The SCOR model allows for identifying the origins of Big Data in supply chain activities. This can initially be applied at elevated levels and drilled down as required; secondly, it will identify potential users of such Big Data and their information requirements. Finally, the originating and communicating activities will need access to and analysis of Big Data sets. Finally, the approach specifies the type of IT infrastructure required to process such data, i.e., whether a batch or real-time streaming capability will be required or a combination of these two.

Additionally, the framework can accommodate the enterprise potential of augmented reality (AR) and virtual reality (VR). These efforts increasingly intersect with opportunities made possible by the Internet of Things (IoT) technology (e.g., sensors and connected devices) that help build a more integrated and extended digital and physical environment. Finally, the framework can help define a typical Big Data strategy for the whole supply chain.

This framework allows the problem of utilizing Big Data for supply chain management to be viewed from the perspective of vital managerial components of business logistics and distinct categories of stakeholders. It can encompass all essential supply chain business foundations such as forecasting, inventory management, transportation, and human recourses management. Different stakeholders, such as manufacturers, carriers, and manufacturers, can then identify ways to benefit from the supply chain Big Data.

FUTURE RESEARCH DIRECTIONS

Business analytics (BA) design and development face numerous future success challenges, such as implementation cost and complexity. BA software systems often have multiple elements that do not integrate well, including best-of-breed components from different commercial vendors. In addition, the biggest challenge is the users' ability to determine how to act based on the results of BA analysis in a manufacturing business. In addition, traditional BA applications have been slow at gathering and analyzing data, making short-term and day-to-day decision-making unsuitable.

Consequently, real-time data gathering, storing, and processing are fundamental challenges. Therefore, the best manufacturing supply chains will be those that can quickly analyze enormous amounts of distributed data and disseminate business insights to decision-makers as close to real-time as possible. In future, the current research will focus on the above issues and the usability of the BA software systems.

CONCLUSION

This chapter provides a systematic approach for using frameworks such as SCOR for supply chains, identifying the potential of Big Data within them, and identifying practical analytics applications for manufacturing supply chain management. The current business environment pushes manufacturing enterprises to optimize business processes and costs. At the same time, the extensive availability of data makes customers increasingly knowledgeable about products, prices, and other essential information. At first, manufacturing supply chain networks appear to easily manage abstract operational frameworks such as SCOR's Plan, Source, Make, Deliver and Return. However, upon deeper examination, manufacturing chains need real-time business process integration, coordination, and collaboration to deliver a higher level of corporate performance.

The emergence of Big Data Analytics opens excellent opportunities in manufacturing business communities. Also, current Information and Communication Technology (ICT) architectures enabled by Big Data can help optimize manufacturing supply chains with shared real-time data, coordination and altering capabilities. Additionally, this architecture needs to accommodate the enterprise potential of augmented reality (AR) and virtual reality (VR) that help build a more integrated and extended digital and physical environment.

REFERENCES

Angeles, R. (2005). RFID technologies: Supply-chain applications and implementation issues. *Information Systems Management*, *22*, 51–65.

Baesens, A. (2014). *Analytics in a big data world: The essential guide to data science and its applications*. John Wiley & Sons.

Ballou, R. H. (2007). *Business Logistics/supply Chain Management, 5/E (With Cd)*. Pearson Education India.

Barhhan, P. (2003). *International Trade Growth and Development: Essays*. Wiley-Blackwell.

Bowman, P., Ng, J., Harrison, M., Lopez, S., & Illic, A. (2009). Sensor based condition monitoring. *Building Radio frequency IDentification for the Global Environment, (Bridge)*. Euro RFID project.

Choi, S., Kim, B. H., & Noh, S. D. (2015). A diagnosis and evaluation method for strategic planning and systematic design of a virtual factory in smart manufacturing systems. *International Journal of Precision Engineering and Manufacturing*, *16*, 1107–1115.

Clark, L. (2013). Tesco Uses Supply Chain Analytics to Save £100 m a Year. *Computer Weekly*. https://www.computerweekly.com/news/2240182951/Tesco-uses-supply-chain-analytics-to-save-10-m-a-year

Cooke, J. A. (2013). Three trends to watch in 2013, Perspective. *Supply chain Quarterly, 1*, 11.

Davenport, T. H. (2006). *Competing on analytics, Harvard Business Review*. Harvard Business Press.

Davenport, T. H., Barth, P., & Bean, R. (2012). How Big Data is different. *MIT Sloan Management Review*, (Fall), 22–24.

Davenport, T. H., & Harris, J. G. (2007). *Competing on analytics – the new science of wining*. Harvard Business School Publishing Corporation.

Davenport, T. H., Harris, J. G., & Morison, R. (2010). *Analytics at work – smart decisions, better results*. Harvard Business Press.

Davenport, T. H., Harris, J. G., & Morison, R. (2010). *Analytics at work – smart decisions, better results*. Harvard Business Press.

Davenport, T. H., & O'Dwyer, J (2011). Tap into the Power of Analytics. *Supply Chain Quarterly*, 28-31.

Davenport, T. H., & Prusak, L. (2000). *Working knowledge: how organizations manage what they know*. Harvard Business Press.

Davis-Sramek, B., Germain, R., & Iyer, K. (2010). Supply Chain Technology: The Role of Environment in Predicting Performance. *Journal of the Academy of Marketing Science, 38*(1), 42–55. doi:10.100711747-009-0137-1

Decker, C., Berchtold, M. L., Chaves, W. F., Beigl, M., Roehr, D., & Riedel, T. (2008). Cost benefit model for smart items in the supply chain. In The Internet of Things, 155–172. Springer.

Deloitte & MHI. (2014). *The 2014 MHI Annual Industry Report – Innovation the driven supply chain*. MHI.

Deloitte & MHI (2016). The 2016 MHI Annual Industry Report – Accelerating change: How innovation is driving digital, always-on. *Supply Chains*.

Demirkan, H., & Delen, D. (2012). Levering the Capabilities of Service-oriented Decision Support Systems: Putting Analytics and Big Data in Cloud. *Decision Support Systems, 55*(1), 412–421. doi:10.1016/j.dss.2012.05.048

Dietrich, B., Plachy, E. C., & Norton, M. F. (2014). *Analytics across the enterprise: How IBM realize business value from big data and analytics*. IBM Press Books.

Fang, C., Liu, X., Pardalos, P. M., & Pei, J. (2016). Optimization for a three-stage production system in the Internet of Things: Procurement, production and product recovery, and acquisition. *International Journal of Advanced Manufacturing Technology, 83*, 689–710.

Fang, J., & Ma, A. (2020). IoT application modules placement and dynamic task processing in edge-cloud computing. *IEEE Internet of Things Journal, 8*(6), 12771–12781.

Fang, J., Qu, T., Li, Z. G., Xu, G., & Huang, G. Q. (2013). Agent-based gateway operating system for RFID-enabled ubiquitous manufacturing enterprise. *Robotics and Computer-integrated Manufacturing*, *29*, 222–231.

Forslund, H., & Jonsson, P. (2007). The Impact of Forecast Information Quality on Supply Chain Performance. *International Journal of Operations & Production Management*, *27*(1), 90–107. doi:10.1108/01443570710714556

Gereffi, G. (1999). International trade and industrial upgrading in the apparel commodity chain. *Journal of International Economics*, *48*(1), 37–70. doi:10.1016/S0022-1996(98)00075-0

Gubbi, J., Buyya, R., Murusic, S., & Palaniswami, M. (2013). Internet of Things (IoT): A vision, architectural elements, and future directions. *Future Generation Computer Systems*, *29*(7), 1645–1660.

Handfield, R. (2006). *Supply Market Intelligence: A Managerial Handbook for Building Sourcing Strategies*. Taylor & Francis. doi:10.4324/9780203339527

Handfield, R., & Nichols, E. Jr. (2004). Key Issues in Global Supply Base Management. *Industrial Marketing Management*, *33*(1), 29–35. doi:10.1016/j.indmarman.2003.08.007

Jiang, J., Li, Z., Tian, Y., & Al-Nabhan, N. (2020). A review of techniques and methods for IoT applications in collaborative cloud-fog environment. *Security and Communication Networks*, 1–15.

Kohli, R., & Grover, V. (2008). Business Value of IT: An Essay on Expanding Research Directions to Keep up with the times. *Journal of the Association for Information Systems*, *9*(1), 23–39. doi:10.17705/1jais.00147

Lavalle, S., Lesser, E., Shockey, R. H., & Crosthwait, N. M. (2011). Big Data, Analytics and the Path from Insight to Value. *MIT Sloan Management Review*, *52*(2), 21–32.

Li, B., Yang, C., & Huang, S. (2014). Study on supply chain disruption management under service level dependent demand. *Journal of Networking*, *9*, 1432–1439.

Li, D. (2016). Perspective for smart factory in petrochemical industry. *Computers & Chemical Engineering*, *91*, 136–148.

Li, D. (2016). Perspective for smart factory in petrochemical industry. *Computers & Chemical Engineering*, *91*, 136–148.

Liberatore, M., & Luo, W. (2010). The Analytics Movement. *Interface: a Journal for and About Social Movements*, *40*(4), 313–324. doi:10.1287/inte.1100.0502

Liukkonen, M., & Tsai, T. N. (2016). Toward decentralized intelligence in manufacturing: Recent trends in automatic identification of things. *International Journal of Advanced Manufacturing Technology*, *87*, 2509–2531.

Mayika, J. M., Chui, B., Brown, J. Bughin, R., Dobbs, Roxburgh, C., & Byers, A. (2011). Big Data: The Next Frontier for Innovation, Competition, and Productivity. *McKinsey Report*.

McAfee, A., & Brynjolfsson, E. (2012). Big data: The management revolution. *Harvard Business Review*, *90*(10), 61–68. PMID:23074865

McAfee, A., & Brynjolfsson, E. (2012). Big data: The management revolution. *Harvard Business Review*, *90*(10), 61–68. PMID:23074865

Mithas, S., Ramasubbu, N., & Sambamurthy, V. (2011). How Information Management Capability Influences Firm Performance. *Management Information Systems Quarterly*, *35*(1), 237–256. doi:10.2307/23043496

Ng, I. C., & Wakenshaw, S. Y. (2017). The Internet-of-Things: Review and research directions. *International Journal of Research in Marketing*, *34*(1), 3–21.

O'dwyer, J., & Renner, R. (2011). The Promise of Advanced Supply Chain Analytics, Supply Chain. *Management Review*, *15*, 32–37.

Oliva, R., & Watson, N. (2011). Cross-Functional Alignment in Supply Chain Planning: A Case Study of Sales & Operations Planning. *Journal of Operations Management*, *29*(5), 434–448. doi:10.1016/j.jom.2010.11.012

Paksoy, T., Karaoğlan, I., Gökçen, H., Pardalos, P. M., & Torğul, B. (2016). Experimental research on closed loop supply chain management with internet of things, Journal of Economy. *Bibliograph.*, *3*, 1–20.

Pal, K. (2016). *Supply Chain Coordination Based on Web Service, Supply Chain Management in the Big Data Era, Hing Kai Chan, Nachiappan Subramanian, and Muhammad Dan-Asabe Abdulrahman (edited Book Chapter)*. IGI Publication.

Pal, K. (2017). Supply Chain Coordination Based on Web Services, in H K. Chan, N. Subramanian, & M. D. Abdulrahman (Eds), Supply Chain Management in the Big Data Era, 137-171. IGI Global Publishing, Hershey PA, USA.

Pal, K. (2018). A Big Data Framework for Decision Making in Supply Chain, in P K Gupta, T. Oren, & M. Singh (Edited). Predictive Intelligence Using Big Data and the Internet of Things, 51-76. IGI Global Publication.

Pal, K. (2019). Quality Assurance Issues for Big Data Applications in Supply Chain Management, in P K Gupta, T. Oren, & M. Singh (Eds.), Predictive Intelligence Using Big Data and the Internet of Things, 51-76. IGI Global Publication.

Pal, K. (2020). Information sharing for manufacturing supply chain management based on blockchain technology, I. Williams (Edited), in Cross-Industry Use of Blockchain Technology and Opportunities for the Future, 1-17. IGI Global Publication.

Pal, K. (2021). Applications of Secured Blockchain Technology. In S. K. Pani, B. Patnaik, S. Lun, & X Liu (Edited), Manufacturing Industry, in Blockchain and AI Technology in the Industrial Internet of Things. IGI Global Publication.

Pal, K. (2021). Privacy, Security and Policies: A Review of Problems and Solutions with Blockchain-Based Internet of Things Applications in Manufacturing Industry. *Procedia Computer Science*, *191*, 176–183.

Pal, K. (2022). A Decentralized Privacy-Preserving Healthcare Blockchain for IoT, Challenges, and Solutions. In M. D. Borah, P. Zhang, & G. C. Deka (Edited), Prospects of Blockchain Technology for Accelerating Scientific Advancement in Healthcare, 158-188. IGI Global Publication.

Pal, K. (2023). Security Issues and Solutions for Resource-Constrained IoT Applications Using Lightweight Cryptography, in S. Verma, V. Vyas, & K. Kaushik (eds.) Cybersecurity Issues, Challenges, and Solutions in the Business World, 158-188. IGI Global Publication.

Pal, K., & Yasar, A. (2020). Internet of Things and blockchain technology in apparel manufacturing supply chain data management. *Procedia Computer Science*, *170*, 450–457. doi:10.1016/j.procs.2020.03.088

Pal, K., & Yasar, A. U. H. (2020). Internet of Things and Blockchain Technology in Apparel Manufacturing Supply Chain Data Management. *Procedia Computer Science*, *170*, 450–457.

Poluha, R. G. (2007). *Application of the SCOR Model in Supply Chain Management*. Youngstown.

Porter, M. E. (1985). *Competitive Advantage: Creating and Sustaining Superior Performance*. The Free Press.

Power, D. J. (2007). A Brief History of Decision Support Systems. *DSS Resource*. http://DSSResource.COM/history/dsshistory.html

Prahalad, C. K., & Mashelkar, R. A. (2010). Innovation's Holy Grail. *Harvard Business Review*, (July-August), 2010.

SAS. (2012). *Supply Chain Analytics: Beyond ERP and SCM*. SAS.

Sathi, A. (2012). Big data analytics: Disruptive technologies for changing the game. MC Press Online.

Shapiro, J. (2010). Advanced Analytics for Sales & Operations Planning. *Analytics Magazine*, (May-June), 20–26.

Shin, T. H., Chin, S., Yoon, S. W., & Kwon, S. W. (2011). A service-oriented integrated information framework for RFID/WSN-based intelligent construction supply chain management. *Automation in Construction*, *20*, 706–715.

Siegel, E. (2013). *Predictive analytics: The power to predict who will click, buy, lie or die*. John Wiley & Sons Inc.

Stiglitz, J. (2017). *Globalization and its Discontents Revisited: Anti-Globalization in the Era of Trump*. Penguin.

Stock, J. R. (2013). Supply chain management: A look back, a look ahead. *Supply Chain Quarterly*, *2*, 22–26.

Stock, T., & Seliger, G. (2016). Opportunities of sustainable manufacturing in industry 4.0. *Procedia*, *40*, 536–541.

Supply Chain Council. (2010). *Homepage*. ASCM. http://supply-chain.org/f/down-load/726710733/SCOR10.pdf

Svilvar, M., Charkraborty, A. & Kanioura, A. (2013). Big data analytics in marketing. *OR/MS Today*.

Tan, J., & Koo, S. (2014). A survey of technologies in internet of things, in IEEE. *Computers & Society*, 269–274.

The Economist. (2010). Data, Data Everywhere. *The Economist.* https://www.economist.com/node/15557443

Trkman, P., McCormack, K., de Oliveira, M. P. V., & Ladeira, M. B. (2010). The Impact of Business Analytics on Supply Chain Performance. *Decision Support Systems*, *49*(3), 318–327. doi:10.1016/j.dss.2010.03.007

Turban, E., & Sepehri, M. (1986). Applications of Decision Support and Expert Systems in Flexible Manufacturing Systems. *Journal of Operations Management*, *6*(34), 433–448. doi:10.1016/0272-6963(86)90015-X

Turban, E., Sharda, R., Delen, D., & King, D. (2011). *Business Intelligence: A Managerial Approach* (2nd ed.). Prentice-Hall.

Verdouw, C. N., Wolfert, J., Beulens, A., & Rialland, A. (2016). Virtualization of food supply chains with the internet of things. *Journal of Food Engineering*, *176*, 128–136.

Veza, I., Mladineo, M., & Gjeldum, N. (2015). Managing innovative production network of smart factories. *IFAC-PapersOnLine*, *48*, 555–560.

Viswanathan, N., & Sadlovska, V. (2010). *Supply Chain Intelligence: Adopt Role-based Operational Business Intelligence and Improve Visibility*. Aberdeen Group.

Vosburg, J., & Kumar, A. (2011). Managing Dirty Data in Organizations Using ERP: Lessons from a Case Study. *Industrial Management & Data Systems*, *101*(1), 21–31. doi:10.1108/02635570110365970

Wang, L., Laszewski, G. V., Young, K. M., & Tao, J. (2010). Cloud Computing: A Perspective Study. *New Generation Computing*, *28*(2), 137–146.

Wang, T., Zhang, Y., & Zang, D. (2016) Real-time visibility and traceability framework for discrete manufacturing shopfloor. In *Proceedings of the 22nd International Conference on Industrial Engineering and Engineering Management*,(pp. 763–772). IEEE.

Watson, H. J. (2009). Tutorial: Business Intelligence – Past, Present, and Future. *Communications of the Association for Information Systems*, *39*(25). doi:10.17705/1CAIS.02539

Watson, M., Lewis, S., Cacioppi, P., & Jayaraman, J. (2013). Supply chain network design – applying optimization and analytics to the global supply chain. *FT Press.*

Xing, B., Gao, W. J., Battle, K., Nelwamondo, F. V., & Marwala, T. (2012). e-RL: the Internet of things supported reverse logistics for remanufacture-to-order. International Conference in Swarm Intelligence: Advances in Swarm Intelligence, (pp. 519–526). Springer.

Yan, B., & Huang, G. (2009). Supply chain information transmission based on RFID and internet of things in Computing. In *Communication, Control, and Management*, (pp. 166–169). ISECS International Colloquium.

Yuvaraj, S., & Sangeetha, M. (2016). Smart supply chain management using internet of things (IoT) and low power wireless communication systems. In *Wireless Communication, Signal Processing and Networking, International Conference*, (pp. 555-558). IEEE.

Zawadzki, P., & Zywicki, K. (2016). Smart product design and production control for effective mass customization in the industry 4.0 concept. *Management of Production Engineering Review*, *7*, 105–112.

KEY TERMS AND DEFINITIONS-

Augmented Reality (AR): It is a modern technology that involves the overlay of computer graphics on real-world applications.

Big Data Analytics (BDA): Analytics is the discovery, interpretation, and visualization of meaningful patterns in Big Data. In order to do this, analytics use data classification and clustering mechanisms.

Decision Making Systems: A decision support system (DSS) is a computer-based information system that supports business or organizational decision-making activities, typically ranking, sorting, or choosing from among alternatives. Decision support systems can be either fully computerized, human-powered or a combination of both.

Supply Chain Management (SCM): A supply chain consists of a network of key business processes and facilities involving end-users and suppliers that provide products, services, and information.

Internet of Things (IoT): The Internet of things (IoT) is the inter-networking of physical devices, vehicles (also referred to as "connected devices" and "smart devices"), buildings, and other items; embedded with electronics, software, sensors, actuators, and network connectivity that enable these objects to collect and exchange data.

Neural Network: Neural network is an information processing paradigm inspired by how biological nervous systems, such as the brain, process information. It uses a classification mechanism that is modelled after the brain and operates by modifying the input through weights to determine what it should output.

Radio Frequency Identification (RFID): This wireless technology is used to identify tagged objects in certain vicinities. Generally, it has three main components: a tag, a reader, and a back-end. A tag uses the open air to transmit data via a radio frequency (RF) signal. However, it is also weak in computational capability. Finally, RFID automates information collection regarding an individual object's location and actions.

Virtual Reality (VR): It is a term used for computer-generated three-dimension (3D) environments that allow the user to enter and interact with synthetic environments. The users can immerse themselves to varying degrees in the artificial computer world, which may either be a simulation of some form of reality or a complex phenomenon.

Chapter 6
Cognitive Load Measurement Based on EEG Signals

Tamanna Tasmi
Bangladesh University of Health Sciences, Bangladesh

Mohammad Parvez
Brac University, South Korea

Jia Uddin
Woosong University, South Korea

ABSTRACT

Measurement of the cognitive load should be advantageous in designing an intelligent navigation system for visually impaired people (VIPs) when navigating unfamiliar indoor environments. Electroencephalogram (EEG) can offer neurophysiological indicators of the perceptive process indicated by changes in brain rhythmic activity. To support the cognitive load measurement by means of EEG signals, the complexity of the tasks of the VIPs during navigating unfamiliar indoor environments is quantified considering diverse factors of well-established signal processing and machine learning methods. This chapter describes the measurement of cognitive load based on EEG signals analysis with its existing literatures, background, scopes, features, and machine learning techniques.

INTRODUCTION

To understand cognitive load, we must first understand the working memory and to understand working memory we have to know what memory is. The memory is the aptitude of brain that deals with encoding, storing and retrieving information as needed. This information is received and transmitted from external environment by sensory nervous system in the form of chemical or physical stimuli and processed by memory in central nervous system. Now memory is classified into two categories: (1) short term or working memory that holds information temporarily (Miyake & Shah, 1999) and allows manipulation

DOI: 10.4018/978-1-7998-7852-0.ch006

Copyright © 2023, IGI Global. Copying or distributing in print or electronic forms without written permission of IGI Global is prohibited.

of stored information for reasoning, decision making, guiding behavior etc., (2) long term memory that can stock data for a long period.

Cognitive load is the quantity of working memory in use. There is another notion called "long-term working memory" that is a set of rescues of parts of long-term memory enabling continuous access to data required for daily activities (Ericsson & Kintsch, 1995). Cognitive load is the basis of problem solving and learning (Sweller, 1998).

Now comes the question, why do we need to measure cognitive load? Cognitive load is associated with learning new things whether be it a study matter or a new skill. So, to design a lesson plan it is important to understand how can a learner learn or memorize it easily and quickly where the measurement of cognitive load comes handy. The constructions and roles of human cognitive built have been used to develop a range of instructional means aiming for the reduction of load of working memory in learners and encouragement of diagram construction (Sweller et al., 1998). Furthermore, there are many other purposes of measuring cognitive load, such as, to know how a disease (neurodegenerative diseases, carcinomas etc.) or it's treatment (chemotherapy, radiotherapy, immunotherapy etc.) affects human cognition. This measurement is also very important for different researches including age related cognitive declines, learning task performances and multiple document handling (Cerdan et al., 2018), Designing navigation aid for blind people (Kalimeri & Saitis, 2016) etc.

Researchers formulated many ways of measuring cognitive load such as subjective scales, task-invoked pupillary response (Paas et al., 2003; Skulmowski & Rey, 2017), EEG signals (Antonenko et al., 2010), fMRI etc. Distraction causes increase in cognitive load (Paas & Sweller, 2012).

Subjective scales vary with varying perception in different individuals; thus, such scales are less reliable. Again, pupillary responses is equivalent to a range of events requiring psychological efforts which may be perceptual, cognitive and/ or response related, thus it is not indicative of cognitive load being linked to task performance (Kramer, 1991). EEG and fMRI give more precise results regarding cognitive load related to task performance although they are expensive and complicated to operate. Compared to fMRI, EEG is easier to perform and read as various software are available now for EEG interpretation.

In this chapter Measurement of cognitive load based on EEG signals is discussed taking references from a study of learning processes of visually impaired people (VIP) while navigating through unfamiliar indoor environment using EEG signal (Afroz et al., 2019).

LITERATURE REVIEW

First, we review the prior research in the area of cognitive load measurement using EEG signals. Much work has been devoted in this area by extracting various features from EEG signals and then machine learning approaches have been applied to quantify cognitive load over years.

Fraser et al. (Fraser et al., 2015) argued that intrinsic cognitive load needs to be adjusted to level of the apprentice, extraneous cognitive load needs to be abridged, and germane load needs to be augmented until the boundaries of working memory are not exceeded.

Gevins et al. (Gevins et al., 1997) studied cortical activity throughout working memory tasks and found that a sluggish (low-frequency), parieto-central, alpha signal lessened as working memory load amplified.

Anderson et al. (Anderson et al., 2011) designed user study performance based on measurement of cognitive load using EEG signals and these trials are used for quantitative evaluation the efficiency of visualizations.

Chandra et al. (Chandra et al., 2015) extracted different features such as root mean square value, sub band energy, power spectral density, and engagement index and then used neural network for classification of the workload.

Kumar et al. (Kumar & Kumar, 2016) postulated that the cognitive load is possible to be quantified by measuring alpha and beta band events in the frontal, temporal and front-central regions of the cortex of cerebellum.

Antonenko et al. (Antonenko et al., 2010) computed Event-related desynchronization (ERD)/ event-related synchronization (ERS) principles for individual partaker's alpha and theta rhythms under respective experimental conditions. Results illustrated that attainment of abstract and physical knowledge was meaningfully better in the lead-facilitated hypertext state.

Fournier et al. (Fournier et al., 1999) illustrated that alpha ERD is insensitive to various workload weights in multi-task situations, it is operative in measuring variances in processing difficulties in the solitary interactive mission condition. Moreover, theta ERS is insensitive to workload and exercising the interactive of multiple task condition.

Klimesch et al. (Klimesch, 1999) illustrated the volume of power of theta and alpha frequency range in EEG is certainly associated with cognitive performance and memory in specific, if a binary dissociation amid absolute and event-related fluctuations in alpha and theta wave is considered.

Ryu et al. (Ryu & Myung, 2005) established technique based on several functional indices for evaluating the mental load all through arithmetic and tracking tasks. The suppression of alpha delivered appropriate data to deduce exertions for math task, but not in case of tracking task. On the other hand, blink intermission and heart rate variability allowed comprehensive readings about workload during tracking task, but not in case of arithmetic task.

Roy et al. (Roy et al., 2013) evaluated psychological exhaustion, ascending from mounting time-on-task (TOT), can pointedly affect the dispersal of the band power features. They exposed contradictory variations in alpha power spreading between Workload (WKL) and TOT settings, and reduction in WKL level discriminability by means of growing TOT in both figure of statistical alterations in band power and cataloguing act.

Krigolson et al. (Krigolson et al., 2015) demonstrated that bigger cognitive load lessens the functional efficiency of a recompence processing structure inside human medial–frontal cortex.

The performance of the quantification of cognitive load using EEG in real time seems to be more challenging due to diverse factors such as different environments, age, sex, social status, therefore, more research will be conducted.

BACKGROUND STUDY

In this section EEG system, event related synchronization/ event related desynchronization (ERS/ERD), as well as different bands from EEG signals are discussed.

EEG System

The foremost International EEG assembly was held in 1947 when it recognized that standard electrodes appointment technique is required for the EEG recording (Oostenveld & Praamstra, 2001). This recognition resulted in the introduction of 10-20 electrode structure by H.H. Jasper in 1958. In this 10-20 system, 21 electrodes are placed utilizing the scalp size conferring to the external landmarks on cranium and their locations at distance of 10% and 20% of measurement of coronal, sagittal and circumference arcs from the nation to inion. Electrodes are marked for identification according to their relative site on scalp. For instance, F, T, C, P and O represents frontal, temporal, central, parietal, and occipital lobes successively. Odd figures mean the electrodes of left side while even figures mean the right-sided ones, whereas the ground electrode is positioned at an unbiased site of the head such as midline of forehead. The two reference electrodes namely A1 and A2 are sited in dynamic zone (ear lobes) on left and right side respectively. The 10–10 system or International 10–20 system is a globally recognized technique for relating and localizing scalp electrodes in the context of EEG based experiments, which was established to guarantee standardized duplicability so that subjective researches as well as subjects could be compared to each other over time. This scheme is founded on the basis of association amid the position of an electrode and its underlying region of cerebral cortex. As for the designation of the scheme, the "10" and "20" actually refer to the real distances among adjacent electrodes which are either 10% or 20% of the entire front–back or right–left dimension of the skull. The 10-10 system, also acknowledged as the 10% system, is not only presented but also recommended by the standard of the American Electroencephalographic Society and the International Federation of Societies for Electroencephalography and Clinical Neurophysiology (Oostenveld & Praamstra, 2001) where electrodes' situation occur at interval of 10% of measured coronal, sagittal and circumference arcs amid the two points explicitly nation and inion. The other stretched variety that is 10-5 system allows more than 300 electrodes to be placed (Jurcak et al., 2007). Intracranial source of EEG signals is applied to facilitate the high-density EEG applications.

Event Related Synchronization/ Event Related Desynchronization

Cognitive load theory (CLT) refers to a hypothetical agenda which is founded upon human reasoning construction of long-term and short term (working memory) built (Sweller et al., 1998). CLT is a theory about information processing that relates working memory manacles to the effectiveness of instruction. Learning procedure is performed in the working memory which has restricted in both terms of capacity and duration as it can hold only up to 7±2 portions of data at a specified time while novel data can be stored within about 15 to 30 seconds (Miller, 1956). There are the three commonest categories of cognitive load, specifically intrinsic, extraneous and germane and these are defined by present CLT (Van Merrienboer & Sweller, 2005). Intrinsic load is mainly imposed according to the intrinsic intricacy of the mission, while extraneous and germane loads are forced in accordance with the methods by which things are needed to be learnt. event-related synchronization and desynchronization (ERSD) are measured to estimate the experimental participants' cognitive load. Antonenko et al. (Antonenko et al., 2010) demonstrated that the alpha band power is augmented in event-related synchronization (ERS) and diminished in event-related desynchronization (ERD) of the task interlude by means of baseline intermission. Hence, cognitive index (CI) of ERSD is designed via the following equation:

$$\partial = \left(\frac{\in_b - \in_t}{\in_b} \right) * 100 \quad (1)$$

where ∂ is cognitive index, $\in_b$ is the base-line intermission of band power, and $\in_t$ is the task internal of band power.

Different Bands Signals

In typical scenarios, the breadth of clinical EEG signals is from 10 to 100 μv as the frequency ranges from 1 to 100 Hz. EEG signal is categorized into five broad rhythmic categories conferring to their frequency bands as clarified underneath:

Delta waves (δ): The frequency range is below 4 Hz with amplitude ranging from 20-200 μv. It arises at the time of deep sleep, infancy, and grave organic brain disorders (He, 2013). It is documented from posterior brain in children and from front brain in grown-ups.

Theta waves (θ): The frequency ranges from 4 to 7 Hz as it is prominent in situations like sleep, psychological strain, and awakening in case of both children and grownups. It can be documented from both temporal and parietal regions of the scalp with a breadth ranging from 5-10 μv (He, 2013).

Alpha waves (α): It is periodic wave that is found in fit grownups during wakefulness, relaxation with eyes kept closed. Its frequency ranges from 8 to 13 Hz with usual voltage array of about 20-200 μv. It disappears during disease conditions like coma or normally in sleep.

Beta waves (β): Its frequency ranges from 13-30 Hz, although their breadths are lesser, ranging from 5-10 μv. It surfaces in situations like excessive excitement of the central nervous system when upsurge in alertness and watchfulness occur. It substitutes the alpha wave if cognitive damage ensues for any reason. It can be recorded placing electrodes on parietal and frontal regions of the scalp.

Gamma waves (γ): Frequency of gamma waves arrays from 30 to 100 Hz. It can be documented from the somatosensory cortex in instances of cross model sensory processing, short term memory processing to identify objects, sounds, palpable sensation and in some pathologies including cognitive deterioration, mainly when it relates to the θ band.

COGNITIVE LOAD MEASUREMENT

Measurement of cognitive load is typically consisting of bands extraction, features extraction, calculation of ERS/ERD, and classification (see in Fig. 1).

EEG Signals

Nine VIPs with different degree of vision lose partaken this experiment wandering through a complex route of the University building which included students' units, class rooms, reading rooms, a book store and two restaurants (Kalimeri & Saitis, 2016). The route was about 200 meters in length whereas the walking distance was roughly 5 minutes. Various environmental condition such as door, elevator, moving people and object, open and narrow space, as well as stairs were considered.

The study got its approval from the National Bioethics Committee of Iceland and data set was anonymized prior to analysis. The data set comprises of EEG recordings from nine healthy VIPs (6 female; average duration of visual damage=30 yrs, range=2-52 yrs) with different degrees of eyesight loss (see Table. 1) as they strolled separately through a multifaceted course in an educational institute. The participants were asked to abstain from smoking, consuming coffee, and sugar roughly 1 hour before the initiation of the experiment. The partakers were instructed to walk through the charted route thrice (i.e., trial 1, trial 2, and trial 3) as drills. Directions were provided for the first time only (i.e., trial 1) to aid the VIPs acquaint with the course. They were also prohibited from needless head activities and hand movements. They also asked to avoid using their O&M instructor except in case of emergencies.

Figure 1. Basic diagram for cognitive load measurement using EEG signals and machine learning approach.

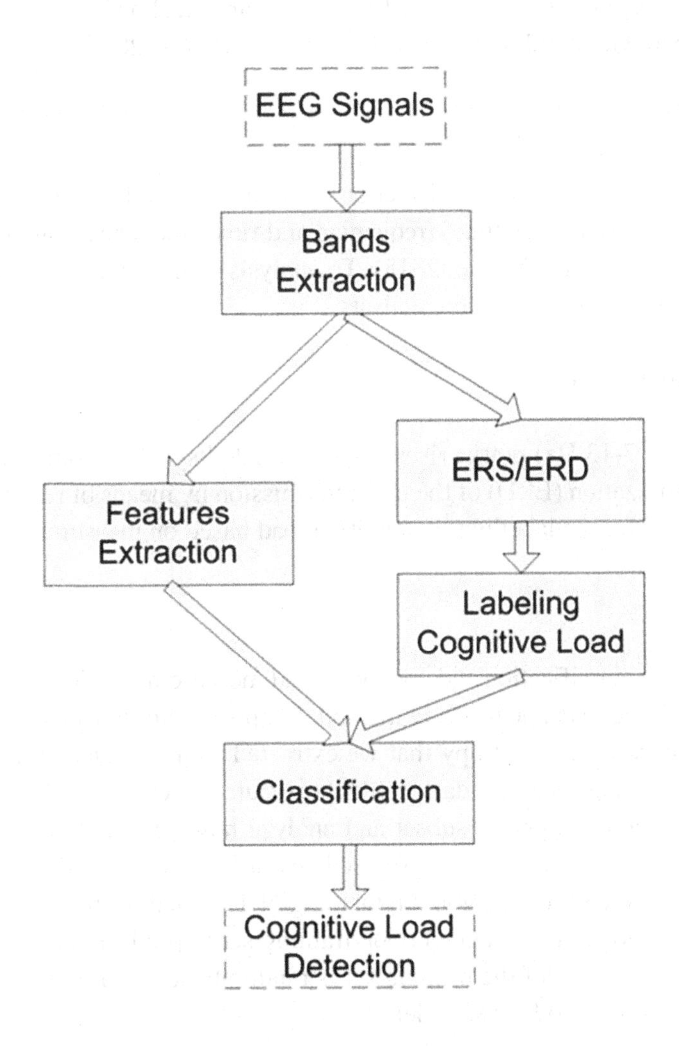

EEG data was attained by the Emotive EPOC+ system which is a portable headset containing 16 dry electrodes with 128 Hz sampling rate registering capacity over the 10-20 system locations specifically

AF3, F7, F3, FC5, T7, P7, O1, O2, P8, T8, FC6, F4, F8, AF4, GYROX and GYROY (i.e. information about how the head accelerates during leaning sessions according to x-axis and y-axis, respectively). This system encompasses a sum of interior signal training phases. Analogue signals are at first filtered by high-pass filter with a 0.16 Hz cut-off, pre-amplified, then low-pass filtered with an 83 Hz cut-off and tried at 2048 Hz. Digital signals are subsequently notch-filtered at 50/60 Hz and down-sampled to 128 Hz before broadcast. The EEG signal of each participant are firstly smothered by his own base signal and then bandpass butter worth filter is used on the signal for mining of gamma (30 - 60 Hz), beta (13 - 30 Hz), alpha (7 - 13 Hz), theta (4 - 7 Hz), and delta (0.5 - 4 Hz) bands respectively.

Bands Extraction

Different bands like gamma, beta, alpha, theta, and delta are extracted to get appropriate features. The power of alpha band is grown in event-related activation and declined in event-related deactivation considering task interval with baseline interval (Skulmowski & Rey, 2017).

Features Extraction

Features extract play an important role for detection of cognitive load based on EEG signals. Different feature extraction techniques such as time, frequency, and time-frequency domains are considered for features extraction (Hernández & Trujillo, 2018). To analysis of the cognitive load, we have extracted entropy feature based on frequency domain analysis.

Labelling of Cognitive Load

Though alpha band (i.e., 7-13 Hz) power rises in event-related synchronization (ERS) and lessens in event-related desynchronization (ERD) of the task intermission by means of reference line interval. So, alpha band power is used for the labelling of cognitive load based on measurement of cognitive index.

Classification

In this section, we discuss classification and cognitive load measurement. Classifier intent to categorize the EEG signals by applying machine learning method. Support vector machine (SVM) is implemented on spatio-temporal feature namely entropy that we extracted using frequency domain analysis. Classification processed revolving cross validation which execute its technique involving the test set and training set of data into complimentary subset and analyze how accurately the predicted model will perform (Parvez & Paul, 2016). In our case, we used five-fold cross validation (i.e., N=5). One set is randomly chosen and reserved for testing and remaining N-1 is used for training and then average the results. five-fold cross validation is operated upon training set to produce an ideal model of the SVM classifier. N-1 which means 80% of the training set is randomly selected to establish the SVM model and remaining 20% is observed to fit the model.

RESULTS AND DISCUSSION

Entropy features extracted from EEG signals using 16 electrodes and different bands. The performance metrics such as sensitivity, specificity, and accuracy are considered to evaluate and measure the performance (Parvez & Paul, 2014). The formulation of sensitivity, specificity, and accuracy are as follows:

$$Psen = \frac{\tau_p}{\tau_p + \varphi_n} \quad (2)$$

$$Pacc = \frac{\tau_p + \tau_n}{\tau_p + \tau_n + \varphi_p + \varphi_n} \quad (3)$$

where τ_p is the true positive, τ_n is the true negative, φ_p is the false positive, and φ_n is the false negative.

Table 1. Calculate performance using signal processing feature and machine learning approach for cognitive load measurement in the indoor environment based different band from EEG signals

Brain Location	Sensitivity	Specificity	Accuracy
Entire Brain	58.37	76.23	73.08
Left hemisphere	56.44	74.32	72.31
Right hemisphere	54.72	73.64	71.45
Left hemisphere and frontal lobe	56.98	72.61	71.64
Right hemisphere and frontal lobe	55.46	72.25	71.45
Left hemisphere and theta and alpha bands	70.11	71.35	70.65
Right hemisphere and theta and alpha bands	56.15	72.20	71.29

To observe the measurement of cognitive load, the performance metrices are considered. In the Table 1, the results demonstrated that accuracy is high when considering entire brain compared to consider partial brain locations.

CONCLUSION

In this study the chief aim of the measurement of cognitive load with emphasis on working memory load was to design a smart motion system for the visually impaired people (VIPs) for their navigation. To achieve this goal an experiment consisting of different phases such as bands extraction, features extraction, labelling of cognitive index using well-established metric, and finally use of the machine learning techniques was conducted to measure cognitive load. The performance of the prediction and/or detection of cognitive load using EEG signals in real time is more challenging, therefore, further research should be conducted for improved performance.

REFERENCES

Afroz, S., Shimanto, Z. H., Jahan, R. S., & Parvez, M. Z. (2019). Exploring the cognitive learning process by measuring cognitive load and emotional states. *Biomedical Engineering: Applications, Basis and Communications*, *31*(04), 1950032.

Anderson, E. W., Potter, K. C., Matzen, L. E., Shepherd, J. F., Preston, G. A., & Silva, C. T. (2011). A user study of visualization effectiveness using EEG and cognitive load. *Computer Graphics Forum*, *30*(3), 791–800. doi:10.1111/j.1467-8659.2011.01928.x

Antonenko, P., Paas, F., Grabner, R., & van Gog, T. (2010). Using electroencephalography to measure cognitive load. *Educational Psychology Review*, *22*(4), 425–438. doi:10.100710648-010-9130-y

Cerdan, R., Candel, C., & Leppink, J. (2018). Cognitive load and learning in the study of multiple documents. *Frontiers in Education.*, *3*, 59. doi:10.3389/feduc.2018.00059

Chandra, S., Sharma, G., Verma, K.L., Mittal, A., & Jha, D. (2015). EEG based cognitive workload classification during NASA MATB-II multitasking. *International Journal of Cognitive Research in Science, Engineering and Education, 3*(1).

Ericsson, K. A., & Kintsch, W. (1995). Long-term working memory. *Psychological Review*, *102*(2), 211–245. doi:10.1037/0033-295X.102.2.211 PMID:7740089

Fournier, L. R., Wilson, G. F., & Swain, C. R. (1999). Electrophysiological, behavioral, and subjective indexes of workload when performing multiple tasks: Manipulations of task difficulty and training. *International Journal of Psychophysiology*, *31*(2), 129–145. doi:10.1016/S0167-8760(98)00049-X PMID:9987059

Fraser, K. L., Ayres, P., & Sweller, J. (2015). Cognitive load theory for the design of medical simulations. *Simulation in Healthcare*, *10*(5), 295–307. doi:10.1097/SIH.0000000000000097 PMID:26154251

Gevins, A., Smith, M. E., McEvoy, L., & Yu, D. (1997). High-resolution EEG mapping of cortical activation related to working memory: Effects of task difficulty, type of processing, and practice. *Cerebral Cortex*, *7*(4), 374–385. doi:10.1093/cercor/7.4.374 PMID:9177767

Hernández, D. E., & Trujillo, L. (2018). Detecting Epilepsy in EEG Signals Using Time, Frequency and Time-Frequency Domain Features. Computer Science and Engineering—Theory and Applications.

Jurcak, V., Tsuzuki, D., & Dan, I. (2007). 10/20, 10/10, and 10/5 systems revisited: Their validity as relative head-surface-based positioning systems. *NeuroImage*, *34*(4), 1600–1611. doi:10.1016/j.neuroimage.2006.09.024 PMID:17207640

Kalimeri, K., & Saitis, C. (2016). Exploring multimodal biosignal features for stress detection during indoor mobility. *Proceedings of the 18th ACM international conference on multimodal interaction*, 53-60. 10.1145/2993148.2993159

Klimesch, W. (1999). EEG alpha and theta oscillations reflect cognitive and memory performance: A review and analysis. *Brain Research. Brain Research Reviews*, *29*(2-3), 169–195. doi:10.1016/S0165-0173(98)00056-3 PMID:10209231

Kramer, A.F. (1991). Physiological metrics of mental workload: A review of recent progress. *Multiple-task performance*, 279-328.

Krigolson, O.E., Hassall, C.D., Satel, J., & Klein, R.M. (2015). The impact of cognitive load on reward evaluation. *Brain Research, 1627*, 225-32.

Kumar, N., & Kumar, J. (2016). Measurement of cognitive load in HCI systems using EEG power spectrum: An experimental study. *Procedia Computer Science*, *84*, 70–78. doi:10.1016/j.procs.2016.04.068

Miller, G. A. (1956). The magical number seven, plus or minus two: Some limits on our capacity for processing information. *Psychological Review*, *63*(2), 81–97. doi:10.1037/h0043158 PMID:13310704

Miyake, A., & Shah, P. (1999). *Models of working memory. Mechanisms of active maintenance and executive control*. Cambridge University Press. doi:10.1017/CBO9781139174909

Oostenveld, R., & Praamstra, P. (2001). The five percent electrode system for high-resolution EEG and ERP measurements. *Clinical Neurophysiology*, *112*(4), 713–719. doi:10.1016/S1388-2457(00)00527-7 PMID:11275545

Paas, F., & Sweller, J. (2012). An evolutionary upgrade of cognitive load theory: Using the human motor system and collaboration to support the learning of complex cognitive tasks. *Educational Psychology Review*, *24*(1), 27–45. doi:10.100710648-011-9179-2

Paas, F., Tuovinen, J. E., Tabbers, H., & van Gerven, P. W. (2003). Cognitive load measurement as a means to advance cognitive load theory. *Educational Psychologist*, *38*(1), 63–71. doi:10.1207/S15326985EP3801_8

Parvez, M. Z., & Paul, M. (2014). Epileptic seizure detection by analyzing EEG signals using different transformation techniques. *Neurocomputing*, *145*, 190–200. doi:10.1016/j.neucom.2014.05.044

Parvez, M. Z., & Paul, M. (2016). Seizure prediction using undulated global and local features. *IEEE Transactions on Biomedical Engineering*, *64*(1), 208–217. doi:10.1109/TBME.2016.2553131 PMID:27093309

Roy, R. N., Bonnet, S., Charbonnier, S., & Campagne, A. (2013). Mental fatigue and working memory load estimation: interaction and implications for EEG-based passive BCI. *2013 35th Annual International Conference of the IEEE Engineering in Medicine and Biology Society (EMBC)*, 6607-6610.

Ryu, K., & Myung, R. (2005). Evaluation of mental workload with a combined measure based on physiological indices during a dual task of tracking and mental arithmetic. *International Journal of Industrial Ergonomics*, *35*(11), 991–1009. doi:10.1016/j.ergon.2005.04.005

Skulmowski, A., & Rey, G. D. (2017). Measuring cognitive load in embodied learning settings. *Frontiers in Psychology*, *8*, 1191. doi:10.3389/fpsyg.2017.01191 PMID:28824473

Sweller, J. (1998). Cognitive Load During Problem Solving: Effects on Learning. *Cognitive Science*, *12*(2), 257–285. doi:10.120715516709cog1202_4

Sweller, J., van Merrienboer, J. J., & Paas, F. G. (1998). Cognitive Architecture and Instructional Design. *Educational Psychology Review*, *10*(3), 251–296. doi:10.1023/A:1022193728205

Van Merrienboer, J. J., & Sweller, J. (2005). Cognitive load theory and complex learning: Recent developments and future directions. *Educational Psychology Review*, *17*(2), 147–177. doi:10.100710648-005-3951-0

Chapter 7
A Study on Deep Learning Methods in the Concept of Industry 4.0

Mehmet Ali Şimşek
Tekirdag Namik Kemal University, Turkey

Zeynep Orman
https://orcid.org/0000-0002-0205-4198
Istanbul University-Cerrahpasa, Turkey

ABSTRACT

Nowadays, the main features of Industry 4.0 are interpreted to the ability of machines to communicate with each other and with a system, increasing the production efficiency, and development of the decision-making mechanisms of robots. In these cases, new analytical algorithms of Industry 4.0 are needed. By using deep learning technologies, various industrial challenging problems in Industry 4.0 can be solved. Deep learning provides algorithms that can give better results on datasets owing to hidden layers. In this chapter, deep learning methods used in Industry 4.0 are examined and explained. In addition, data sets, metrics, methods, and tools used in the previous studies are explained. This study can lead to artificial intelligence studies with high potential to accelerate the implementation of Industry 4.0. Therefore, the authors believe that it will be a handbook and very useful for researchers who want to do research on this topic.

INTRODUCTION

Industry 4.0 is defined as the fourth phase of the industrial revolution. The invention and use of steam engines led to Industry 1.0; the development of electric and batch production devices led to Industry 2.0; the use of robots and computers on production lines led to Industry 3.0, and lastly, the big industrial revolution today named as Industry 4.0 continues to impact our lives. The desire to create super-smart societies and not to make any mistakes will be referred to as Industry 5.0.

DOI: 10.4018/978-1-7998-7852-0.ch007

Copyright © 2023, IGI Global. Copying or distributing in print or electronic forms without written permission of IGI Global is prohibited.

Industry 4.0 is a process that aims to reduce the costs by decreasing the human impact in the production to a minimum through artificial intelligence methods, internet of things (IoT), and big data. The purposes of Industry 4.0 include the transition of traditional factories into digital factories, fast and reliable productions, development of smart cities, creating data storages with instant notifications and obtaining more meaningful knowledge from the data, minimum use of energy, and reducing the environmental damage to a minimum.

Industry 4.0 is classified in multiple categories based on the improvements in the hardware of electronic improvement cards, sensor architectures and structures, cloud services and wireless communication systems, and electronic systems. These categories are; Internet of Things (IoT), Cyber-Physical Systems, Cloud Technology, Augmented Reality, Autonomous Robots, Layered Production, Cyber Security, Big Data and Analysis, System Integration (Süzen & Kayaalp, 2019; Tekin & Karakuş, 2018).

The main characteristics of Industry 4.0 include communication of machines with each other and a system, an increase of production efficiency, and improvement of the mechanisms of robots. These situations are required to use some new analytical algorithms for Industry 4.0. Various industrial challenging problems in Industry 4.0 can be solved by using deep learning-based technologies.

Deep learning is a machine learning method that uses large amounts of data to perform supervised or unsupervised activities including feature extraction, transformation, and categorization through mimicking the skills of observation, analysis, learning, and decision making that human brain uses for complex problems (Süzen & Kayaalp, 2018a). It is a machine learning method developed by expanding traditional neural networks.

Today's industrial environments produce large amounts of regular or irregular data. Processing the data generated with the right methods and strategies would allow for reduced costs, the ability to calculate the life of materials used, easy use of autonomous systems, taking measures for potential mistakes, and efficient energy usage.

This study is to examine and explain the Industry 4.0 concepts and deep learning methods in terms of the methods used, the success of recommended systems, the success of the metrics, and the data libraries used in the related studies in the literature. Therefore, a literature review will be provided with classifications by interpreting the findings of related studies in the field. The study will include the following sections; background on Industry 4.0 and concepts related, deep learning and methods, metrics used to measure the success of deep learning methods, literature review, findings, and discussion.

BACKGROUND

In order to understand the literature review section better, this section will focus on the introduction of Industry 4.0, deep learning and success metrics of the research. The research strategy of the study is also mentioned.

Research Strategies

We have used the SMS (Systematic Mapping Study) method to examine the studies in the literature related to deep learning methods that can be used for Industry 4.0. SMS is a literature review method that aims to classify by interpreting the outputs from the relevant studies to find answers to the research questions identified.

The keywords that were determined for searching in online libraries are "deep learning, industry 4.0". The search was carried out by connecting the two keywords with the "and" conjunction. It was observed that when the searches were connected with the "or" conjunction, quite a lot of publications unrelated to the subject were listed. Searches were made in IEEE Xplore, ScienceDirect, Google Scholar and Wiley online databases and a total of 60 results were found. Publications encountered in online databases are given in Table 1. Within the scope of this study, a total of 27 publications were used after the SMS "inclusion and exclusion criteria" process.

Table 1. Publications encountered by the online database

	Conferences	Journals	Early Access Articles	Magazines	Total
IEEE Xplore	32	11	3	3	49
ScienceDirect	0	3	0	0	3
Google Scholar	3	5	0	0	8
Wiley	0	0	0	0	0
Total	**35**	**19**	**3**	**3**	**60**

Relevant publications have been examined according to criteria such as industry 4.0 domain, deep learning models used, whether traditional machine learning models were used, online data set, success metrics used, year of publication, and type of implementation tool.

Industry 4.0

Industry 4.0 is divided into some subtitles in parallel with the developments in electronic development cards, sensor architectures, and structure, cloud services and wireless communication systems, in the hardware that make up the electronic systems. These are the Internet of Things (IoT), Cyber-Physical Systems, Cloud Technology, Augmented Reality, Autonomous Robots, Layered Production, Cyber Security, Big Data and Analysis, System Integration subheadings (Süzen & Kayaalp, 2019; Tekin & Karakuş, 2018). Some subtopics of Industry 4.0 are described below.

Internet of Things (IoT): Communication between different devices has become easier due to the developments in internet technology, internet protocols, and information processing systems in recent years. These developments, in which billions of devices enter the Internet, have enabled the development of the Internet of Things concept. The purpose of the Internet of Things (IoT) is to develop a smarter environment and a simplified lifestyle by saving time, energy and money (Mahdavinejad et al., 2018).

Autonomous robots: Autonomous robots, also called intelligent robots, are robots that can make desired operations by entering dangerous areas where people will have trouble reaching. They are robotic systems that have the feature of knowing their own positions, can use artificial intelligence techniques thanks to the embedded systems they contain, can produce results according to the information they receive from the external environment with the help of sensors, and can take decisions according to the result (Akyol & Uçar, 2019; Süzen & Kayaalp, 2019). Autonomous robots also have the ability to work together as a herd and collect data.

Augmented Reality: A combination of virtual and real-world created by combining and enriching the real environment with virtual objects (Doğan, Sönmez, & Cankül, 2020). Augmented reality is the 3D positioning of a camera, computer, and practical world elements; enrichment of real-world spaces and objects by producing virtual items in the computer environment (Tekin & Karakuş, 2018). Augmented reality applications, a new method in human-computer interaction, are frequently encountered in the field of education, in the field of medicine (Joda, Gallucci, Wismeijer, & Zitzmann, 2019), in the areas of games and sports.

Big data: Over the years, regular or irregular data sets are accumulated that consist of content such as companies in the business world, social network users, articles gathered by authors on blogs opened on the internet, video & photography produced by content producers (Aktan, 2018; Tekin & Karakuş, 2018). The process of producing meaningful and useful information as a result of some methods from these datasets is called big data.

Deep Learning

Machine learning can be used in processes such as content filtering in web searches and social networks, deciding suggestions on e-commerce websites, identifying objects, converting speech to text, matching news content, publications or products with users' interests and showing matched results. Such practices can also be done with deep learning techniques recently (Lecun, Bengio, & Hinton, 2015). Deep learning is a machine learning technique that mimics the abilities of the human brain such as observation, analysis, learning, and decision making for complex problems, and can perform operations such as supervised or unsupervised features such as feature extraction, transforming and classifying using large amounts of data (Süzen & Kayaalp, 2018b). In its simplest form, deep learning is a machine learning technique created by expanding traditional neural networks.

Depending on the type of the problem and the data set, the deep learning methods to be applied also differ. Some of these are convolutional neural networks, recurrent neural networks, long-short term memory networks, restricted Boltzmann machines, deep autoencoders (Pekmezci, 2012; Süzen & Kayaalp, 2018b).

Convolutional neural networks (CNN): They are often used with image datasets. It is an algorithm that can distinguish objects by clustering the objects on the given image. Some filters are applied to recognize objects on images and detection of objects on the image is provided.

Recurrent neural network (RNN): In traditional artificial neural networks, all inputs and outputs are independent of each other. The recurrent neural network is an artificial neural network model in which the output from the previous step is given as an input to the current step. Recurrent neural networks can be used as working logic as single input-single output, single input-multiple output, multiple input-single outputs, multi-row input-multi-row output and multi-synchronous input-multi-synchronous output(Eşref, 2019; Süzen & Kayaalp, 2018a).

The long short-term memory (LSTM): It is a deep learning algorithm that reduces the error flow of recurrent neural networks. It determines how much of the input data should be forgotten, thanks to Sigmoid and forget gates(Graves & Schmidhuber, 2005).

Restricted Boltzmann machines (RBM): Limited Boltzmann Machines is a random neural network that can learn probability distribution on the input set. It consists of the first two layers (Sertkaya, 2018). The first layer is called the visible or input layer, and the second layer is called the hidden layer. Nodes in the hidden layer are where calculations are made for output.

Deep auto-encoders: It is a deep learning algorithm consisting of three layers that produces the data as output, which they receive as input. It is used to convert an N-dimensional vector into a smaller-sized vector with little loss. They are not used for the classification process. They can be used for data compression and feature extraction processes (Süzen & Kayaalp, 2018a).

Measurement of Model Success (Success Metrics)

It is required to measure whether the prepared models are more successful than or lag behind similar studies. The methods used for this are called metrics. The metrics encountered as a result of the literature research conducted within the scope of this study are mentioned in this section. Some metrics applied in machine learning or deep learning algorithms are given and explained below.

1. Classification Accuracy
2. Logarithmic Loss
3. Confusion Matrix
4. F1 Score
5. The area under Curve- ROC Curve
6. Mean Absolute Error
7. Mean Squared Error
8. Root Mean Squared Error

Classification Accuracy: It is the ratio of the number of correct estimates to the input sample. It is seen that it does not work efficiently in unbalanced data sets. It gives a percentage result. The accuracy rate is shown in Formula 1.

$$Accuracy\ Rate = \frac{Number\ of\ correct\ estimates}{Number\ of\ total\ estimates\ made} \tag{1}$$

Logarithmic Loss - Log Loss: It is an evaluation criterion that works with the method of punishing false estimates. It works well for multi-class classifiers. It determines the probability that the estimate belongs to each class. The logarithmic loss range is between [0, ¥]. Formula 2 shows a logarithmic loss.

$$Log\ Loss = \frac{-1}{N}\sum_{i=1}^{N}\sum_{j=1}^{M} yij * \log(Pij) \tag{2}$$

Confusion Matrix: Also called the confusion matrix or error matrix, this metric gives a matrix as a result to understand the full performance of the model. There are 4 important terms for this metric. These are;

True Positive: the situations we estimated as "yes" and the real output is "yes".
True Negatives: the situations we estimated as "no" and the real output is "no".
False Positives: the situations we estimated as "yes" and the real output is "no".

False Negatives: the situations we estimated as "no" and the real output is "yes". When these situations are evaluated, a matrix structure is created, as shown in Table 2.

Table 2. Confusion matrix table

		Predicted	
		Positive	**Negative**
Actual	Positive	True Positive (TP)	False Negative (FN)
	Negative	False Positive (FP)	True Negative (TN)

Success criteria such as Accuracy, Error rate, Precision, Recall, F1-Score and MCC are found by using this table (Bulut, 2016). The formula for the accuracy metric is given in Formula 3, the calculation of the error rate metric is given in Formula 4, the formula for the precision metric is given in Formula 5, the formula for the Recall metric is given in Formula 6 (Coşkun & Baykal, 2011).

$$Accuracy = \frac{TP + TN}{Total\ Sample} \tag{3}$$

$$Error\ rate = \frac{FP + FN}{Total\ Sample} \tag{4}$$

$$Precision = \frac{TP}{TP + FP} \tag{5}$$

$$Recall = \frac{TP}{TP + FN} \tag{6}$$

F1 Score: This metric, also called the F1 score, is used to measure the accuracy of a model. The F1 score range is between [0-1]. It is the harmonic average of precision, recall performance evaluation criteria. It allows evaluating both criteria together. In Formula 7, the calculation formula of the performance of the F1 score is given.

$$F_1 = \frac{2 * precision * recall}{precision + recall} \tag{7}$$

Area Under Curve-AUC and Receiver Operating Characteristic Curve-ROC: The area under the curve is used for the binary classification problem. AUC scale is used to characterize the overall performance of a classification solution.

True Positive Rate (Sensitivity) (TPR): True Positive Rate is defined as TP / (FN + TP). True Positive Rate corresponds to the ratio of positive data points that are considered positive relative to all positive data points.

False Positive Rate (Specificity) (FPR): False Positive Rate is defined as FP / (FP + TN). False Positive Rate corresponds to the rate of negative data points that are mistakenly considered positive compared to all negative data points.

By placing these two metrics on the TPR and FPR x and y axes, the calculation of the area under the line is AUC.

Mean Absolute Error -MAE: The mean absolute error is the average of the difference between the actual values and the predicted values. It measures how far the estimates are from the original output. However, they do not give us any idea about the direction of the error, i.e., whether we anticipate the data or overestimate the data. Mathematical representation of the mean absolute error is given in Formula 8.

$$\text{Mean Absolute Error} = \frac{1}{N}\sum_{j=1}^{N}\left|yj - \widehat{yj}\right| \tag{8}$$

Mean Squared Error-MSE is very similar to the Mean absolute error; the only difference is that the MSE averages the square of the difference between the original values and the predicted values. The advantage of MSE is that it is easier to calculate the gradient, whereas Mean Absolute Error requires complex linear programming tools to calculate the gradient. Because we've resolved the error, the effect of larger errors becomes more distinct and gets caught by smaller errors, so the model can now focus on larger errors. MSE formula is given in Formula 9.

$$\text{Mean Squared Error} = \frac{1}{N}\sum_{j=1}^{N}\left(yi - \widehat{yi}\right)^2 \tag{9}$$

Root Mean Squared Error (RMSE) is a quadratic scoring rule that measures the mean error size. RMSE is the most popular evaluation criterion used in regression problems. The difference between the estimate and the observed values is taken on each square and then centered on the sample. Both evaluation tools can be used together to diagnose variations of errors in a series of estimates. RMSE will always be greater than or equal to MAE. The bigger difference between them is that the variance of individual errors in the sample is larger (Aydın, 2018). RMSE formula is given in Formula 10.

$$RMSE = \sqrt{\frac{\sum_{i=1}^{n}\left(P_i - O_i\right)^2}{n}} \tag{10}$$

Except for the above - mentioned performance metrics, there are other metrics used. Gain and Lift Charts, Kolmogorov Smirnov Chart, Gini Coefficient, Concordant - Discordant Ratio are some of these.

LITERATURE REVIEW

In a study by Maggipinto, (2018) et al., a deep learning-based system was developed in the production environment to reduce costs and increase efficiency by providing production optimization. This system has been run on a real industrial dataset. This data set contains optical emission spectroscopy data. The reason for choosing deep learning algorithms in this study was that the feature extraction process for the developed model could be done by the system. Since the data set consists of visual data, CNN and FNN (Feedforward Neural Networks) were used to create this model. The success of the designed system was measured with MSE, R2 and MAE metrics. The deep Convolutional Neural Network used enables efficient use of plasma OES (Optical Emission Spectroscopy) data without requiring a precise feature extraction. Keras and TensorFlow libraries were used to develop the model.

In a study conducted by Yu, (2019) et al., a study was conducted to estimate the remaining usage times of the devices by making use of the sensor data produced by the devices used in the field of health. C-MAPSS (https://data.nasa.gov/widgets/xaut-bemq), which is an open-source data set, was used for this study. CNN, due to its feature extraction ability, and LSTM deep learning algorithms due to time series operations, were used in this study, which deals with smart manufacturing, cyber-physical systems and big data systems, which are among the concepts of Industry 4.0. The data obtained as a result of the CNN-LSTM combined model was compared with the traditional machine learning model data created from another publication. The reason why the designed system works well is that it is said that the data collected in the real environment cause different distributions between the training set and the test set due to the disturbances in the working environment. This affects generalization in traditional machine learning models. RMSE metric was used to measure the performance of this study.

The study by Yue, (2018) et al. proposes an end-to-end method based on deep learning and transfer learning for industrial fault diagnosis. Industrial data are time series with multiple features collected from different sensors, and an excellent feature presentation is crucial for accurate results. CNN-LSTM combined model combined with CNN deep learning algorithms due to feature extraction ability and LSTM due to time series operations were used. The model created on the data set with the blade icing problem in wind turbines was applied, and some good results were obtained. The model created was compared with LR (Logistic Regression), SVM (Support Vector Machine), 1, 2 and 3-layer LSTM, 1, 2 and 3-layer CNN and CNN-SVM / RNN / GRU models, and it was reported to give the best results with 0.9465 accuracy. These model evaluation criteria are accuracy, RMSE and the score calculated by the score function in the competition description that reflects the precision and recall.

In the study conducted by Pillai (2018) et al., an effort was made in Industry 4.0 factories to reduce the estimated production cost and waste of product quality. Due to the unbalanced behavior of traditional machine learning methods for quality prediction, the proposed new model has recommended a framework that combines the feature extraction capabilities of CNN and the field information features of fuzzy systems. It has been observed that there is progress compared to existing methods. The success of the system was measured with G-Mean, ROC-AUC, MCC, F1-score. In this study, adult dataset (https://archive.ics.uci.edu/ml/index.php) and semiconductor industry dataset, which are accessible to everyone, are used.

In a study conducted by Bodkhe (2019) et al., a system called *Blockchain Effective Intelligent Tourism and Accommodation Management* (BloHosT) is proposed. BloHosT is a system that comes with Industry 4.0 and proposes the collection of digital payments that are increasing day by day under one roof. Through the LSTM deep learning algorithm, experiences are obtained from the comments made by tourists about the places they have visited before. Since LSTM has many hidden layers, it has the ability to extract different features from the data. The data set has been trained with LSTM classifiers to create rating points. These experiences are produced for BloHosT. It is stated that BloHosT has a high return on investment compared to traditional methods in the tourism sector.

The study conducted by Miškuf and Zolotová (2016) focused on Industrial systems, cloud architecture and data analytics, which will be the main pillars of Industry 4.0. Using the UCI repository datasets (https://archive.ics.uci.edu/ml/index.php), H20 and Microsoft Azure (MA) service for deep learning were compared. The classification was made on the selected data set, and accuracy, sensitivity and MSE values were calculated with the help of complexity matrix. This comparison process is among some traditional methods (Neural Network, Decision Forest, Decision Jungle, Logistic Regression, SVM) using H20 deep learning and MA services. In experiments with H20 deep learning, it was observed that the accuracy was higher, and the error rate was lower. It is emphasized in the study that such cloud services will play a more prominent role in future data analytics.

In a study by Vafeiadis (2019) et al., a data analysis platform for less waste was presented from the data obtained by a waste management company through sensors. The platform, created using the Internet of Things devices, aims to provide intelligent solutions to the waste management company that will provide planning support, ease of monitoring and managing processes. A web-based tool has been developed for the platform. Deep learning algorithms are used in the price estimation of products such as paper, metal, plastic, etc., which are materials in waste management. LSTM, which is one of the deep learning algorithms, has been used since price prediction is difficult in the machine learning algorithm. The data of the related company for 2016-2018 has been a data set for the price estimation of the developed tool.

In a study by Hao (2019) et al., an artificial intelligence model called *Privacy Enhanced Federated Learning (PEFL)* was proposed to solve various industrial challenging problems in Industry 4.0. PEFL was created with a federated learning approach that allows the system to be learned in collaboration with different participants without disclosing the local data obtained. IoT devices collect and encrypt encrypted data via a cloud server. Systems such as Microsoft Azure Machine Learning, Google Cloud ML Engine have been proposed for cloud service. MNIST (http://yann.lecun.com/exdb/mnist/) data set was used to test the proposed system, and performance evaluation was done with CNN. The accuracy rate of the system, which was designed with a collision rate of 0.05 to 0.1, was shown as 95.7% and 95.5%. It is stated that the collision rate has a bad result above 0.5. It is stated that studies will be carried out with more massive data sets in future studies. It is also emphasized that it provides data privacy since it is a federated learning-based system.

In a study by Calderisi (2019) and colleagues, an approach that combines different networks and techniques has been proposed to detect defects that occur during weaving on a fabric. This study consists of 71878 images taken with high-resolution vision systems. In the study, 3 different deep learning approaches were used. First, the system was created with CNN. The second is the combination of CNN and feed-forward neural network (CNN + META). In the third, variable influence on projection (VIP) value is taken as <1. The success of the system was measured with Classification Accuracy. These approaches have been tested in Kode, mobeNet, inception, resNet and denseNet models.

Another study involving the concept of augmented reality and the Internet of Things, which are components of Industry 4.0 for smart factories, has been conducted by Subakti and Jiang (2018). Mobile augmented reality system design has been designed to quickly visualize and interact with the machines used in smart factories. Through the received images, the image classification library mobileNets and TensorFlow, industrial machine parts are marked on the screen of the mobile device. For this, a cloud

SCADA server was used. This process used CNN because of its success in mobile classification and image classification. The data set used for this study relates to a Wire Electrical Discharge Machine (WEDM) located in a mechanical workshop at National Central University.

Zabiński (2019) et al. mentioned the idea of applying prototype industrial platform computational intelligence methods for condition monitoring in Industry 4.0 production systems. As an example, CNC milling set head mechanical imbalance prediction system developed with the use of deep learning is explained. The model was built with CNN, and 99.8% success was achieved.

In the study conducted by Yan (2018) et al., an algorithm based on deep-compensating auto-coder (DDA) and regression operant have been proposed to estimate the remaining life of industrial equipment. The proposed system consists of 2 DAA architectures and linear regression analysis. The error between the system estimated and the actual value is about 20%. It has been stated that Industry 4.0 demonstrates its applicability to speed up implementation. Caffe was used for the training of the developed model. It is stated that the proposed concept and algorithm combines the typical industrial scenario and high artificial intelligence with the potential to accelerate the implementation of Industry 4.0.

A study by Duan (2019) et al. is on smart device monitoring systems, one of the benefits of Industry 4.0. In this study, it is suggested to determine the internal thermal failure of the transformer with CNN. The transformer, which monitors temperature and speed images under error conditions, was simulated using the lattice Boltzmann Method (LBM) and feature extraction was made on the images. There is insufficient data because the transformers are not damaged frequently. This problem was tried to be solved by transfer learning. Then, image segmentation was performed to extract the features of the error fields and to simplify the data volumes. CNN was used to make recurrent error localization. After the image segmentation, the accuracy was calculated using the information obtained by the sensors, the mean error localization accuracy decreased from 97.95% to 94.42%, while the data volume was reported to decrease to almost 1% of the original. It was stated that the average calculation time decreased by 8.816% and the loss value decreased by 37.68%. Architecture has been tried in different neural networks and it has been stated that the best performance value has been reached by using GoogLeNet.

In a study by Saravanan (2019) et al., an indoor interactive robot was designed using the Internet of Things, CNN and Q-learning. This robot designed with IoT features, is built on having an arm that can reach the desired target and perform the appropriate action in the indoor environment. To facilitate navigation in a space, a minimum number of grid matches (4x4) were made, and it was aimed to learn how to go to the desired object in the most appropriate way. CNN is used for object detection. The created models are sent to Raspberry Pi, which acts as fog node in industry 4.0 environments. Testing and development processes were done here. Values such as temperature, humidity read by the robot can be transmitted to Ericsson's IOT Accelerator platform and indoor plant monitoring system can be realized. The movement of the robot was carried out with Q-learning, reinforcement learning (RL) and CNN. The success of the system has been examined in three different cases; (i) 100% when only RL is used, (ii) 50% when only CNN is used, and (iii) 50% when both CNN and RL are used. Keras library was used to conduct this study.

In a study by Richter (2017) et al., it was stated that not only data were produced by industry 4.0, but also a large amount of data was started to be stored, and this data has the potential to help develop the control machines. Instead of manual control of printed circuit boards (PCBs) on all production lines, today, automatic optical inspection (AOI) is used. In this study, an architectural structure integrated with deep learning techniques is proposed to AOI manufacturers. It has been stated that surface mounted devices can be preferred due to their success in feature extraction of deep convolutional neural networks, although there are studies with traditional machine learning approaches to find defects in wipe joints. It has been stated that the data set required for the training and testing of the developed system can be obtained from the devices coming to the repair stations and the better the data received from the devices of different brands, the better the proposed architecture can give results. It was stated in the study that the model proposed in the future is to make the first version. It is stated that the proposed AOI system can use deep convolutional neural networks (CNN) to simplify the optical inspection process. It is stated that tools such as tensorflow, theano, caffee, which have become widespread in recent years, can be selected for the implementation of the system. It is stated that a huge data set is needed to realize such a system. It is stated that there is no such database for now and it is necessary to work for it. A new system has been proposed for AOI, which is included in the Industry 4.0 concept, and it has been stated that it will be developed.

A study by Tsai and Chang (2018) focused on a deep learning-based application of the coil level adjustment system used in sheet metal production/shaping. An automated encoder neural network was used to implement this system, which was recommended for smart manufacturing. The developed model was first evaluated with 200 samples, but when it was observed that it did not work well, it was increased to 500 samples. It has been increased to 700 samples for the system to reach a stable state. The system has achieved 80% success. It is stated that future studies will be to try different deep learning methods to compare similar applications and improve the current solution.

In a study by Ferrari (2019) et al., performance evaluation of full-cloud and edge-cloud architectures for the determination of anomaly based on deep learning in the concept of the Internet of objects, which is one of Industry 4.0 concepts, was performed. LSTM is used in architecture as it is suitable for the online time series anomaly approach. LSTM is a deep learning algorithm that can represent the relationship between current events and previous events and addresses time-series events. Siemens controller and Microsoft Azure platform cloud systems are also used in system architecture. It has been shown that full-cloud architecture can outperform edge-cloud architecture.

Collaborative production ecosystems provide a high amount of data collected from factories and generally related to machine data, sensor measurements and production processes. In a study by Nizamis (2018) et al., it was explained how these data are modeled and stored using ontologies, how deep learning tools are obtained and analyzed by continuous learning algorithms, and ultimately how they are returned to the semantic framework and how they increase the efficiency of a semantic mapper respectively. The purpose of using deep learning tools is to imitate prices for raw material needs and therefore provide sustainable estimates for raw materials. RNN and LSTM are used. The system has been tested with the data obtained from London metal exchange including 1400 price samples selected from 2012-2017 (https://www.quandl.com/data/LME-London-Metal-Exchange). Mean Squared Error was used to measure the performance of the system.

One of the important components in the realization of Industry 4.0 is unmanned aerial vehicles. In a study by Li (2019) et al., a system was developed that simultaneously counts and detects vehicles from images from unmanned aerial vehicles. CNN is used for this system; feature extraction layer and object count layer are created. CARPK data set (https://web.inf.ufpr.br/vri/databases/parking-lot-database/), VisDrone2018 data set (http://www.aiskyeye.com) and UAVDT were used with almost 90000 tools for training of this system. The applied methods were tested with RMS and RMSE. MatConvNet was used to implement the system.

In a study by Siddiqui (2019) et al., a new model is presented for determining evolutionary deep learning models for time series analysis. The B5M data set in the internet traffic data set (Cortez, Rio,

Rocha, & Sousa, 2012) was used for the training and testing of this model. CNN, one of the deep learning methods, was used to create the model, and keras and tensorflow were used for its realization. MSE metric was used for the evaluation of the success of the model. The proposed model has been found to achieve 99.88% success. The developed system is published with open access (https://github.com/shoaibahmed/TSViz-Core).

A system for early error detection in the steel industry has been proposed by Karagiorgou et al., (2019) et al. Produced from sensor data using deep learning techniques, this model offers some early results on modeling and predicting the complex and dynamic behavior in the manufacturing environment, one of the indispensable concepts of Industry 4.0. This system consists of a perception module, detection module and prediction module. LSTM was used to model the dynamic structure of manufacturing systems. The reason why LSTM is preferred is that it is successful in collecting the information of the sequential data and predicting the future ones. The success of the system was measured by RMSE and MAE metrics. It was stated that the results were promising. The sampling frequency was obtained from 3 sensors with 0.2 Hz (1 reading in 10 seconds). Acceleration, speed and maximum amplitude measurement were obtained by creating the information for 1 month. Creating a dataset from sensor data is also associated with the IoT concept.

In a study by Tsutsui and Matsuzawa (2019) the Virtual Metrology Model, which performs Optical emission spectroscopy (OES) data with deep learning methods, is proposed. In this study, real data were studied, and CNN was used. The proposed system (OESNet) has been compared with AlexNet, ZFNet, GoogLeNet, ResNet and SENet and it has been stated that it gives successful results. R-Squared value and sMAPE are used as comparison metrics.

In a study by Chen and Guhl (2018), object learning based on deep learning and industrial robot control design was mentioned. This robot allows the UR5 to detect, find and interact with different objects such as tools and office supplies. Using a stereo view camera, it is possible to obtain both RGB and depth data of the surrounding robots and work area. This data is fed into the Faster-RCNN Network than deep learning algorithms to realize the recognition and localization of existing objects from 50 different classes. Appropriate operations can be planned and implemented with the information obtained. The model has been successfully trained and has been able to recognize and find with 68% accuracy.

Industry 4.0 and deep learning have also been the subject of research/review/compilation studies. The structures involving the problems faced by intelligent production lines such as collecting big industrial data obtained from smart factories, modeling smart product groups based on ontology, forecasting diagnostics, and prediction methods based on the deep neural network were examined. It is stated that one of the topics suggested to deal with these problems is deep learning and deep neural network (Xu & Hua, 2017). Another research study explores the latest technology available to automate, strengthen and combine systems used in industry 4.0 concepts, such as smart cities, autonomous vehicles, and energy efficiency, in smart manufacturing and healthcare. Although these systems apply big data and deep learning areas, they focus on the areas that are missing (McKee, Clement, Almutairi, & Xu, 2018). In a study by Liu (2017) et al., it was stated that Industry 4.0, Industrial Internet, unmanned aerial vehicles, smart driving, 3D printing are only products obtained from the development and budding of the motor nervous system of the Internet. The relationship between artificial intelligence and the internet has been examined in terms of brain science. The creation of a new artificial intelligence system model with the Internet and brain science was discussed.

SOLUTIONS AND RECOMMENDATIONS

In this study, it has been investigated how much deep learning and deep learning methods are used in industry 4.0 concept. The evaluation metrics of deep learning methods, datasets and sub-concepts of Industry 4.0 are also included in the evaluation criteria. It was observed that no data was specified for the classification created in some publications.

As a result of the literature research, the existence of deep learning practices was found in the Industry 4.0 concept. Although these two concepts have entered our lives in recent years, they have become very popular. However, the findings show that it is not at the desired level. The graphic of the publication of the relevant studies by years is given in Figure 1. According to the graphic, it has been observed that new studies have been carried out within the industry 4.0 concept with deep learning methods. The fact that the first publication was made in 2016 and the increase in publications in other years shows that the research area may still be more popular in the coming years.

Figure 1. Number of publications by year in the related field

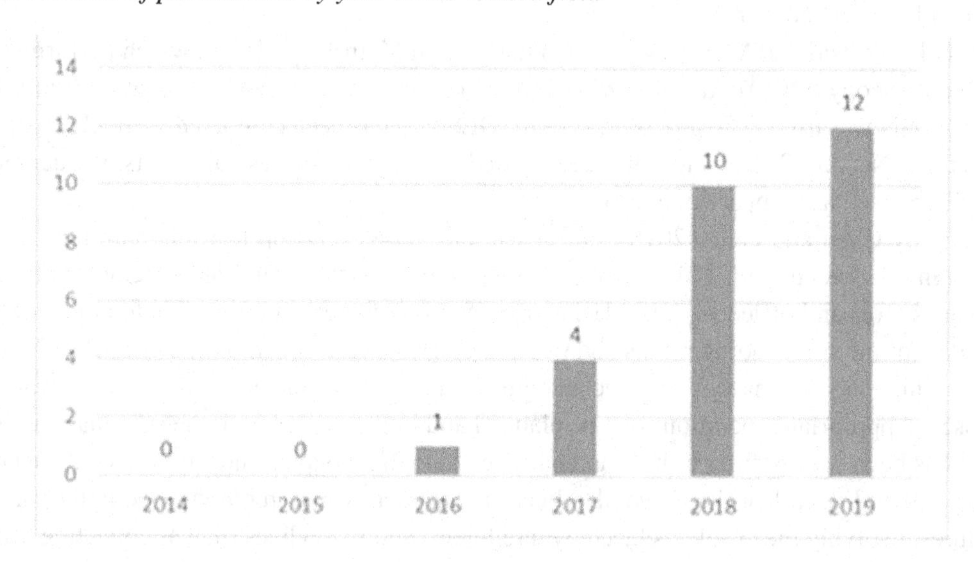

When evaluated according to the type of publication, the graphic of the publication of the relevant studies according to the publication types is given in Figure 2. It is seen that the related publications are mostly the conference publications. However, there are publications published as original articles. In 12 original articles, 2 different articles are studies that suggest a concept, not an experimental study on these fields. It gives ideas on what can be done thanks to deep learning methods in Industry 4.0. Some experimental studies are presented in Table 3 as conference papers. Some publications are presented in Table 3 as the original articles. Original articles mostly consist of industry 4.0 subfields, production, manufacturing and big data. It is observed that CNN is the most used one among the deep learning methods.

Figure 2. Publication types

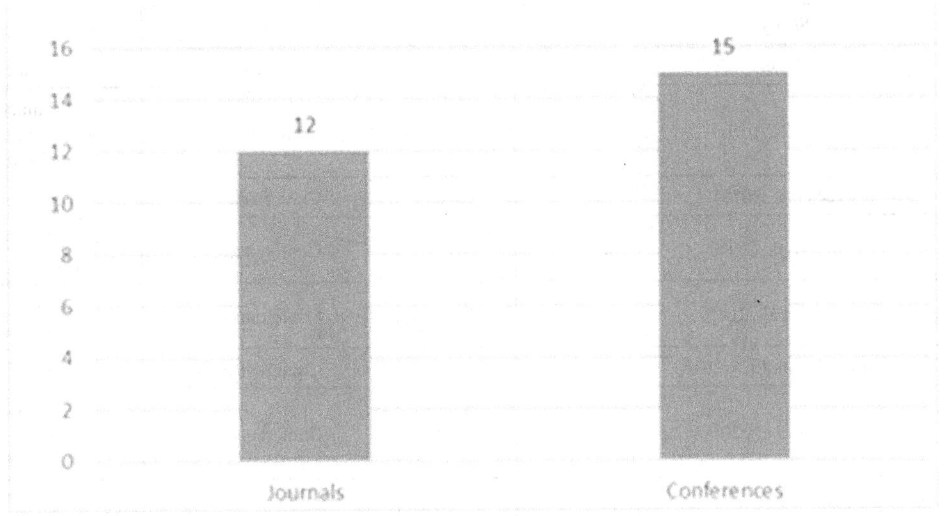

When the literature study was evaluated according to regions, it shows that there were no publications from Turkey. It has been observed that the studies related to the relevant field were conducted mostly in China.

Evaluation by Industry 4.0 areas: Industry 4.0 concept is divided into 9 main areas in total and publications are classified under these 9 main areas. These areas were determined as manufacturing, big data, internet of things (IoT), machine-machine interaction-automated tools- (MTM), cyber-physical systems, cloud systems, augmented reality, process control, security.

Distribution of publications according to Industry 4.0 fields is given in Table 3. The most studies were conducted on manufacturing, big data and the internet of things. Some publications are publications containing more than one industry 4.0 field. As Industry 4.0 requires smart factories, positive use of accumulated data and machine interaction it is natural to work in these areas. That these systems will lead to efficiency and change in many different sectors.

Table 3. Industry 4.0 field of publications

Study	Years	Publication type	Industry 4.0 field(s)
(Maggipinto, et al. 2018)	2018	Journals	Manufacturing
(Yu etal., 2019)	2019	Conferences	Manufacturing, Cyber-physical systems, Big data
(Yue et al., 2018)	2018	Conferences	Manufacturing
(Pillai et al., 2018)	2018	Conferences	Manufacturing, Big data, Process control
(Bodkhe et al., 2019)	2019	Conferences	Cloud system, Big data
(Miškuf & Zolotová, 2016)	2016	Conferences	Big data
(Vafeiadis et al., 2019)	2019	Conferences	IoT, Big data, MTM
(Hao et al., 2019)	2019	Journals	IoT, Big data, Atifial intellice
(Calderisi et al., 2019)	2019	Conferences	Manufacturing,
(Subakti & Jiang, 2018)	2019	Conferences	IoT, Big data, Augmented reality, Manufacturing
(Xu & Hua, 2017)	2017	Journals	Big data, Manufacturing, Cyber-physical systems
(Yan et al., 2018)	2018	Journals	Big data, Manufacturing, Cyber-physical systems
(Duan et al., 2019)	2019	Journals	Security
(Saravanan et al., 2019)	2019	Conferences	IoT
(Richter et al., 2017)	2017	Conferences	Manufacturing
(Tsai & Chang, 2018)	2018	Conferences	Manufacturing
(Ferrari et al., 2019)	2019	Conferences	IOT, Cloud system
(Nizamis et al., 2018)	2018	Conferences	Manufacturing
(Li et al., 2019)	2019	Journals	MTM
(McKee et al., 2018)	2018	Journals	Cyber-physical systems, IoT
(Siddiqui et al., 2019)	2019	Journals	MTM
(Karagiorgou et al., 2019)	2019	Conferences	IoT, Manufacturing
(Alenazi, Niu, Wang, & Savolainen, 2018)	2018	Conferences	Cyber-physical systems
(Tsutsui & Matsuzawa, 2019)	2019	Journals	Manufacturing
(Liu et al., 2017)	2017	Journals	IoT, Big data
(Chen & Guhl, 2018)	2018	Journals	MTM
(Zabiński et al., 2019)	2019	Journals	Manufacturing

Deep learning is basically divided into 5 main methods and subclasses are created according to the use of these methods. These methods are CNN, RNN, LSTM, Auto encoder, and Boltzmann method (LBM). CNN is the most used method among related studies. It is stated that CNN is successful in feature extracting and classifying operations on pictures. Deep learning methods used are shown in Table 4. The LSTM method is another preferred method. It is stated that LSTM is preferred due to its success in time series operations and having many hidden layers.

Table 4. Deep learning methods, tools and metrics used in the studies

Study	Years	Tool(s)	Deep learning method(s)	Metric(s)
(Maggipinto et al., 2018)	2018	Keras, Tensorflow	CNN	MSE, R2, MAE
(Yu et al., 2019)	2019	-	CNN-LSTM (combination)	RMSE
(Yue et al., 2018)	2018	-	CNN-LSTM (combination)	RMSE
(Pillai et al., 2018)	2018	-	CNN	G-Mean, ROC-AUC MCC, F1-score
(Bodkhe et al., 2019)	2019		LSTM	
(Miškuf & Zolotová, 2016)	2016	H20, Microsoft Azure	-	MSE
(Vafeiadis et al., 2019)	2019	-	LSTM	-
(Hao et al., 2019)	2019	-	CNN	-
(Calderisi et al., 2019)	2019	-	CNN	Classification Accuracy
(Subakti & Jiang, 2018)	2018	Tensorflow	CNN	
(Zabiński et al., 2019)	2019	-	CNN	-
(Yan et al., 2018)	2018	Caffe	AUTO ENCODER	
(Duan et al., 2019)	2019	GoogLeNet	-	Accuracy
(Saravanan et al., 2019)	2019	Keras	CNN	-
(Richter et al., 2017)	2017	Tensorflow, Theano, caffee (offer)	CNN	-
(Ferrari et al., 2019)	2019	Microsoft Azure platform	LSTM	-
(Nizamis et al., 2018)	2018	-	RNN, LSTM	MSE
(Li et al., 2019)	2019	MatConvNet	CNN	RMS, RMSE
(Siddiqui et al., 2019)	2019	Keras, tensorflow	-	MSE
(Karagiorgou et al., 2019)	2019	-	LSTM	RMSE, MAE
(Tsutsui & Matsuzawa, 2019)	2019	-	CNN	R2, sMAP
(Chen & Guhl, 2018)	2019	-	Faster-RCNN	-

Studies that use LSTM and CNN in combination are also seen (Yu, Wu, Zhu, & Pecht, 2019; Yue, Ping, & Lanxin, 2018). It is seen that CNN and hybrid algorithms are used on visual images and LSTM is used in the evaluation of sensor data. It has been seen that LSTM is used for estimation processes and CNN + LSTM is used together to determine the remaining lifetimes of the devices.

It was observed that open-source deep learning libraries and cloud computing methods were used in the development of deep learning applications. In some studies, data sets were created for the study but it was seen that they were not shared. In some studies, it was observed that more than one tool was used. It was observed that the most used tools are tensorflow and keras, and Microsoft Azure, GoogleNet, H2O machine learning is used for cloud computing (Table 4).

Only the data set used in 7 of the studies was specified. you can see these studies in Table 5. Some studies have worked on the data they produce, and the data set has not been shared with the reader. It can be seen that some studies can be done about this field using shared data sets.

Table 5. Databases used in the studies

Study	Years	Database(s)
(Yu et al., 2019)	2019	C-MAPSS (https://data.nasa.gov/widgets/xaut-bemq)
(Pillai et al., 2018)	2018	https://archive.ics.uci.edu/ml/index.php
(Miškuf & Zolotová, 2016)	2016	https://archive.ics.uci.edu/ml/index.php
(Hao et al., 2019)	2019	http://yann.lecun.com/exdb/mnist/
(Nizamis et al., 2018)	2018	https://www.quandl.com/data/LME-London-Metal-Exchange
(Li et al., 2019)	2019	CARPK dataset: (https://web.inf.ufpr.br/vri/databases/parking-lot-database/), VisDrone2018 dataset: (http://www.aiskyeye.com)
(Siddiqui et al., 2019)	2019	(Cortez et al., 2012)

Considering the success metrics used in the studies, it has been seen that metrics such as RMSE, MAE, sMAPE, R-Squared are used more than others.

FUTURE RESEARCH DIRECTIONS

This section introduces some useful, practical, and theoretical suggestions for potential research fields to researchers. This literature study showed that the application of deep learning methods in the field of industry 4.0 is quite a new subject. Studies on this field will remain popular for a while.

In general, it is seen that the metrics used in machine learning are used in studies in this field. But it was not encountered much, which shows that it can be used in future studies.

Datasets used in only 7 of the publications were shared. When these datasets are analyzed, it is seen that new studies can be done by using industry 4.0 data with different metrics and methods.

When the realization tools used are examined, it is seen that they are used in cloud systems together with tensorflow and keras tools. Since real-time applications are needed in Industry 4.0, it is seen that more research can be done on cloud computing methods.

Most of the studies conducted a comparison between classical machine learning methods and deep learning methods.

It was seen that much work has been done on the remaining lifetime. It has been seen that this kind of work can be done for many devices.

In some studies, it is thought that data sets cannot be shared due to data privacy. Therefore, studies can be conducted on federated learning approaches for being able to share outputs of models without sharing data on Industry 4.0 concepts.

CONCLUSION

This section reviews the articles using deep learning approaches and reviews the Industry 4.0 and Industry 4.0 field in a comprehensive way. These studies have grouped objectives such as industry 4.0 field, deep learning methods, metrics used, databases. The literature study conducted also guided other studies in this area.

It has been observed that the development of decision-making systems required in smart production and manufacturing environments, the communication of the devices, and the calculation of the remaining lifetimes of the devices are facilitated through deep learning methods.

ACKNOWLEDGMENT

This research received no specific grant from any funding agency in the public, commercial, or not-for-profit sectors.

REFERENCES

Aktan, E. (2018). Büyük Veri: Uygulama Alanları [Big Data: Application Areas]. *Analitiği ve Güvenlik Boyutu.*, *1*(1), 1–22.

Akyol, S., & Uçar, A. (2019). Rp-Lidar ve Mobil Robot Kullanılarak Eş Zamanlı Konum Belirleme ve Haritalama [Simultaneous Positioning and Mapping Using Rp-Lidar and Mobile Robot]. *Fırat Üniversitesi Mühendislik Bilimleri Dergisi*, *31*(1), 137–143.

Alenazi, M., Niu, N., Wang, W., & Savolainen, J. (2018). Using obstacle analysis to support SysML-based model testing for cyber physical systems. *Proceedings - 2018 8th International Model-Driven Requirements Engineering Workshop, MoDRE 2018*, (pp. 46–55). 10.1109/MoDRE.2018.00012

Aydın, C. (2018). Makine Öğrenmesi Algoritmaları Kullanılarak İtfaiye İstasyonu İhtiyacının Sınıflandırılması [Classification of Fire Station Needs Using Machine Learning Algorithms]. *European Journal of Science and Technology*, (14), 169–175. doi:10.31590/ejosat.458613

Bodkhe, U., Bhattacharya, P., Tanwar, S., Tyagi, S., Kumar, N., & Obaidat, M. S. (2019). BloHosT: Blockchain Enabled Smart Tourism and Hospitality Management. *2019 International Conference on Computer, Information and Telecommunication Systems (CITS)*. IEEE. 10.1109/CITS.2019.8862001

Bulut, F. (2016). Sınıflandırıcı Topluluklarının Dengesiz Veri Kümeleri Üzerindeki Performans Analizleri [Performance Analyzes of Classifier Communities on Unbalanced Datasets]. *Bilişim Teknolojileri Dergisi, 9*(2), 153–159. doi:10.17671/btd.81137

Calderisi, M., Galatolo, G., Ceppa, I., Motta, T., & Vergentini, F. (2019). Improve image classification tasks using simple convolutional architectures with processed metadata injection. *Proceedings - IEEE 2nd International Conference on Artificial Intelligence and Knowledge Engineering, AIKE 2019*, (pp. 223–230). IEEE. 10.1109/AIKE.2019.00046

Chen, X., & Guhl, J. (2018). Industrial robot control with object recognition based on deep learning. *Procedia CIRP, 76*, 149–154. doi:10.1016/j.procir.2018.01.021

Cortez, P., Rio, M., Rocha, M., & Sousa, P. (2012). Multi-scale Internet traffic forecasting using neural networks and time series methods. *Expert Systems: International Journal of Knowledge Engineering and Neural Networks, 29*(2), 143–155.

Coşkun, C., & Baykal, A. (2011). *Veri Madenciliğinde Sınıflandırma Algoritmalarının Bir Örnek Üzerinde Karşılaştırılması. Akademik Bilişim 2011 [Comparison of Classification Algorithms in Data Mining on a Sample. Academic Informatics 2011]*. Akademik Bilişim.

Doğan, A., Sönmez, B., & Cankül, D. (2020). Yiyecek-İçecek İşletmelerinde İnovasyon Ve Artırılmış Gerçeklik Uygulamaları [Innovation And Augmented Reality Applications in Food And Beverage Companıes]. *Journal of Business Research - Turk, 10*(3), 576–591. doi:10.20491/isarder.2018.488

Duan, J., He, Y., Du, B., Ghandour, R. M. R., Wu, W., & Zhang, H. (2019). Intelligent Localization of Transformer Internal Degradations Combining Deep Convolutional Neural Networks and Image Segmentation. *IEEE Access: Practical Innovations, Open Solutions, 7*, 62705–62720. doi:10.1109/ACCESS.2019.2916461

Eşref, Y. (2019). *Türkçe Dizi Etiketleme İçin Sinir Ağ Modelleri [Neural Network Models for Sequence Labeling in Turkish]*. Hacettep Üniveristesi.

Ferrari, P., Rinaldi, S., Sisinni, E., Colombo, F., Ghelfi, F., Maffei, D., & Malara, M. (2019). Performance evaluation of full-cloud and edge-cloud architectures for Industrial IoT anomaly detection based on deep learning. *IEEE International Workshop on Metrology for Industry 4.0 and IoT, MetroInd 4.0 and IoT 2019 - Proceedings*, (pp. 420–425). 10.1109/METROI4.2019.8792860

Graves, A., & Schmidhuber, J. (2005). Framewise Phoneme Classification with Bidirectional LSTM and other Neural Network Architectures. *Neural Networks, 18*(5–6), 602–610. doi:10.1016/j.neunet.2005.06.042 PMID:16112549

Hao, M., Li, H., Luo, X., Xu, G., Yang, H., & Liu, S. (2019). Efficient and Privacy-enhanced Federated Learning for Industrial Artificial Intelligence. *IEEE Transactions on Industrial Informatics, 3203*, 1–11. doi:10.1109/TII.2019.2945367

Joda, T., Gallucci, G. O., Wismeijer, D., & Zitzmann, N. U. (2019). Augmented and virtual reality in dental medicine: A systematic review. *Computers in Biology and Medicine, 108*, 93–100. doi:10.1016/j.compbiomed.2019.03.012 PMID:31003184

Karagiorgou, S., Vafeiadis, G., Ntalaperas, D., Lykousas, N., Vergeti, D., & Alexandrou, D. (2019). Unveiling Trends and Predictions in Digital Factories. *15th Annual International Conference on Distributed Computing in Sensor Systems, DCOSS 2019*, (pp. 326–332). IEEE. 10.1109/DCOSS.2019.00073

Lecun, Y., Bengio, Y., & Hinton, G. (2015). Deep learning. *Nature*, *521*(7553), 436–444. doi:10.1038/nature14539 PMID:26017442

Li, W., Li, H., Member, S., Wu, Q., Chen, X., & Ngan, K. N. (2019). *Simultaneously Detecting and Counting Dense Vehicles From Drone Images.*, *66*(12), 9651–9662.

Liu, F., Shi, Y., & Li, P. (2017). Analysis of the Relation between Artificial Intelligence and the Internet from the Perspective of Brain Science. *Procedia Computer Science*, *122*, 377–383. doi:10.1016/j.procs.2017.11.383

Maggipinto, M., Terzi, M., Masiero, C., Beghi, A., & Susto, G. A. (2018). A Computer Vision-Inspired Deep Learning Architecture for Virtual Metrology Modeling with 2-Dimensional Data. *IEEE Transactions on Semiconductor Manufacturing*, *31*(3), 376–384. doi:10.1109/TSM.2018.2849206

Mahdavinejad, M. S., Rezvan, M., Barekatain, M., Adibi, P., Barnaghi, P., & Sheth, A. P. (2018, August 1). Machine learning for internet of things data analysis: A survey. *Digital Communications and Networks*, *4*(3), 161–175. doi:10.1016/j.dcan.2017.10.002

McKee, D. W., Clement, S. J., Almutairi, J., & Xu, J. (2018). Survey of advances and challenges in intelligent autonomy for distributed cyber-physical systems. *CAAI Transactions on Intelligence Technology*, *3*(2), 75–82. doi:10.1049/trit.2018.0010

Miškuf, M., & Zolotová, I. (2016). Comparison between multi-class classifiers and deep learning with focus on industry 4.0. *2016 Cybernetics and Informatics, K and I 2016 - Proceedings of the 28th International Conference*. IEEE. 10.1109/CYBERI.2016.7438633

Nizamis, A., Vergori, P., Ioannidis, D., & Tzovaras, D. (2018). Semantic Framework and Deep Learning Toolkit Collaboration for the Enhancement of the Decision Making in Agent-Based Marketplaces. *Proceedings - 2018 5th International Conference on Mathematics and Computers in Sciences and Industry, MCSI 2018*, (pp. 135–140). IEEE. 10.1109/MCSI.2018.00039

Pekmezci, M. (2012). *Kısıtlanmış Boltzman Makinesi ile Zaman Serilerinin Tahmini, Y [Estimation of Time Series with Restricted Boltzman Machine, Y]*. Lisans Tezi, Bilgisayar Mühendisliği Anabilim Dallı. Maltepe Üniversitesi.

Pillai, S., Punnoose, N. J., Vadakkepat, P., Loh, A. P., & Lee, K. J. (2018). An Ensemble of fuzzy Class-Biased Networks for Product Quality Estimation. *23rd IEEE International Conference on Emerging Technologies and Factory Automation, ETFA IEEE International Conference on Emerging Technologies and Factory Automation, ETFA, 2018-Septe*, (pp. 615–622). IEEE. 10.1109/ETFA.2018.8502492

Richter, J., Streitferdt, D., & Rozova, E. (2017). On the Development of Intelligent Optical Inspections. *7th Annual Computing and Communication Workshop and Conference, CCWC 2017*. IEEE. 10.1109/CCWC.2017.7868455

Saravanan, M., Satheesh Kumar, P., & Sharma, A. (2019). IoT enabled indoor autonomous mobile robot using CNN and Q-learning. *2019 IEEE International Conference on Industry 4.0, Artificial Intelligence, and Communications Technology, IAICT 2019*, (pp. 7–13). IEEE. 10.1109/ICIAICT.2019.8784847

Sertkaya, M. E. (2018). *Derin Öğrenme Tekniklerinin Biyomedikal İmgeler Üzerine Uygulamaları [Applications of Deep Learning Techniques on Biomedical Images]*. Fırat Üniversitesi.

Siddiqui, S. A., Mercier, D., Munir, M., Dengel, A., & Ahmed, S. (2019). TSViz: Demystification of Deep Learning Models for Time-Series Analysis. *IEEE Access: Practical Innovations, Open Solutions*, *7*, 67027–67040. doi:10.1109/ACCESS.2019.2912823

Subakti, H., & Jiang, J. R. (2018). Indoor Augmented Reality Using Deep Learning for Industry 4.0 Smart Factories. *42nd IEEE International Conference on Computer Software & Application*, *2*, (pp. 63–68). IEEE. 10.1109/COMPSAC.2018.10204

Süzen, A. A., & Kayaalp, K. (2018a). Derin Öğrenme ve Türkiye'deki Uygulamaları [Deep Learning and its Applications in Turkey] (1st ed.). ResearchGate. https://www.researchgate.net/publication/327666072_Derin_Ogrenme_ve_Turkiye'deki_Uygulamalari

Süzen, A. A., & Kayaalp, K. (2018b). Derin Öğrenme Yöntemleri İle Sıcaklık Tahmini: Isparta İli Örneği [Temperature Estimation with Deep Learning Methods: Example of Isparta Province]. *International Academic Research Congress INES 2018*, (December).

Süzen, A. A., & Kayaalp, K. (2019). Endüstri 4.0 ve Adli Bilişim [Industry 4.0 and Forensic Informatics]. *Bilişim ve Teknoloji Araştırmaları*, *2019*, 23–31.

Tekin, Z., & Karakuş, K. (2018). Gelenekselden Ak ı ll ı Üretime Spor Endüstrisi 4 . 0-. *İnsan Ve Toplum Bilimleri Araştırmaları Dergisi,* [Sports Industry 4.0 from Conventional to Smart Production] *7*(3), 2103–2117.

Tsai, S. Y., & Chang, J. Y. J. (2018). Parametric Study and Design of Deep Learning on Leveling System for Smart Manufacturing. *2018 IEEE International Conference on Smart Manufacturing, Industrial and Logistics Engineering, SMILE 2018, 2018-Janua*, (pp. 48–52). IEEE. 10.1109/SMILE.2018.8353980

Tsutsui, T., & Matsuzawa, T. (2019). Virtual metrology model robustness against chamber condition variation using deep learning. *IEEE Transactions on Semiconductor Manufacturing*, *32*(4), 428–433. doi:10.1109/TSM.2019.2931328

Vafeiadis, T., Nizamis, A., Pavlopoulos, V., Giugliano, L., Rousopoulou, V., Ioannidis, D., & Tzovaras, D. (2019). Data Analytics Platform for the Optimization of Waste Management Procedures. *15th Annual International Conference on Distributed Computing in Sensor Systems, DCOSS 2019*, (pp. 333–338). IEEE. 10.1109/DCOSS.2019.00074

Xu, X., & Hua, Q. (2017). Industrial Big Data Analysis in Smart Factory: Current Status and Research Strategies. *IEEE Access: Practical Innovations, Open Solutions*, *5*, 17543–17551. doi:10.1109/ACCESS.2017.2741105

Yan, H., Wan, J., Zhang, C., Tang, S., Hua, Q., & Wang, Z. (2018). Industrial Big Data Analytics for Prediction of Remaining Useful Life Based on Deep Learning. *IEEE Access: Practical Innovations, Open Solutions*, *6*, 17190–17197. doi:10.1109/ACCESS.2018.2809681

Yu, S., Wu, Z., Zhu, X., & Pecht, M. (2019). A Domain Adaptive Convolutional LSTM Model for Prognostic Remaining Useful Life Estimation under Variant Conditions. *Proceedings - 2019 Prognostics and System Health Management Conference, PHM-Paris 2019*, (pp. 130–137). 10.1109/PHM-Paris.2019.00030

Yue, G., Ping, G., & Lanxin, L. (2018). An End-to-End model base on CNN-LSTM for Industrial Fault Diagnosis and Prognosis. [IEEE.]. *Proceedings of IC-NIDC, 2018*, 274–278.

Zabiński, T., Maoczka, T., Kluska, J., Madera, M., & Sęp, J. (2019). Condition Monitoring in Industry 4.0 Production Systems - The Idea of Computational Intelligence Methods Application. *Procedia CIRP*, *79*, 63–67. doi:10.1016/j.procir.2019.02.012

KEY TERMS AND DEFINITIONS

CNN: Convolutional Neural Network is a sub-branch of deep learning and is often used to analyze visual information.

Deep Learning: Deep Learning is an artificial neural network technique with hidden layers.

Industry 4.0: Industry 4.0 is the subset of the fourth industry-related industrial revolution. It covers all smart systems.

IoT: The Internet of Things is to equip devices with the ability to transfer data to each other over a network.

LSTM: The Long Short-Term Memory is an artificial repetitive neural network architecture used in deep learning.

Metrics: Metrics are the scales used to evaluate and compare of a performance system.

RNN: Recurrent neural network is a class of artificial neural networks in which the connections between the nodes form a guided graphic across a temporary array.

SMS: Systematic Mapping Studies is providing an overview of a research area, and identify the quantity and type of research and results available within it.

SVM: Support Vector Machine can be defined as a vector space-based machine learning method that finds a decision boundary between two classes that are farthest from any point in the training data.

Chapter 8
Smart Manufacturing:
Post-Pandemic and Future Trends

Khalid H. Tantawi
https://orcid.org/0000-0002-2433-6815
University of Tennessee at Chattanooga, USA

Ismail Fidan
Tennessee Tech University, USA

Yasmin Musa
Motlow State Community College, USA

Anwar Tantawy
Smart Response Technologies, Canada

ABSTRACT

In this chapter, the current state and future trends in smart manufacturing (SM) and its technologies are presented with the perspective of economic growth and evolution of policies and strategies that steer its growth. The long-term effect of the COVID-19 pandemic on manufacturing is investigated. As a result of the COVID-19 pandemic, a long-lasting effect on manufacturing is foreseen, particularly in the supply chain dependency. To overcome future supply chain disruptions, attention is expected to shift towards incorporating industrial and service robotics, additive manufacturing, and augmented and virtual reality. Additive manufacturing will continue to play an increased role in customized product manufacturing. More demand is expected in the long term of additive manufacturing to counter future supply chain interruption.

DOI: 10.4018/978-1-7998-7852-0.ch008

Copyright © 2023, IGI Global. Copying or distributing in print or electronic forms without written permission of IGI Global is prohibited.

INTRODUCTION

In 2011 the term "Industry 4.0" (Rüßmann, et al., 2015) (Erol, Schumacher, & Sihn, 2016) Came to define the 4th Generation of industry (Digitale Wirtschaft und Gesellschaft, n.d.) after the Hannover fair in Germany. In the years that follow, Germany and the European Union adopted policies and plans that rely on Smart Manufacturing (SM) technologies to increase their national manufacturing production. *The Wall Street Journal* calls it the "New Industrial Revolution", and the Huffington Post called it a "bullet train" that "propels the manufacturers that climb on board" (Kennell, 2015).

However, before the adoption of policies and plans for "Industry 4.0" in Europe, in the year 2010, China surpassed the United States, for the first time in modern history, as the world's largest manufacturer and it continues to widen its lead (see *Table 1*) (China Solidifies Its Position as the World's Largest Manufacturer, 2015). For the past four decades, China adopted strict and fruitful industrial policies, that resulted in the successful transformation of China's economy to a leading industrialized one.

The term "Intelligentization" was first introduced by the Chinese State Council in its National Artificial Intelligence Strategy (Webster, Creemers, Triolo, & Kania, 2017) translated as "Next Generation Artificial Intelligence Development Plan" to refer to the next generation of intelligent systems, in which decision making is a key capability of the AI machines.

Table 1. Top 10 countries that provide value-added manufacturing ranked by the percentage of the world's manufacturing services provided, 2015 data

Rank	Country	Percent		Rank	Country	Percent
1	China	23.2%		6	Italy	3.1%
2	United States	17.2%		7	France	2.4%
3	Japan	7.8%		8	Russia	2.4%
4	Germany	7.3%		9	Brazil	2.3%
5	Korea	6.3%		10	United Kingdom	2.1%

The spread of the Corona Virus Disease of 2019 (CoVid-19) caused a turning point in the manufacturing industry and intensified the need for "intelligentization" and "networkization" such as the use of augmented reality for remote troubleshooting and collaborative robots in both service and industrial applications. As of September 2021 the confirmed cases of the Corona Virus Disease of 2019 (CoVid-19) reported by the World Health Organization exceeded 220 million, including more than 4.5 million deaths (Coronavirus disease (COVID-19) pandemic, 2021).

The spread of the Coronavirus pandemic affected almost every aspect of the human civilization. By the end of April 2020, the pandemic resulted in a 52% decline in the automotive sales in the U.S. (Collie, et al., 2020) and a paralyzed aviation industry that shrunk at an unprecedented level of 94% (IATA, 2020) (Tantawi, Literature Review: Rethinking BioMEMS in the Aftermath of CoViD-19, 2020). Almost every manufacturing industry faced supply chain disruptions, however, on the long term, the automotive industry and other industries are expected to experience an increase in demand, with more individuals indicating that they will likely switch mode of transportation from public transportation to walking or private scooter or car (Bert, et al., 2020).

The pandemic also exposed the need for service robots in the medical field, such as in performing surgical operations (Zemmar, Lozano, & Nelson, 2020).

Since they were discovered in 1965, Corona viruses were known to cause approximately 5-15% of upper respiratory tract illnesses (Falsey, et al., 1997), such as the 229E virus (Kahn & McIntosh, 2005), the B814 virus, the NL63, OC43, and HKU1 viruses. Other Coronaviruses cause acute respiratory illnesses such as the Middle East Respiratory Syndrome (MERS-CoV), the Severe Acute Respiratory Syndrome (SARS-CoV), and the SARS-CoV-2 which causes CoViD-19.

However, Zoonotic spillover of viruses (the process by which a virus jumps from wildlife to humans) (Olival, et al., 2017) is not new, some of the zoonotic viruses have been coexisting with humanity for centuries, some are very lethal such as Rabies (originated in bats), and some are uncurable such as HIV (originated in primates) and some are very contagious and lethal such as smallpox (from rodents). Biologists estimate that there are as many as 840,000 unknown virus species in the wildlife that can infect humans (Lu M. C., 2020). With the logarithmic increase in human population and the reduced wildlife natural habitat, chances of contact between humans and these viruses will be inevitable, and therefore, having another pandemic is just a matter of time. Therefore, more preparedness is needed for future pandemics on all levels, such as developing self-sufficiency in the supply chain (Ayati, Saiyarsarai, & Nikfar, 2020).

In the rest of this chapter, the current trends in the major technologies that constitute Smart Manufacturing and the risks associated with SM will be discussed.

BACKGROUND

In 2015, Smart Manufacturing was defined in the United States in Congressional Bill S.1054, and since then, three definitions are widely adopted of Smart Manufacturing as follows (Terry, et al., 2020) (Tantawi, Fidan, Chitiyo, & Cossette, 2021):

- U.S Congress Definition of Smart Manufacturing: The Congressional Bill S.1054 define Smart Manufacturing as "a set of advanced sensing, instrumentation, monitoring, controls, and process optimization technologies and practices that merge information and communication technologies with the manufacturing environment for the real-time management of energy, productivity, and costs across factories and companies" (Malik & A, 2016). This definition is characterized by an easy-to-understand language, and the utilization of non-technical words. The keywords in this definition are "technologies and practices", and therefore it is essential to include both the technologies and the policies and practices that are in place to adopt and implement the technologies.
- The National Institute of Standards and Technology (NIST) definition: Here, Smart Manufacturing systems are defined as those that are "fully-integrated, collaborative manufacturing systems that respond in real time to meet changing demands and conditions in the factory, in the supply network, and in customer needs" (Smart Manufacturing Operations Planning and Control Program, n.d.). The keyword in this technical definition is "collaborative". The word "collaborative" has recently come to define one of the major applications of the era of Artificial Intelligence, in which AI is used for human-machine collaboration such as in the use of collaborative robots.
- The Smart Manufacturing Leadership Coalition (SMLC) defines Smart Manufacturing as: "the ability to solve existing and future problems via an open infrastructure that allows solutions to be

> implemented at the speed of business while creating value-added results" (Davis & Swink, Smart Manufacturing as a Real-Time Networked Enterprise and a Market-Driven Innovation Platform) (Davis, et al., 2015). In this definition, from the SMLC perspective, the open infrastructure is the main feature of Smart Manufacturing.

From the definitions above, one common characteristic of Smart Manufacturing systems is their ability to change in order to answer demands with maximum efficiency (Jung, Morris, Lyons, Leong, & Cho, 2015).

The U.S. Senator Jeanne Shaheen states that consumption of electrical energy can be reduced by as much as $25 billion annually (Shaheen, 2015) by providing Smart Manufacturing technologies to small- and medium-sized manufacturers. In addition to that, Shaheen states that global gross domestic product can be increased by $15 trillion over the next two decades if Smart Manufacturing is employed (Shaheen, 2015). In response to this need, the United States Department of Energy (DOE) created Industrial Assessment Centers (IAC) in order to provide medium-sized and small-sized manufacturers with free assessment of energy and waste management (Energy, 2016) (Terry, Fidan, Zhang, & Tantawi, 2019). These centers only targets companies that have less than 500 employees and annual gross sales of less than $100 million (Malik & A, 2016). However, the IACs do not provide workforce training to utilize and perform SM technologies and techniques.

On the other hand, the SMLC concentrates its efforts on creating a platform for Smart Manufacturing applications and creating educational and training resources for current manufacturing employees (Davis, et al., 2015).

Although, the SMLC efforts demarcate the skill set that is needed for the SM age, and the much-needed application platform. Neither the SMLC nor the IAC provide the needed and trained workforce pool. There still exists a deep need on the national level for a program and/or a strategy that provides a workforce pool trained to handle SM technologies. Furthermore, current data suggest that this skill gap is widening significantly more than what was previously expected.

In a study that was conducted by the Korn Ferry Institute (Michael Franzino, 2020) found that the expected global shortage will reach as many as 85 million skilled personnel by 2030 and that this shortage of human talent will be the biggest challenge by that year.

The Korn Ferry study reports much larger numbers than previous research studies. In 2015 Deloitte and the Manufacturing Institute expected that over the next decade about 2.7 million jobs in advanced manufacturing will not be filled due to retiring workforce (The skills gap in U.S. manufacturing 2015 and beyond, 2015), and that natural business growth will add another 700 thousand jobs (Arnold, Fidan, & Tantawi, 2018). However, only 1.4 million jobs will be filled, and consequently resulting in a shortage of 2 million jobs in advanced manufacturing due to the skill gap (The skills gap in U.S. manufacturing 2015 and beyond, 2015). Furthermore, Accenture reports that in a survey research, 84% indicated that workforce is unprepared to adopt technologies such as IoT (Leathers, 2016) (Growing the Impact Economy in Greater Philadelphia, 2016) (Giffi, et al., 2015) (Desai, 2014).

All these indicators show that strategies on the national levels are needed to create a skilled workforce that is trained to handle the upcoming changes in manufacturing and cover the expanding skill gap.

MAIN FOCUS OF THE CHAPTER

Issues and Problems

In this chapter both aspects of Smart Manufacturing technologies and practices are introduced. The chapter starts with an overview of the major industrial policies in the world that govern the practices in Smart Manufacturing and direct research fundings. After that, the main technologies are presented, with an emphasis on the future trends in these technologies taking into account the challenges and opportunities that resulted from the CoVid-19 pandemic.

Current data show that intelligent industrial robotics market is experiencing a dramatic increase in East Asia, that a "robotics revolution" is already taking place in that market, but lack of safety standards is the main challenge that faces the next-generation of industrial robotics.

The main challenge that manufacturers faced during the CoVid-19 pandemic was the supply chain disruptions that took place; a challenge that will result in a dramatic change in the way it is looked at some technologies, such as robotics and additive manufacturing, and virtual and augmented reality. For example, the continuous travel restrictions have caused manufacturers to speed up incorporating augmented reality in their plants to facilitate remote maintenance of equipment, such that the away personnel can guide the on-ground personnel on troubleshooting equipment using the augmented reality headsets.

MAJOR INDUSTRIAL POLICIES IN THE WORLD

In this section, some of the major industrial strategies and policies that are related Smart Manufacturing in the world will be investigated:

The German "Industrie 4.0" Policy

The term Industry 4.0 (spelled Industrie 4.0 in German, also known as I40) came to existence in Hannover, Germany in 2011. It came to identify the technologies and practices of the fourth generation of industry. Figure 1 summarizes the four generations of industry.

The term Industry 4.0 became widely used due to its unique definition that summarizes the evolution of industry. In the first stage of industry (Industry 1.0) which refers to the period from the beginning of the industrial revolution in the 1700's to the end of the 19th century, the major character of industry in that period is the use of advanced mechanical systems, such as built drives and gear drives run by steam engines. This is referred to as "Mechanization".

The second stage of industry is that which took place after the adoption of electric machines in industry the late 1800's until the beginning of the computer age in the 1960's. This period is referred to as the "Automation" period. Figure 1 shows the first automated assembly line in 1913.

The computer age which extends from the 1960's until the 2010's is best known for the "Computer Integrated Manufacturing" paradigm, and thus is referred to as the "Computerization" stage.

Lastly, the current stage of industry is characterized by the use of Artificial Intelligence.

Figure 1. The four stages of industry

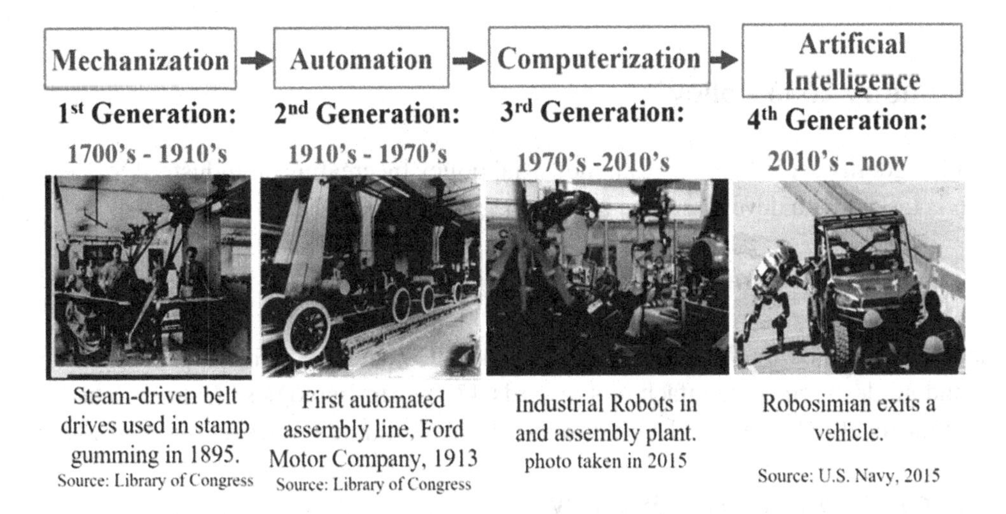

The "Industrie 4.0" policy initiative targets both manufacturers and politicians, the European Commission lists in its Digital Transformation Monitor publication of 2017 that some of the achievements of the policy are the reduction of industry segregation and the development of a reference architecture.

In Germany, the "Industrie 4.0" policy is governed through two ministries:

i. the Ministry of Education and Research
ii. the Ministry of Economic Affairs and Energy

The Chinese "Made in China 2025" Industrial Policy

In January 2018, China started a national campaign for its industrial policy "Made in China 2025", which was launched in 2015. The policy promotes intelligent robotics and a robot revolution in China's manufacturing plants to "turbocharge" its economy (Reuters, 2018) (Bateman, 2018).

China aggressively takes the global lead in its push towards "intelligent robotics", and places it as a priority for its growth by 2030 as stated by China's president Xi Jinping (Faggella, 2018). The Intelligent Industrial Robotics market is expected to grow to \$14.29 billion in 2023 from \$4.94 billion in 2018 (Demaitre, 2018).

The Korean "Manufacturing Innovation 3.0" Policy

This policy was launched in 2014 by the then- Korean president with the three main objectives: enhancing the core industries of Korea, introducing innovation to manufacturing, such as the development of key technologies for Internet of Things and additive manufacturing; and the third objectives of the policy is the establishment of 30,000 smart factory by the end of 2025.

The name 3.0 indicates the method employed in the advancement of the manufacturing industry specific to Korea, with "Manufacturing Innovation 1.0" focusing on the replacement of imported light industries, and "Manufacturing Innovation 2.0" primarily focusing on assembly machines (Liebhart & Hohmann, 2016).

The UK Industry 2050 Policy

The United Kingdom decided to pursue an industrial policy by creating an Industrial Strategy Commission, which is tasked with developing the new strategy (Laing, 2017).

The New Industrial France Initiative

This policy was launched by the then-French president Hollande in 2013. It presents nine industrial solutions, and guides French manufacturers towards 47 "key technologies" (France, The New Face of Industry in France, 2013) (France, New Industrial France Building France's Industrial Future, 2016).

The Japanese "Society 5.0" Strategy

A vision initiative in Japan of the fifth stage of society, that was launched in 2016, this strategy aims at a comprehensive transformation, that leads to the ultimate objective of digital transforming of the society in all levels. The strategy name comes to indicate the fifth stage of society with the five stages of society being: Hunting society, Agrarian society, Industrial society, Information society, and in the fifth stage the "Super Smart" society, and hence, 5.0 (From Industry 4.0 to Society 5.0: the big societal transformation plan of Japan) (Japan).

Figure 2. The Japanese initiative of 2016 defines five stages of society

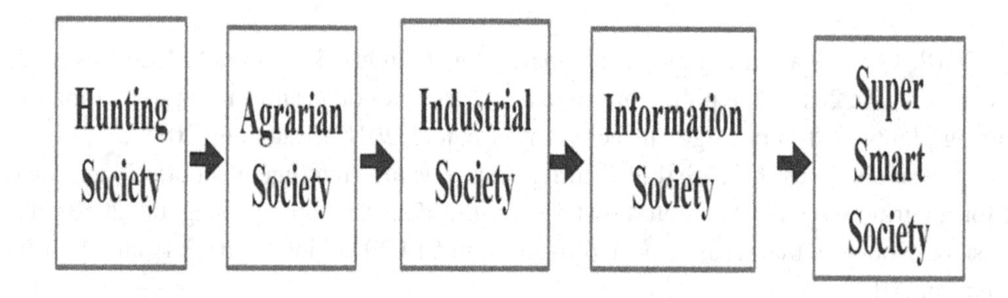

EVOLUTION OF SMART MANUFACTURING

The historical evolution of the different manufacturing paradigms is illustrated in Table 2. In their work, Lu et. al. (Lu, Morris, & Frechette, 2016), distinguish between eight paradigms of manufacturing.

Despite that Computer Integrated Manufacturing (CIM) solved many issues faced by industries in the 1980's era, it was soon realized that CIM models were significantly centralized and as a result would be an obstacle to achieve flexibility, the purpose for which they existed in the first place (Babiceanu & Chen, 2006). To overcome that, Holonic Manufacturing models were introduced.

The Cloud Manufacturing paradigm was introduced by Hu et. al. in China in the late 2000's (Bo-hu, et al., 2010). The main feature of this paradigm was that it was the first to deploy the Internet of Things (IoT) technology into a manufacturing environment.

Table 2. Top 10 Historical Evolution of the Manufacturing paradigms and models

Manufacturing Model / Paradigm	Development Date	Main Features
Lean Manufacturing	1940's (Epply & Nagengast)	Waste management
Flexible Manufacturing Systems	1970's (A History Of Flexible Manufacturing Systems, 2017)	Flexible adjustments in processes
Computer Integrated Manufacturing	1970's (Harrington, 1979)	Computer-controlled processes
Sustainable Manufacturing	1990's (Alting & Jogensen, 1993)	Environment friendly
Holonic Manufacturing	1990's (Giret, 2008)	Autonomy
Agile Manufacturing	1990's (Thilak, Devadasan, & Sivaram, 2015)	Customer satisfaction with reduced cost
Cloud Manufacturing	2000's (Bo-hu, et al., 2010)	Deploys IoT technology
Smart Manufacturing	2010's (Tantawi, Fidan, & Tantawy, Status of Smart Manufacturing in the United States, 2019)	Artificial Intelligence

PROJECTED GROWTHS IN MAIN TECHNOLOGIES OF SMART MANUFACTURING

In this section, the current state and projected trend in the main technologies that constitute SM are investigated (Kang, et al., 2016):

1. **Internet of Things (IoT) and Cyber-Physical Systems (CPS):** the interconnection of devices, machines, and equipment over a communication network (Internet of Things Global Standards Initiative, n.d.). As a result, IoT allows the machines, equipment, products and devices to be monitored and even controlled remotely over the communication network. Products can be improved and large costs can be avoided. **CPS** systems are developed to achieve mass production with a high flexibility for customization in real-time (Kang, et al., 2016). Internet of Things technology is currently the most prominent in the smart manufacturing era with the annual spending on IoT expected to climax at $450 billion in 2023 (Business Insider Intelligence Estimates 2018, 2017, 2018).

The well-known robotics manufacturer FANUC launched in 2016 the FANUC Intelligent Edge Link and Drive (FIELD) industrial internet of things platform. ABB Robotics followed in 2017 with the launching of its "Ability" platform. These systems are developed to achieve mass production with a high flexibility for customization in real-time (Kang, et al., 2016) (Internet of Things Global Standards Initiative, n.d.). **Table 3** below shows some IIoT platforms and their sizes as of 2018 (Francis, 2018) (Webster S. A., 2016) (Terry, et al., 2020) (Tantawi, Fidan, & Tantawy, Status of Smart Manufacturing in the United States, 2019) (Tantawi, Sokolov, & Tantawi, Advances in Industrial Robotics: From Industry 3.0 Automation to Industry 4.0 Collaboration, 2019). The tables show an aggressive growth in the IIoT platforms in a short period of time, reaffirming the trend of a steady and fast growth of this technology.

Table 3. Common Industrial Internet of Things platforms and their sizes as of 2018.

IIoT platform	Owner	Year launched	Number of devices on the network
Ability	ABB Robotics	2017	7000 robots
FIELD	FANUC Robotics	2016	Not reported, but there are at least 18,000 Robots and CNCs connected to the iZDT app of the FIELD platform
Connyun	Kuka Robotics	2016	One million devices
Predix	General Electric	2016	The largest IIoT platform in the world. Number of devices not reported.
MindSphere	Siemens	2017	More than 30 million devices. Second largest in the world.

Sustainable Manufacturing

The U.S. Environmental Protection Agency defines sustainable manufacturing as "the creation of manufactured products through economically-sound processes that minimize negative environmental impacts while conserving energy and natural resources" (Sustainable Manufacturing, n.d.). Environmental impact of manufacturing technologies will steadily continue to be a decisive consideration factor in the developed countries.

Additive Manufacturing

This layer-by-layer production technology is rapidly finding an increased number of applications. Initially, AM saw applications in rapid prototyping, altering production lines and adding a high level of flexibility to manufacturing processes (Attaran, 2017), part repair, and direct manufacturing of parts as well (Liu, Wang, Sparks, Liou, & Newkirk, 2016) (Strickland, 2016) (Kellner, 2017).

However, despite the advances in additive manufacturing, recent research suggests that several challenges still exist in the field, in particular, low accuracy when compared to subtractive manufacturing techniques (Abdulhameed, Al-Ahmari, Ameen, & Mian, 2019),surface roughness of the final product continues to be a key challenge in the field (Fox, Moylan, & Lane, 2016) (Technology, 2012). In addition to that, the lack of established and standardized methods of quality assurance of additively manufactured parts, makes this technology slowly progressing in the manufacturing industry, in which failure of components is minimally accepted (K.Everton, Hirsch, Stravroulakis, K.Leach, & T.Clare, 2016). Up until recently, optical transparency was another major drawback of additive manufacturing, in the work by Kotz et al. it was shown that highly transparent silica glass was 3D printed with resolutions down to tens of micrometers (Kotz, et al., 2017).

From the above analysis, it can be seen that growth of additive manufacturing in the process and manufacturing industry is still in the initial phase experiencing a gradual growth. It currently excels in the industries that produce customer-specified products such as in the aerospace and healthcare industries, with a growth of 2X over the period from 2016 to 2018 (Research, 2016).

2. **Incorporation of Industrial Wireless Technologies:** significant cost cuts can be achieved from utilizing industrial wireless technologies in both the factory level, and in end-products. For example, truck engines have recently been equipped with wireless transmitters that relay engine performance and oil temperature to manufacturers and owners, who in turn, can avoid costly recalls and repairs (Kusiak, 2017). In the factory level, wireless technologies save on installation costs.

Although advanced wireless systems have been available in the market, they are being slowly adopted by manufacturers, mainly due to a perceived low performance (Martinez, Cano, & Vilajosana, 2018). In particular, control operations, which require deterministic and real-time response, as well as the fear of security breaches.

3. Intelligent Industrial Robotics

Over the last six decades, Artificial Intelligence (AI) industry underwent a rapid growth and evolved from simple digitization and "networkization" to "intelligentization" (Webster, Creemers, Triolo, & Kania, 2017). Forbes reports that the Artificial Intelligence industry accomplished a 70% growth rate in 2018 (Coleman, 2018). When Artificial Intelligence is introduced into industrial robotics, the new field of "Intelligent Industrial Robotics" is created. This collaborative robotics had undergone significant progress in the recent years. They can reduce costs effectively for manufacturers by employing artificial intelligence and advanced perception and can be taught movements using a range of methods such as electromyography (Mohamed, et al., 2020). Machine intelligence has already been established in many other technologies and applications, for example, optical scanners with the proper image processing algorithms may be used to troubleshoot faults in printed circuit boards (Kusiak, 2017) (Berggren, Nilsson, & Robinson, 2007).

Although applications that rely on artificial intelligence have been demonstrated in service robots, this technology however, progressed slowly in the field of industrial robotics. This is mainly due to the difficulty of establishing an infrastructure of technicians that are trained to handle AI and the associated preventive and predictive maintenance protocols.

Prior to the pandemic, the Intelligent Industrial Robotics market was expected to grow to $14.29 billion in 2023 from $4.94 billion in 2018 (Tantawi, Sokolov, & Tantawi, Advances in Industrial Robotics: From Industry 3.0 Automation to Industry 4.0 Collaboration, 2019) (Demaitre, 2018). The 2020 report of the International Federation of Robotics (IFR), shows that there are more than 2.7 million industrial robotic units in the world. *Figure 3* lists the top 10 countries in growth of industrial robotics for the year 2016. The data shows that the market size of the top three countries China, Korea, and Japan constitutes more than twice the market size of the next seven countries combined. Furthermore, in 2016, the electronics industry exceeded, for the first time, the automotive industry in demand for industrial robotics in the three Asian markets of China, Japan, and Korea (Coronis & Tantawi, 2019). The top 10 countries in growth of industrial robotics for the year 2019 are shown in *Figure 4*. The data shows that the market size of China alone grew in 2019 to constitute 46% of the top 10 market size. These numbers indicate that there is a "robotics revolution" that is currently taking place in East Asia at the time this chapter is written. By sector, the report indicates that the largest demand for industrial robotics was by the automotive sector at 28% followed by the electronics sector at 24% (*Figure 5*). The data also show that a slight shrink in the world industrial robotics market took place in 2018 and 2019, particularly in the automotive and electronics sectors, that was attributed to a slowdown in the world economy and the trade conflict that involved the U.S., Europe, and China.

Figure 3. Top 10 markets of industrial robotics in 2016

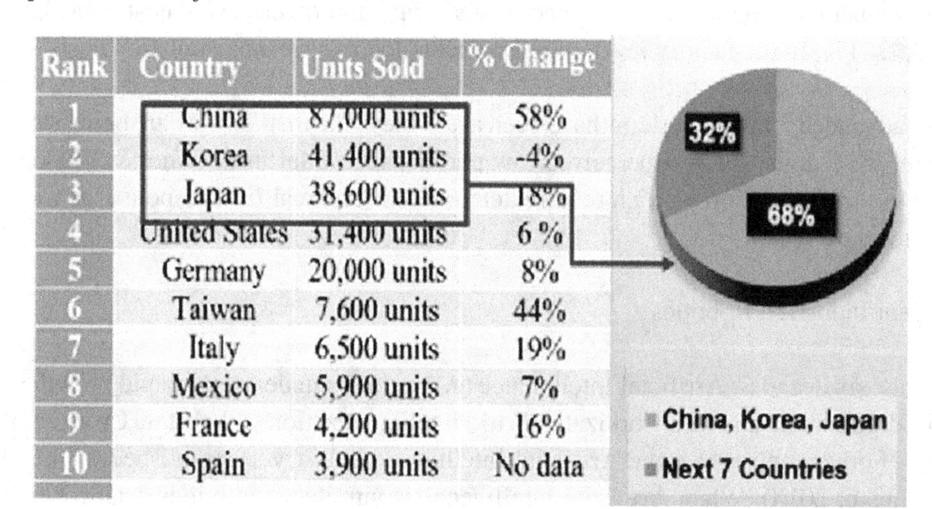

Rank	Country	Units Sold	% Change
1	China	87,000 units	58%
2	Korea	41,400 units	-4%
3	Japan	38,600 units	18%
4	United States	31,400 units	6 %
5	Germany	20,000 units	8%
6	Taiwan	7,600 units	44%
7	Italy	6,500 units	19%
8	Mexico	5,900 units	7%
9	France	4,200 units	16%
10	Spain	3,900 units	No data

Figure 4. Top 10 markets of industrial robotics in 2019, and China's share of the top 10 market

Rank	Country	Units Sold
1	China	140,492 units
2	Japan	49,908 units
3	United States	33,339 units
4	Korea	27,873 units
5	Germany	20,473 units
6	Italy	11,100 units
7	France	6,700 units
8	Taiwan	6,400 units
9	Mexico	4,600 units
10	India	4,300 units

54%
46%
China
Next 9 Countries including the U.S.

Figure 5. Shares of Industrial robotics by industry sector over the period from 2017 to 2019

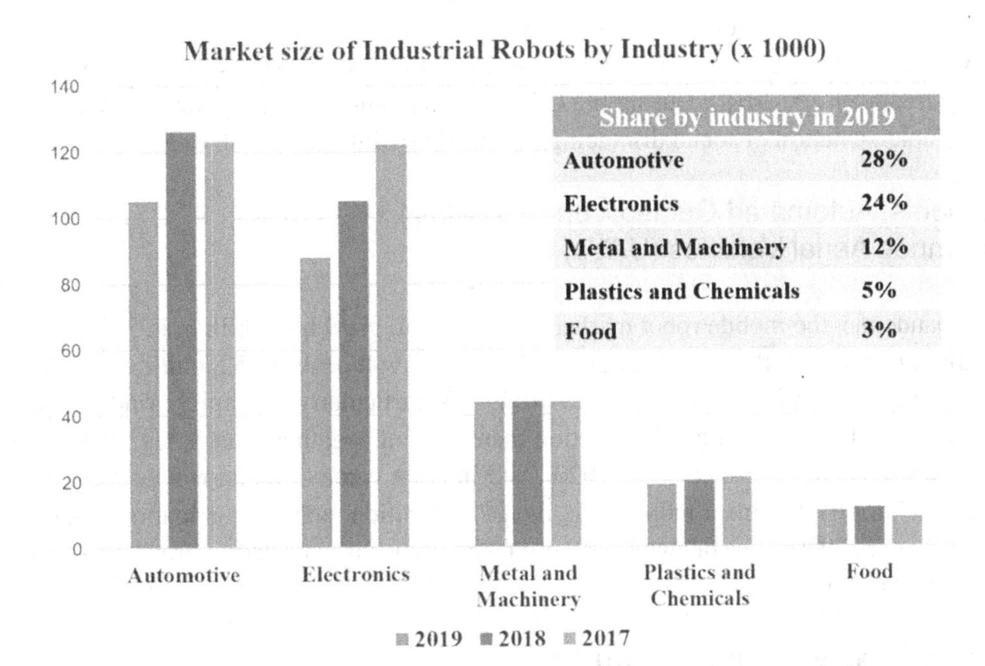

Furthermore, prior to the pandemic, the IFR expected that the total number of industrial robots in 2020 would be double the number in 2014. In the aftermath of the pandemic, the forecast growth of the industrial robotics market is projected to witness a slight decrease by 3% by 2025 (COVID-19 Impact on the Industrial Robotics Market by Type (Articulated, SCARA, Parallel, Cartesian Robots), Industry (Automotive; Electrical and Electronics; Food & Beverages; Pharmaceuticals and Cosmetics), and Region – Global Forecast to 2025, 2020). This decrease is significantly less than that witnessed in other industries. This is due to the increased demand on both industrial and service robots to counter the shortage in human operators, and to function in hospitals and medical centers.

This change in technology requires "targeted training and further education for employees" (Robots double worldwide by 2020, 2018). Moreover, the IFR forecasts that intelligent robots will take the lead in the robotics industry in the upcoming years (The Robot Revolution: The New Age of Manufacturing | Moving Upstream S1-E9, 2018).

Okuda et al. (Okuda, Haraguchi, Domae, & Shiratsuchi, 2016) identify four technological aspects in which artificial intelligence excels in industrial robotics and as a result increases their flexibility and adaptability: Collision avoidance in real time, Performing Human-Robot Cooperative (HRC) tasks (Okuda, Haraguchi, Domae, & Shiratsuchi, 2016) (Pellegrinelli, Orlandini, Pedrocchi, Umbrico, & Tolio, 2017), Three-Dimensional vision and sensing capability, and Force control and inspection.

Collaborative Robotics (Co-Bots)

Based on machine intelligence, the next generation of industrial robotics, Co-bots, are more dynamic and compact in size than traditional industrial robots. They are equipped with an advanced sensory system to sense human presence around them. Reports indicate that deploying co-bots resulted in increased production in some of the major automotive manufacturers (Toyota, Ford, and Mercedes Benz) (Estolatan, Geuna, Guerzoni, & Nuccio, 2018). The IFR forecasts that collaborative robots will take the lead in the robotics industry in the upcoming years (The Robot Revolution: The New Age of Manufacturing | Moving Upstream S1-E9, 2018). During the Covid-19 pandemic, cobots were used in applications to support the implementation of social distancing in factories (Malik, Masood, & Kousar, 2020).

Mobile Robots, Automated Guided Vehicles (AGV), and Unmanned Aerial Vehicles (UAVs)

Prior to the pandemic, the mobile robot market is expected to grow to $7 billion in 2022 from slightly above $1billion in 2017, on the other hand, the UAV market will peak at $52.3 billion in 2025 from its $20 billion market size in 2018 (Demaitre, 2018). UAVs particularly are rapidly finding applications in a multitude of industries such as construction management, weather forecasting, mining, telecommunication, disaster relief (Park & Choi, 2020), and in what came to be known as "Smart Farming" or "Farming 4.0" for environment monitoring, image acquisition, and even irrigation and fertilization (D.Boursianis, et al., 2020). The demand for these technologies is expected to grow even more rapidly in the post-pandemic era.

Augmented Reality in Robotic Control

The augmented reality technology is based on coupling virtual reality with the real world. The Low-cost robotic solutions based on augmented reality are shown to reduce the programming complexity and increase accuracy, and therefore, can significantly reduce training costs and increase productivity in small-to-medium sized enterprises and large manufacturers alike (Stadler, et al., 2016). Robot controllers can be connected to an Augmented reality application to enable the operator to visualize robot intentions, locations, and states such as target location and if selected objects are correct or not. In addition to that, it can be used when communication channels are limited such as the limited controllability of the teaching pendant due to hand gloves, or when physical barriers exist. Thus complete human-robot communication, and controllability can be performed at the network level without having to be in the field with the robot (Maly, Sedlacek, & Leitao, 2016) (Ruffaldi, Brizzi, Tecchia, & Bacinelli, 2016). This capability also allows for cooperative interface between humans and robots rather than performing tasks independently. Studies show that cooperative work is the most optimized form for production when compared to processes in which humans and robots work independently.

POTENTIAL RISKS ASSOCIATED WITH SMART MANUFACTURING

Although the benefits from deploying SM tehnologies can be significant, there are many risks that may arise, and generally, can be characterized to be less in occurrence but can have catastrophic effects.

As mentioned previously, artificial intelligence progresses at a lower pace in industrial robotics than in service robots, due to the need for a well-established infrastructure of trained technicians, maintenance protocols, and safety laws that are necessary to handle the new technologies. This is also true for the other technologoies that constitute smart manufacturing. Hence, one of the main risks associated with

SM is the potential of lacking the personnel with the proper skills to maintain these technologies if they fail, and as a result, may cause an entire production line to halt.

Another risk associated with manufacturing plants that deploy SM is the potential of a cyberattack. With the advancements in the Internet of Things technology, a large number of equipment and machines can be connected to the network, and consequently every connected device can be an access point for potential threats; as a result of "putting all the eggs in once basket", a successful cyber attack may have a catastrophic and long-lasting effect.

A third risk that may result, is that from the commercial incorporation of UAVs, that can include loss of control, misuse of technology, collision, and trespassing, which can ultimately result in devastating lawsuits.

As with all new technologies, the current safety standards associated with collaborative robots lack sufficient documented accident data in an industrial setting (Matthias, Oberer-Treitz, Staab, Schuller, & Peldschus, 2010). The non-shielded working area of a collaborative robot means that any person can get at a close proximity to the robot without the presence of a physical barrier, thence, the risk potential essentially increases, with personal safety solely relying on the effectiveness of the robotic sensory system. Thus safety standards for working with collaborative robots, particularly mobile collaborative robots are still evolving, and can differ significantly by year and location. This raises a challenge for international companies, and as a result can slow down the incorporation of these technologies.

The technical standards that relate to collaborative industrial robotics are detailed in the Technical Standard number 15066:2016 of the International Standards Organization (ISO) and in the Technical Report 15.606-2016 of the U.S.-based Robotic Industries Association (RIA) (Gallagher, 2019).

In addition to them, RIA TR15.806-2018 details the testing procedure for the power force limiting systems found on most collaborative robots.

ISO standard 10218-2011 and the RIA standard ANSI/R15.06-2012 detail the technical standards for industrial robotics. Both standards are undergoing an update that is projected to be completed by 2021 and 2022.

Standard ANSI/ITSDF B56.5:2019 addresses safety for automated guided vehicles (AGVs) and automated maned vehicles, that typically follow prescribed paths, and therefore the standard does not apply to industrial robots and mobile collaborative robots (Gerstenberger, 2019).

The European standards EN 1525 and EN 1526, and the ISO standard 3691-4 address driverless industrial trucks. Standard ISO 13482:2014 address the safety requirements of mobile service robots, and explicitly states that industrial robotics are not covered by the standard. However, there is currently no safety standard for mobile collaborative robots, the only work available is a "Guidance for Manufacturers" that was very recently published by the RIA under R15.08. A summary of the technical standards that relate to industrial robotics is shown in *Table 4*.

Table 4. Currently available safety standards that relate to industrial robotics

Safety Standard	Description	Year Published
ISO 15066 ANSI-RIA TR 15.606	Collaborative industrial robotics	2016
RIA TR15.806	Testing procedure for PFL systems	2018
ISO 10218-2011 ANSI/R15.06-2012	Industrial robotics	2011 and 2012
ANSI/ITSDF B56.5:2019	Automated Guided Vehicles (AGVs)	2019
EN 1525 EN 1526 ISO 3691-4	Driverless industrial trucks	
ISO 13482:2014	Service robots	2014
RIA R15.08 (incomplete)	Mobile Industrial Robots	2019

According to the standards, four safety methods are utilized in collaborative robots:

1- Safety-rated monitor stops
2- Hand guiding
3- Speed and separation monitoring
4- Power force limiting systems

In conclusion, deploying Smart Manufacturing technologies, can have a dual effect: significant reduction in risky occurrences, however the risks that may occur can have a strong impact. More effort is needed to standardize response plans and protocols across the spectrum of manufacturing industries that minimize the impact of these risks.

SOLUTIONS AND RECOMMENDATIONS

In this chapter, some of the main challenges for industry that came as a result of the CoVid-19 pandemic include the disruption in supply chains. One way to remedy that is to increase incorporation of the next-generation technologies that are based on artificial intelligence, such as collaborative robotics, and increase the "networkization". Additive manufacturing found new applications as a result of the pandemic, and continues to grow steadily.

There is an amounting evidence that the skill gap in handling Artificial Intelligence- based technologies is widening significantly more than was previously expected. To direct research fundings, education, and technical training more efficiently, long-term policies and strategies are needed in place.

Risks associated with Smart manufacturing can be characterized to be low in occurrence but can have a catastrophic effect, thus more effort is needed to develop standard response plans and protocols for the manufacturing industry that minimize the impact of these risks.

FUTURE RESEARCH DIRECTIONS

There is lately a considerable attention by governments to follow suit with the Chinese success story, and adopt official policies that provide guidance and direction for scientific research, technical training and education. A large amount of future trends in research are centered around the incorporation of artificial intelligence in manufacturing. Technical training programs need to cope with the new trend of Intelligentization in manufacturing, particularly after the pandemic, such as the increased use of collaborative robots, augmented reality, and vision systems.

CONCLUSION

In this chapter, current state and future trends in Smart Manufacturing technologies in the aftermath of the COVID-19 Pandemic are presented. All current data suggest that the need in skilled workforce will widen over the next few years in the United States and the world.

Artificial Intelligence continues to be the main characteristic of Smart Manufacturing. One of its main application, intelligent industrial robotics, is expected to take the lion's share of smart manufacturing in the upcoming decade. It is expected that in the next decade the intelligent robotics market will grow aggressively worldwide, with current trends showing that a revolution in industrial robotics is taking place in the East Asian markets. IIoT continues at a steady pace to play a major role in defining the fourth generation of manufacturing (Musa, et al., 2019). The IIoT market is experiencing a rapid growth currently. Other technologies such as additive manufacturing and industrial wireless systems also continue to play an important role, but to a less extent than intelligent industrial robotics and IIoT. The main challenge that faces deploying additive manufacturing in industry is the lack of an established quality assurance method for produced end-products. As a result, this technology is currently suitable for customized end-products such as in the medical and aerospace industries.

One of the main challenges that emerged as a result of the CoVid-19 pandemic was the supply chain disruption. To counter that on the long run, manufacturers will have to increase self-sufficiency. Consequently, that means more demand for intelligent industrial and service robotics and additive manufacturing.

Risks associated with Smart manufacturing can be characterized to be less frequent than those if Smart Manufacturing technologies are not deployed. However, the effect of the risks can have catastrophic impacts. More effort is needed to develop standard response plans and protocols for the manufacturing industry that minimize the impact of these risks.

ACKNOWLEDGMENT

This research was supported by the National Science Foundation [grant numbers 1801120 and 2000685].

REFERENCES

A History Of Flexible Manufacturing Systems. (2017). Retrieved January 20, 2018, from https://www.ukessays.com/essays/information-technology/a-history-of-flexible-manufacturing-systems-information-technology-essay.php

Abdulhameed, O., Al-Ahmari, A., Ameen, W., & Mian, S. H. (2019). Additive manufacturing: Challenges, trends, and applications. *Advances in Mechanical Engineering*, *11*(2), 1–27. doi:10.1177/1687814018822880

Alting, D. L., & Jogensen, J. (1993). The Life Cycle Concept as a Basis for Sustainable Industrial Production. *CIRP Annals*, *42*(1), 163–167. doi:10.1016/S0007-8506(07)62417-2

Arnold, R., Fidan, I., & Tantawi, K. (2018). Transforming Industry towards Smart Manufacturing in the United States. In *NSF-ATE 2018 Conference*. Washington, DC: National Science Foundation.

Attaran, M. (2017). Additive Manufacturing: The Most Promising Technology to Alter the Supply Chain and Logistics. *Journal of Service Science and Management*, *10*(03), 189–205. doi:10.4236/jssm.2017.103017

Ayati, N., Saiyarsarai, P., & Nikfar, S. (2020). Short and long term impacts of COVID-19 on the pharmaceutical sector. *Daru: Journal of Faculty of Pharmacy, Tehran University of Medical Sciences*, *28*(2), 799–805. Advance online publication. doi:10.100740199-020-00358-5 PMID:32617864

Babiceanu, R., & Chen, F. F. (2006). Development and applications of holonic manufacturing systems: A survey. *Journal of Intelligent Manufacturing*, *17*(1), 111–131. doi:10.100710845-005-5516-y

Bateman, J. (2018). *Why China is spending billions to develop an army of robots to turbocharge its economy*. CNBC. Retrieved from https://www.cnbc.com/2018/06/22/chinas-developing-an-army-of-robots-to-reboot-its-economy.html

Berggren, M., Nilsson, D., & Robinson, N. D. (2007). Organic materials for printed electronics. *Nature Materials*, *6*(1), 3–5. doi:10.1038/nmat1817 PMID:17199114

Bert, J., Schellong, D., Hagenmaier, M., Hornstein, D., Wegscheider, A. K., & Palme, T. (2020). *How COVID-19 Will Shape Urban Mobility*. Boston Consulting Group.

Bo-hu, L., Lin, Z., Shi-long, W., Fei, T., Jun-wei, C., Xiao-dan, J., . . . Xu-dong, C. (2010). Cloud manufacturing:a new service-oriented networked manufacturing model. Computer Integrated Manufacturing Systems, 16.

Boursianis, Papadopoulou, Diamantoulakis, Liopa-Tsakalidi, Barouchas, Salahas, . . . Goudos. (2020). Internet of Things (IoT) and Agricultural Unmanned Aerial Vehicles (UAVs) in smart farming: A comprehensive review. *Internet of Things*.

Business Insider. (2015). China Solidifies Its Position as the World's Largest Manufacturer. Manufacturers Alliance for Productivity and Innovation (MAPI).

Coleman, L. (2018). *Inside Trends And Forecast For The $3.9T AI Industry*. Forbes.

Collie, B., Wachtmeister, A., Waas, A., Kirn, R., Krebs, K., & Quresh, H. (2020). *Covid-19's Impact on the Automotive Industry*. Boston Consulting Group.

Coronavirus disease (COVID-19) pandemic. (2021). Retrieved September 6, 2021, from World Health Organization: https://covid19.who.int/

Coronis, A., & Tantawi, K. (2019). Advances in Energy-Efficient Manufacturing using Industrial Robotics. In *NSF-ATE 2019 Conference*. Washington, DC: National Science Foundation.

COVID-19 Impact on the Industrial Robotics Market by Type (Articulated, SCARA, Parallel, Cartesian Robots), Industry (Automotive; Electrical and Electronics; Food & Beverages; Pharmaceuticals and Cosmetics), and Region – Global Forecast to 2025. (2020). Research and Markets.

Davis, J., & Swink, D. (n.d.). *Smart Manufacturing as a Real-Time Networked Enterprise and a Market-Driven Innovation Platform*. Smart Manufacturing Leadership Coalition.

Davis, J., Swink, D., Tran, J., Wetzel, J., Profozich, G., McKewen, E., & Thys, R. (2015). *Smart Manufacturing The Next Revolution in Manufacturing*. California Manufacturing Technology Consulting.

Demaitre, E. (2018). *RBR50 2018 Names the Leading Robotics Companies of the Year*. Robotics Business Review.

Desai, A. (2014). *Economy League's 2014 World Class Summit: Tracking Philadelphia's Progress on Growth and Opportunity*. Global Philadelphia Association. Retrieved from https://globalphiladelphia.org/news/eceonmy-leagues-2014-world-class-summit-tracking-philadelphia%E2%80%99s-progress-growth-and-opportunity

Digitale Wirtschaft und Gesellschaft. (n.d.). Federal Ministry of Education and Research-Germany. Retrieved August 17, 2017, from https://www.bmbf.de/de/zukunftsprojekt-industrie-4-0-848.html

Energy, U. S. (2016). *Industrial Assessment Centers (IACs)*. Retrieved June 12, 2017, from https://energy.gov/eere/amo/industrial-assessment-centers-iacs

Epply, T., & Nagengast, J. (n.d.). *The Lean Manufacturing Handbook* (2nd ed.). Continental Design and Engineering.

Erol, S., Schumacher, A., & Sihn, W. (2016). Strategic guidance towards Industry 4.0 – a three-stage process model. *International Conference on Competitive Manufacturing*.

Estolatan, E., Geuna, A., Guerzoni, M., & Nuccio, M. (2018). *Mapping the Evolution of the Robotics Industry: A Cross Country Comparison*. University of Toronto.

Everton, K., Hirsch, M., Stravroulakis, P., Leach, R. K., & Clare, A. T. (2016). Review of in-situ process monitoring and in-situ metrology for metal additive manufacturing. *Materials & Design*, *95*, 431–445. doi:10.1016/j.matdes.2016.01.099

Faggella, D. (2018, May 29). *Global Competition Rises for AI Industrial Robotics*. Retrieved September 24, 2018, from https://www.techemergence.com/global-competition-rises-ai-industrial-robotics/

Falsey, A. R., McCann, R. M., Hall, W., Criddle, M. M., Formica, M. A., Wycoff, D., & Kolassa, J. E. (1997). The "Common Cold" in Frail Older Persons: Impact of Rhinovirus and Coronavirus in a Senior Daycare Center. *Journal of the American Geriatrics Society*, *45*(6), 706–711. doi:10.1111/j.1532-5415.1997.tb01474.x PMID:9180664

Fox, J. C., Moylan, S. P., & Lane, B. M. (2016). Effect of process parameters on the surface roughness of overhanging structures in laser powder bed fusion additive manufacturing. *Procedia CIRP*, *45*, 131–134. doi:10.1016/j.procir.2016.02.347

France, G. o. (2013). *The New Face of Industry in France*. Author.

Francis, S. (2018). *ABB claims to have connected 7,000 of its industrial robots to its IIoT platform*. Retrieved from https://www.i-scoop.eu/industry-4-0-society-5-0/

Gallagher, R. M. (2019, August 30). *New Safety Standards for Collaborative Robots*. Retrieved from https://www.engineering.com/AdvancedManufacturing/ArticleID/19403/New-Safety-Standards-for-Collaborative-Robots.aspx

Gerstenberger, M. (2019). Industrial Mobile Robots Safety Standard Update. *National Robot Safety Conference*.

Giffi, C., Dollar, B., Drew, M., McNelly, J., Carrick, G., & Gangula, B. (2015). *The Skills Gap in U.S. Manufacturing 2015 and Beyond*. Deloitte Development LLC.

Giret, A. (2008). Holonic Manufacturing Systems. In *A Multi-Agent Methodology for Holonic Manufacturing* (pp. 7-20). Springer. Retrieved from http://impactphl.org/wp-content/uploads/2016/07/ELGP-BFT-Impact-Report.pdf

Harrington, J. (1979). *Computer Integrated Manufacturing*. Krieger Pub Co.

IATA. (2020). *Air Passenger Market Analysis*. International Air Transport Association (IATA).

Internet of Things Global Standards Initiative. (n.d.). Retrieved 6 16, 2017, from https://www.itu.int/en/ITU-T/gsi/iot/Pages/default.aspx

Japan, G. o. (n.d.). *Realizing Society 5.0*. Author.

Jung, K., Morris, K., Lyons, K. W., Leong, S., & Cho, H. (2015). Mapping Strategic Goals and Operational Performance Metrics for Smart Manufacturing Systems. *Procedia Computer Science*, *44*, 184–193. doi:10.1016/j.procs.2015.03.051

Kahn, J. S., & McIntosh, K. (2005). History and Recent Advances in Coronavirus Discovery. *The Pediatric Infectious Disease Journal*, *24*(11), S223–S227. doi:10.1097/01.inf.0000188166.17324.60 PMID:16378050

Kang, H. S., Lee, J. Y., Choi, S., Kim, H., Park, J. H., Son, J. Y., . . . Noh, S. D. (2016). Smart Manufacturing: Past Research, Present Findings, and Future Directions. *International Journal of Precision Engineering and Manufacturing-Green Technologies, 3*, 111-128.

Kellner, T. (2017). GE, CFM Expect $15 Billion. In *New Business In Paris; New LEAP Engines Are Giving A Lift To The Aviation Industry*. General Electric.

Kennell, B. (2015). *Smart Manufacturing: A Path to Profitable Growth.* Retrieved from https://www.huffingtonpost.com/brian-kennell/smart-manufacturing-a-pat_b_7314828.html

Kotz, F., Arnold, K., Bauer, W., Schild, D., Keller, N., Sachsenheimer, K., Nargang, T. M., Richter, C., Helmer, D., & Rapp, B. E. (2017). Three-dimensional printing of transparent fused silica glass. *Nature, 544*(7650), 337–339. doi:10.1038/nature22061 PMID:28425999

Kusiak, A. (2017). Smart manufacturing must embrace big data. *Nature, 544*(7648), 23–25. doi:10.1038/544023a PMID:28383012

Laing, T. (2017). *UK Industrial Strategy: Navigating a changing world.* University of Cambridge Institute for Sustainability Leadership.

Leathers, M. L. (2016). How to Prepare Your Workforce for Smart Manufacturing. *Industry Week.*

Liebhart, R., & Hohmann, L. (2016). *Korea: Evolution of manufacturing industry.* Maschinen Markt International.

Liu, R., Wang, Z., Sparks, T., Liou, F., & Newkirk, J. (2016). Aerospace Applications of Laser Additive Manufacturing. In Laser Additive Manufacturing: Materials, Design, Technologies, and Applications (pp. 351-353). Woodhead Publishing-Elsevier.

Lu, M. C. (2020). Future pandemics can be prevented, but that'll rely on unprecedented global cooperation. *The Washington Post.*

Lu, Y., Morris, K., & Frechette, S. (2016). Current Standards Landscape for Smart Manufacturing Systems. *National Institute of Standards and Technology NISTIR 8107.*

Malik, A. A., Masood, T., & Kousar, R. (2020). Repurposing factories with robotics in the face of COVID-19. *Science Robotics, 5*(43), eabc2782. doi:10.1126cirobotics.abc2782 PMID:33022618

Malik, N., & A, J. (2016, January). US expects energy savings through smart manufacturing. *MRS Bulletin, 41*(1), 10-11.

Maly, I., Sedlacek, D., & Leitao, P. (2016). Augmented Reality Experiments with Industrial Robot in Industry 4.0 Environment. *IEEE International Conference on Industrial Informatics*, 176-181. 10.1109/INDIN.2016.7819154

Martinez, B., Cano, C., & Vilajosana, X. (2018). *A Square Peg in a Round Hole: The Complex Path for Wireless in the Manufacturing Industry.* Cornell University.

Matthias, B., Oberer-Treitz, S., Staab, H., Schuller, E., & Peldschus, S. (2010). *Injury Risk Quantification for Industrial Robots in Collaborative Operation with Humans.* ROBOTIK.

Michael Franzino, A. G. (2020). *The $8.5 Trillion Talent Shortage*. Korn Ferry. Retrieved from https://www.kornferry.com/insights/this-week-in-leadership/talent-crunch-future-of-work

Mohamed, E., Tantawi, K. H., Pemberton, A., Pickard, N., Dyer, M., Hickman, E., . . . Nasab, A. (2020). Real Time Gesture-Controlled Mobile Robot using a Myo Armband. In *Proceedings of the 2nd African International Conference on Industrial Engineering and Operations Management* (pp. 2432-2437). Harare, Zimbabwe: IEOM Society International.

Musa, Y., Tantawi, O., Bush, V., Johson, B., Dixon, N., Kirk, W., & Tantawi, K. (2019). Low-Cost Remote Supervisory Control System for an Industrial Process using Profibus and Profinet. 2019 IEEE SoutheastCon, 1-4.

Okuda, H., Haraguchi, R., Domae, Y., & Shiratsuchi, K. (2016). Novel Intelligent Technologies for Industrial Robot in Manufacturing - Architectures and Applications. *Proceedings of ISR 2016: 47st International Symposium on Robotics.*

Olival, K. J., Hosseini, P. R., Zambrana-Torrelio, C., Ross, N., Bogich, T. L., & Daszak, P. (2017). Host and viral traits predict zoonotic spillover from mammals. *Nature*, *546*(7660), 646–650. doi:10.1038/nature22975 PMID:28636590

Park, S., & Choi, Y. (2020). Applications of Unmanned Aerial Vehicles in Mining from Exploration to Reclamation: A Review. *Minerals (Basel)*, *10*(8), 663. doi:10.3390/min10080663

Pellegrinelli, S., Orlandini, A., Pedrocchi, N., Umbrico, A., & Tolio, T. (2017). Motion planning and scheduling for human and industrial-robot collaboration. *CIRP Annals*, *66*(1), 1–4. doi:10.1016/j.cirp.2017.04.095

Research, F. &. (2016). *Global Additive Manufacturing Market, Forecast to 2025*. Frost & Sullivan MB74-10.

Reuters. (2018, June 25). *Facing US blowback, Beijing softens its 'Made in China 2025' message*. Retrieved September 24, 2018, from https://www.cnbc.com/2018/06/25/facing-us-blowback-beijing-softens-its-made-in-china-2025-message.html

Robotics, I. F. (2017). *How robots conquer industry worldwide*. IFR Press Conference.

Robotics, I. F. (2018). Industrial robot sales increase worldwide by 31 percent. International Federation of Robotics.

Ruffaldi, E., Brizzi, F., Tecchia, F., & Bacinelli, S. (2016). Third Point of View Augmented Reality for Robot Intentions Visualization. *International Conference on Augmented Reality, Virtual Reality and Computer Graphics*, *9768*. 10.1007/978-3-319-40621-3_35

Rüßmann, M., Lorenz, M., Gerbert, P., Waldner, M., Justus, J., Engel, P., & Harnisch, M. (2015). *Industry 4.0: The Future of Productivity and Growth in Manufacturing Industries*. Boston Consulting Group.

Shaheen, J. (2015, April 22). *Shaheen Introduces Bill to Emhance Innovation, Energy Efficiency and Economic Competitiveness of Nation's Manufacturers*. Retrieved June 2017, from https://www.shaheen.senate.gov/news/press/shaheen-introduces-bill-to-enhance-innovation-energy-efficiency-and-economic-competitiveness-of-nations-manufacturers

Smart Manufacturing Operations Planning and Control Program. (n.d.). National Institute of Standards and Technology. Retrieved 6 16, 2017, from https://www.nist.gov/sites/default/files/documents/2017/05/09/FY2014_SMOPAC_ProgramPlan.pdf

Stadler, S., Kain, K., Giuliani, M., Mirnig, N., Stollnberger, G., & Tscheligi, M. (2016). Augmented reality for industrial robot programmers: Workload analysis for task-based, augmented reality-supported robot control. *25th IEEE International Symposium on Robot and Human Interactive Communication*. 10.1109/ROMAN.2016.7745108

Strickland, J. D. (2016). Applications of Additive Manufacturing in the Marine Industry. *Proceedings of PRADS2016*.

Sustainable Manufacturing. (n.d.). United States Environmental Protection Agency. Retrieved June 17, 2017, from https://www.epa.gov/sustainability/sustainable-manufacturing

Tantawi, K. (2020). Literature Review: Rethinking BioMEMS in the Aftermath of CoViD-19. *Biomedical Journal of Scientific & Technical Research*, *31*(1), 23944–23946. doi:10.26717/BJSTR.2020.31.005053

Tantawi, K., Fidan, I., Chitiyo, G., & Cossette, M. (2021). Offering Hands-on Manufacturing Workshops Through Distance Learning. *ASEE Annual Conference*.

Tantawi, K., Fidan, I., & Tantawy, A. (2019). Status of Smart Manufacturing in the United States. 2019 IEEE SoutheastCon.

Tantawi, K., Sokolov, A., & Tantawi, O. (2019). Advances in Industrial Robotics: From Industry 3.0 Automation to Industry 4.0 Collaboration. *4th Technology Innovation Management and Engineering Science International Conference (TIMES-iCON)*. 10.1109/TIMES-iCON47539.2019.9024658

Technology, N. I. (2012). Measurement Science Roadmap for Metal-Based Additive Manufacturing. Academic Press.

Terry, S., Fidan, I., Zhang, Y., & Tantawi, K. (2019). *Smart Manufacturing for Energy Conservation and Savings*. NSF-ATE Conference.

Terry, S., Lu, H., Fidan, I., Zhang, Y., Tantawi, K., Guo, T., & Asiabanpour, B. (2020). The Influence of Smart Manufacturing Towards Energy Conservation: A Review. *Technologies*, *8*(2), 31. doi:10.3390/technologies8020031

The Robot Revolution: The New Age of Manufacturing | Moving Upstream S1-E9. (2018). Wall Street Journal.

The skills gap in U.S. manufacturing 2015 and beyond. (2015). The Manufacturing Institute.

Thilak, V., Devadasan, S., & Sivaram, N. (2015). A Literature Review on the Progression of Agile Manufacturing Paradigm and Its Scope of Application in Pump Industry. *TheScientificWorldJournal.*

Webster, G., Creemers, R., Triolo, P., & Kania, E. (2017). *Full Translation: China's 'New Generation Artificial Intelligence Development Plan*. New America.

Webster, S. A. (2016). *FANUC Launches New IoT System for Smart Manufacturing Era*. Society of Manufacturing Engineering.

Zemmar, A., Lozano, A. M., & Nelson, B. J. (2020). The rise of robots in surgical environments during COVID-19. *Nature Machine Intelligence*, *2*, 566–572.

KEY TERMS AND DEFINITIONS

Additive Manufacturing: A process by which an object is constructed from a three-dimensional model, in a layer-by-layer manner under computer control.

Artificial Intelligence: A computerized system that has the ability to perceive its environment.

Augmented Reality: The enhancement of real objects by a computerized system that is accomplished by a set of sensory systems.

Automated Guided Vehicle: A mobile robotic system that uses guides of different forms for navigation such as floor line tracks, wires, or vision systems.

Industrial Robot: A robot that is used in the manufacturing industry.

Intelligentization: A term that emerged in China to refer to the use of artificial intelligence with decision making capability.

Smart Manufacturing: The United States Congress defines Smart Manufacturing as a set of advanced sensing, instrumentation, monitoring, controls, and process optimization technologies and practices that merge information and communication technologies with the manufacturing environment for the real-time management of energy, productivity, and costs across factories and companies.

Chapter 9
Social Perspective of Suspicious Activity Detection in Facial Analysis:
An ML-Based Approach for Digital Transformation

Rohit Rastogi
https://orcid.org/0000-0002-6402-7638
ABES Engineering College, India

Parul Singhal
ABES Engineering College, India

ABSTRACT

Technology is demanded on to curb crimes, especially image recognition, which can be used to detect suspicious activities. Image, object, and face recognition along with speech identification can be used as great tools to achieve this target. The machine lerning algorithm gave immense capabilities to detect faces, objects, and speech to identify malicious activities, and with several epochs, the accuracy can be enhanced. The chapter applies the various ML algorithms on real-time video data to increase the accuracy and gets satisfactory results in this social cause of utmost importance.

INTRODUCTION

Suspicious Activity

The chapter describes initially the basics of initial crime status at world, face recognition and its tells and techniques, machine learning and data analytics and its tool and techniques. For ease of audiences, it also gives a brief review of Data mining, Regressions, AI and CPS. In Next section of the literature review, the famous authors and gist of their work on the content have been enlisted. Then in the application

DOI: 10.4018/978-1-7998-7852-0.ch009

Copyright © 2023, IGI Global. Copying or distributing in print or electronic forms without written permission of IGI Global is prohibited.

section, the latest classification techniques of support SVM, Dlib, CNN and RNN has been introduced and their application on the data set is reflected. In the result section, the different emotions and object have been recognized and their accuracy has been discussed. Then in the later part the recommendations, novelty, application, limitations have of the research work is explained followed by concluding remarks.

Face Recognition

The world is witnessing an unprecedented growth of cyber-physical systems (CPS), which are foreseen to revolutionize our world via creating new services and applications in a variety of sectors such as environmental monitoring, mobile health systems, and intelligent transportation systems and so on. The information and communication technology (ICT) sector is experiencing significant growth in data traffic, driven by the widespread usage of smartphones, tablets and video streaming, along with the significant growth of sensors deployments that are anticipated soon (Onsen Toyger et al., 2003) (Viola, P. et al., 2004).

Machine Learning

An agent is said to learn from experience (E) for some class of tasks(T) performance measure(P), if its Performance at tasks T, as measured by P, improves with experience. E.g. Playing checkers game, Mailing system (Tom Mitchell 1997).

There are different categories of m/c learning

1. Supervised learning-learn an input and output map (classification: categorical output, regression: continuous output).
2. Unsupervised learning-discover patterns in the data(clustering: cohesive grouping, --association: frequent co-occurrence)
3. Reinforcement learning-learning control

Data Analysis

This is the technique used for extracting useful, relevant, and meaningful information from the huge amount of data in a systematic manner. For the purpose, Parameter estimation (inferring the unknowns), Model development and prediction (forecasting), Feature identification and classification, Hypothesis testing and Fault detection

Tools of Data Analysis: Weka, R, Python

Python is an object-oriented high-level programming language and widely used with semantic dynamic, used for general-purpose programming. It is interpreted programming language. It is used for: web development (server-side), software development. The way to run a python file is like this on the command line:

```
helloworld.py
print ("Hello, World!")
```

Weka and R

It is a freely available s/w package containing a collection of machine learning algorithms under the GNU (General Public License). It is an open-source. The algorithms present in Weka are all coded in java and they can be used by calling them from their java pod. However with also provides a graphical user interface from which the algorithms can directly be applied to data sets.

R software: R is the programming language. It is freely available s/w. It is used for, statistical and analysis, data manipulation, graphic display. Effective data handling and storage of o/p is possible.

> 2 + 2=4 (Tripathi, R. et al., 2014).

DATA MINING AND REFERENTIAL TECHNIQUES

Data Mining (DM)

To extract potentially useful previously unknown information from a large amount of data is called data mining (DM). E.g. - databases text, web, and images. The key properties of Data Mining are: Predictions of likely outcomes, automatic discovery of patterns, focuses on a large database and data sets.

Tracking Patterns

One of the best techniques in data mining is learning to recognize patterns from the data sets.

Classification

Learn a method to predict the frequency class with pre-labelled (classified) examples.

Figure 1. Examples of classification

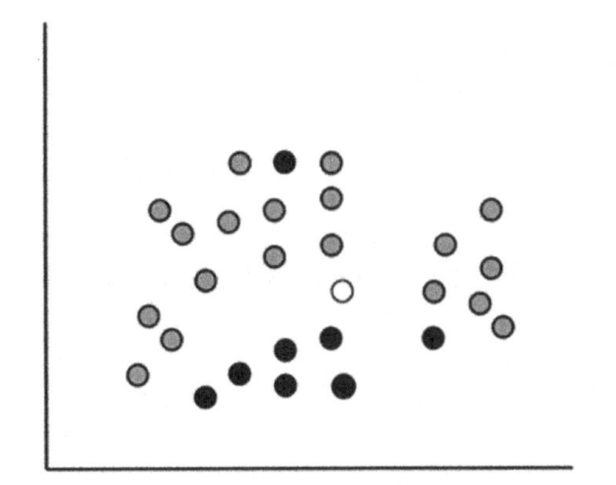

Many approaches: Statistics, Decision Trees, Neural Networks,

...

Supervised learning maps learning input to output and follows two techniques.

1. Classification-categorical output
2. Regression –continuous output

The task of classification is generally used to measure error (As per Figure 1).

Association(Dependency Modeling)

Looking at a set of records, each of which has a few numbers from the collection given;

- Production of dependency rules which will predict the occurrence of an object based on the incidence of other items (As per Figure 2).

Figure 2. Production of dependency rules

TID	Items
1	Bread, Coke, Milk
2	Beer, Bread
3	Beer, Coke, Diaper, Milk
4	Beer, Bread, Diaper, Milk
5	Coke, Diaper, Milk

Rules Discovered:
{Milk} --> {Coke}
{Diaper, Milk} --> {Beer}

Outlier Detection (Anomaly/ Deviation Detection)

The verification or identification of particular data records that might be the data errors or interesting that requires further analysis (As per Figure 3).

Clustering

Figure 3. Example of clustering

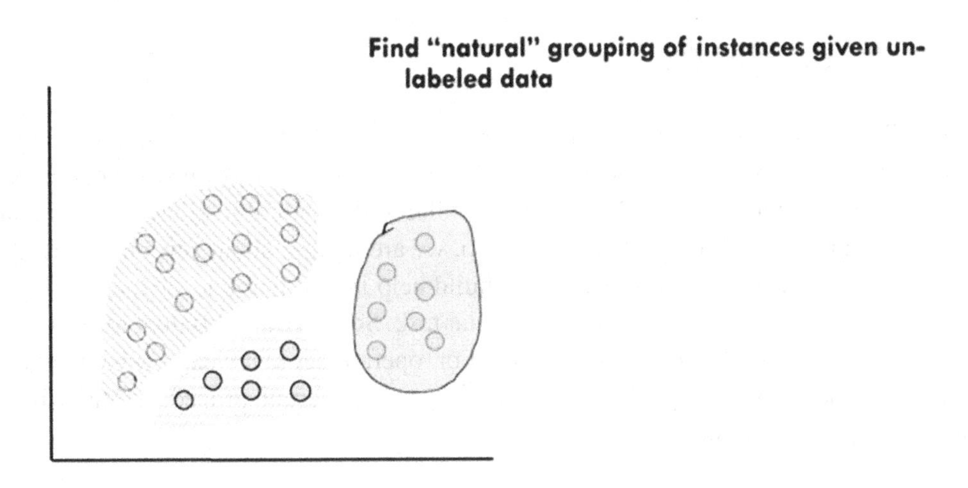

Find "natural" grouping of instances given un-
labeled data

Regression

Try to find a function that models the data with the minimum inaccuracy. A predictor is constructed that predicts a continuous-value function, or ordered value, as opposed to a categorical label.

Regression is a static methodology that is most often used for numeric prediction.

Rule-Based Inferences

This is a very important method: There are different ways to handle uncertainty. The different ways of uncertainty: one can be probabilities where specify the likelihood of a conclusion, apply Bayesian reasoning. And the certainty factor is an estimate of confidence in conclusions. It is an estimate provided by the experts. It is not as mathematically precise as is the case of probabilities but it is also very effective. e.g. - MYCIN is a medical expert system for analysis of meningitis infections. MYCIN uses a backward chaining rule-based expert system.

Application of Data Mining

1. Data Mining in agriculture.
2. Mass surveillance /Surveillance
3. NSA (National security agency)
4. Customer analytic
5. Educational data mining
6. Retail and services
7. Manufacturing
8. Finance and insurance

9. Telecommunication and utilities
10. Transport industry
11. Healthcare and medical industry

ARTIFICIAL INTELLIGENCE

Artificial Intelligence (AI) is a branch of computer science that deals with the study and the creation of computer systems that exhibit some form of intelligence. So, we are talking about computer systems exhibiting some form of intelligence which is very similar to the natural intelligence of human beings. AI is heavily based on the concept of knowledge. So, we are talking about several pieces of data that can lead to information, and then information can build help in building knowledge. So, knowledge is the information that can be used to perform a particular task. So, there are different forms of knowledge particularly in the context of AI. There is procedural or, operational knowledge, which talks about the procedures that will have to be adopted too; the knowledge about certain procedures that will have to be adopted to come up with a particular problem.

CYBER PHYSICAL SYSTEMS

Cyber-physical systems (CPSs) also often known as smart systems and these are co-engineered interacting networks of physical and computational components. Interacting networks of physical components and computational. So, computational is the cyber one-and physical is the physical world in which these systems are operating, physical world. Cyber-physical systems are embedded systems. So, these are cyber-physical systems are you can think of the conceptually as a cyber-physical system, conceptually as an embedded system plus the physical system together with you what you get is the cyber-physical system.

Cyber-physical systems (CPSs) are large scale, geographically dispersed, federated, heterogeneous, life-critical systems that comprise sensors, actuators and control, and networking components. First responder situational awareness systems, pervasive health care systems, smart grids, and unmanned aircraft systems are some examples of CPSs. These systems have multiple control loops, strict timing requirements, predictable network traffic, legacy components, and possibly wireless network segments. CPSs fuse cyber (comprising network components and commodity servers) and physical (comprising sensors and actuators) domains (Khatoun, R. 2017).

Modern technological development is fueled by simultaneous advances in software, data science and AI, communications, computation, sensors, actuators, materials, and their combinations such as 3D printing, batteries, and augmented reality. Different perspectives on these advances have led to the creation of many terms—CPS, the IoT, Industry 4.0, and the Swarm—to represent new classes of technologically enabled systems. We focus on CPS as a more general notion. CPS was introduced in 2006 in the U.S. to characterize "the integration of physical systems and processes with networked computing" for systems that "use computations and communication deeply embedded in and interacting with physical processes to add new capabilities to physical systems". In this context, the word cyber alternatively refers to the dictionary deðnition of "relating to or involving computers or computer networks" or more general feedback systems as in the field of cybernetics pioneered by Wiener. We consider both interpretations of CPS to be valid and, unless otherwise noted, use the term cyber to refer to computing or software parts of a CPS.

LITERATURE SURVEY AND PREVIOUS WORK

The third stage includes object detection. Pradhan, A. and Team (2012) mention CNN-based deep learning techniques. He focused on the fact that CNN is trained per line from the raw pixel level to the final object categories. He also told about various challenges, which need to be removed in the form of partial or complete obstruction, separating the state of light, position, scale, etc. The reader also said that with the image map activation function is decorated, which is treated when the feature maps are generated with pooling layers. The process is repeated until the desired result is produced. He also mentioned about important datasets for object detection such as Microsoft COCO, CIFAR-10, and CIFAR-100, CUB-200-2011, Caltech-256, ILSVRC, PASCAL VOC Challenge dataset.

Pradhan, A. (2012) has discussed various approaches for object detection in his work. 'You Only Look Once' (YOLO) is one such framework for object detection in real-time. The improved model YOLO9000 can categorize into 9000 object categories. According to him, YOLO9000 combines a distinct dataset using a joint training approach. (Sharma and Rameshan, 2017).

They have discovered the need for near-sensor data to detect real-time objects and have uncovered this problem for integrating a centralized and powerful processor to process data from different servers because the large datasets are deep to learn necessary. Pathak concluded that there is a scope of providing "Object Detection as a Service" in a variety of applications. The last stage involves speech recognition. Andy and Allen have suggested an approach to learning the conversation in the Indonesian language using the recurrent neural network and long-short-term memory. He used to use words in vocabulary with a vector-based model.

According to Miles, B. (2018), the conversation system is categorized into two groups: retrieval-based which picks the best answer from possible answers and a generative model in which the answers are loaded previously. Their work exposes deficiencies that are the difficulty in training, errors of grammar, and the challenge of finding the best dataset for a generative model. He explained that word formatting can be implemented in a statistical method in vector.

Miles, B. (2018) also discovered a method to deep learn a conversation in which the model learns from both the input and output sequence of words. Initial processing involves finding and filtering. Initially, data (news data over five years from Compass) is collected from two sources. Crawled data is then filtered by removing punctuation and split paragraphs in one sentence per line. Then processed data is used for intensification of learning words to create vectors, which learns a conversation using the Dual Encoder LSTM. They simultaneously searched for two pre-trained models: Conversation models and word representation vectors to learn deeper conversations.

The malicious use of AI increases the speed and success rate and augments the capabilities of attacks. Information and communication technologies (ICTs) and AI expand the opportunities to commit a crime and form a new threat landscape in which new criminal tactics can take place (Brundage, 2018). In the report on malicious AI, the authors warned about the changing threat landscape by the malicious uses of

AI technologies. The AI field is broadly distinguished between the rule-based techniques and the machine learning (ML)-based techniques, which allow computer systems to learn from a large amount of data.

Cybercriminals learn to use AI technologies-enhanced learning approaches to their advantage and weaponize them by automating the attack process. The shift to AI technologies with learning capability, such as deep learning reinforcement learning (RL), support vector machines, and genetic algorithms, has potentially unintended consequences, such as facilitating criminal actions more efficiently (Goodfellow, L. et al., 2016).

Therefore, awareness of new trends in cybercrime is becoming significantly more important to drive appropriate defensive actions. Based on the way the crime is committed, we can classify it as a computer crime when it is carried out with the use of a computer and as a cybercrime when it is carried out with the use of a network. Along with cybercrime, AI can support cybercriminal activities without human intervention through, for example, automating fraud and data-based learning. In the context of CPS, recent works discussed advanced threats against CPS from a different level of sophistication: an indirect self-learning attack on well-hardened computing infrastructure (CI) by compromising the cyber-physical control systems, whereas another study presented a framework to build cyber-physical botnets attacking a water distribution system but without learning aspects involved in this attack model. Therefore, CPS is a potentially fruitful area for committing artificial intelligence crimes (AICs) due to the decision-making and interconnectivity features (Antonioli, D. et al., 2018).

Traditional CPS are systems that seamlessly integrate sensing, control, networking, and computational algorithms into physical components, connecting them to the Internet and each other. CPS has applications in energy, infrastructure, communication, healthcare, manufacturing, the military, robotics, physical security, building systems, and transportation. Integrating networked computational resources with physical systems that control various sensors and actuators impacts the environment. Advances in connectivity enable the evolution of scraps. The term sCPS refers to a new generation of embedded systems, which are increasingly interconnected, and their operations are dependent on software, such as industrial IoT. They are becoming more sophisticated with increased capabilities, which collect data from various sources to address real-world problems, such as traffic management (King, T C. et al., 2019).

Bures et al. defined sCPS as follows: "Smart Cyber-Physical Systems (sCPS) are modern CPS systems that are engineered to seamlessly integrate a large number of computation and physical components; they need to control entities in their environment smartly and collectively to achieve a high degree of effectiveness and efficiency."

A key to the "smartness" of those systems is their ability to handle complex tasks through the features of self-awareness, self-optimization, and self-adaptation. The feature of smartness becomes apparent from sCPS being highly connected, having cooperative behaviour with others, and being able to make effective decisions automatically. Bures et al. said that "most of the smartness is implemented in software, which makes the software one of the most complex and most critical constituents of sCPS." An outcome relates to highly sophisticated capabilities aimed at providing some degree of automation. Emerging technologies can be used to perform increasingly sophisticated functions in various sCPS, including smart healthcare systems, smart grids, smart buildings, autonomous automotive systems, autonomous ships, robots, smart homes, and intelligent transportation systems, with little or no human oversight. They represent the areas of innovation that are integrated into many CPS components to improve the quality of our lives (Khatoun, R. 2017).

METHODOLOGY- SVM, DLIB, CNN, RNN

The methodology is that to detect anything suspicious we first need to identify the facial expressions. To find the facial expressions, we first detect faces using LBP and Haar-cascade algorithms (Vishnupriya et al., 2018). Haar Cascade is a machine learning-based approach that trains cascade features with many positive and negative images. It requires many positive images (face images) and negative images (no faces).

For each feature calculation, we need to find the sum of the pixels under the white and black rectangles. The integral image was introduced to solve this. This makes it easy to calculate the number of pixels. This is the maximum number of pixels for an operation that has only 4 pixels associated with it. The best of 160000+ features is selected and it is achieved using Adaboost.In the LBP approach, the algorithm is first trained and a dataset is used with people's face images. In the greyscale, an image 8*3 grid is obtained. A threshold value is decided, if it's greater than the threshold value then it is assigned 1 else 0. Now, we have a binary matrix, concatenate them in a clockwise direction, convert to decimal form, and assign it to the central pixel (Alaa Eliyan, 2017).

As we have an image in the grayscale, each histogram will contain only 255 locations (0 to 255), which represent the events of each pixel intensity. Then, we have to change each histogram to create a new,w, and big histogram. The face is identified. Each histogram created is used to represent each image from the training set.LBP is fast (some times faster) but less accurate (less than 10-20% Haar). Therefore, if a face is not detected using LBP then we again perform face detection using Haar to prevent false results.

RESULT AND DISCUSSION

The result and discussion section elaborate the product perspective where the author team is presenting the website and snapshots of the app which will be used for suspicious activity detection.

Module 1: Face detection

Face detection is a computer technology being used in a variety of applications that identifies human faces in digital images. Face detection also refers to the psychological process by which humans locate and attend to faces in a visual scene.

Video Slicing

A cut is mostly referred to when removing a section of the video clip, so you cut in two places and lift out the middle section and join the leftover video back together. If you have a very long video, you might just cut it into sections to work with separately, and then you're splitting your video.

Figure 4. Frames extracted from the video

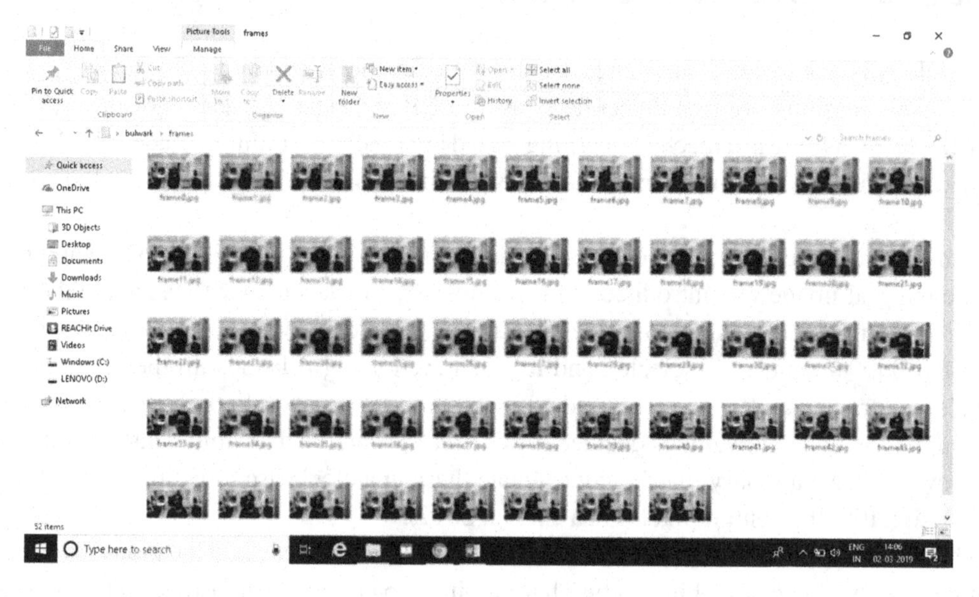

Detecting Faces

It resulted in 52 frames as every 6 frames were taken. From every frame, the face was detected. The module detected faces for 45 frames (didn't detect faces for frame numbers 6, 7, 8, 10, 14, 15, 16). Produced44 correct cases and 1 false case (for frame number 17). This was the best result found when the scale factor and min neighbours parameters for LBP were adjusted to 1.2 and 5 respectively. And for haar cascade, scale factor and min neighbours parameters were adjusted to 1.2 and 3 respectively (As per Figure 4 and Figure 5).

Figure 5. Faces detected from the extracted frames

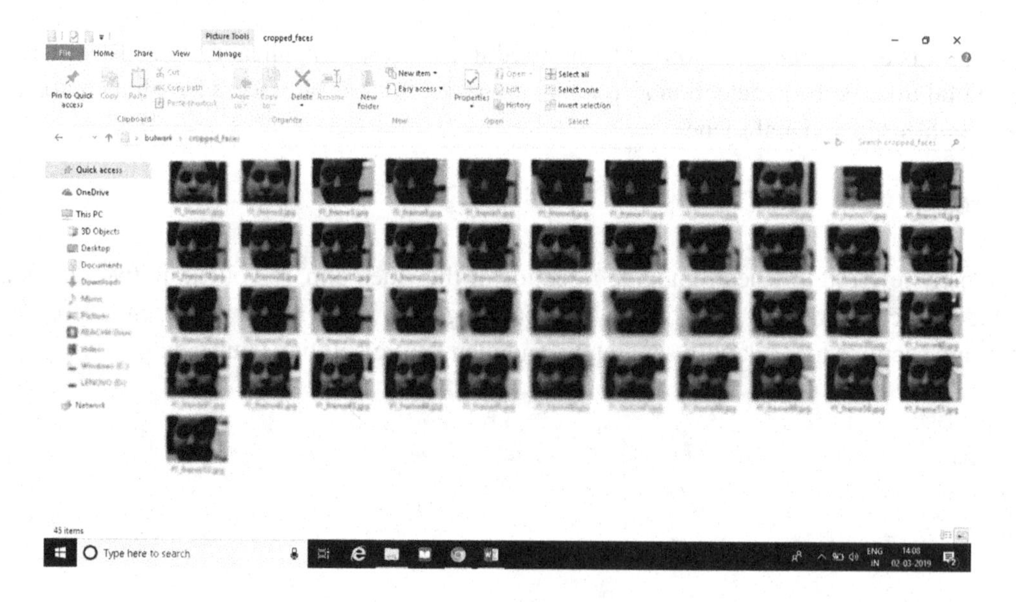

Module 2: Face Expression Recognition

Facial expression recognition software is a technology that uses biometric markers to detect emotions in human faces. More precisely, this technology is a sentiment analysis tool and can automatically detect the six basic or universal expressions: happiness, sadness, anger, surprise, fear, and disgust.

Raw dataset

For module 2 which is intended to recognize facial expression from the faces detected by module

Cohn Kanade's CK+ dataset is being used to train the model for prediction of face expression (As per Figure 6).

Each folder consists of similar kinds of images for the chon Canada data set(As per Figure 7).

Figure 6. Cohn Kanade dataset

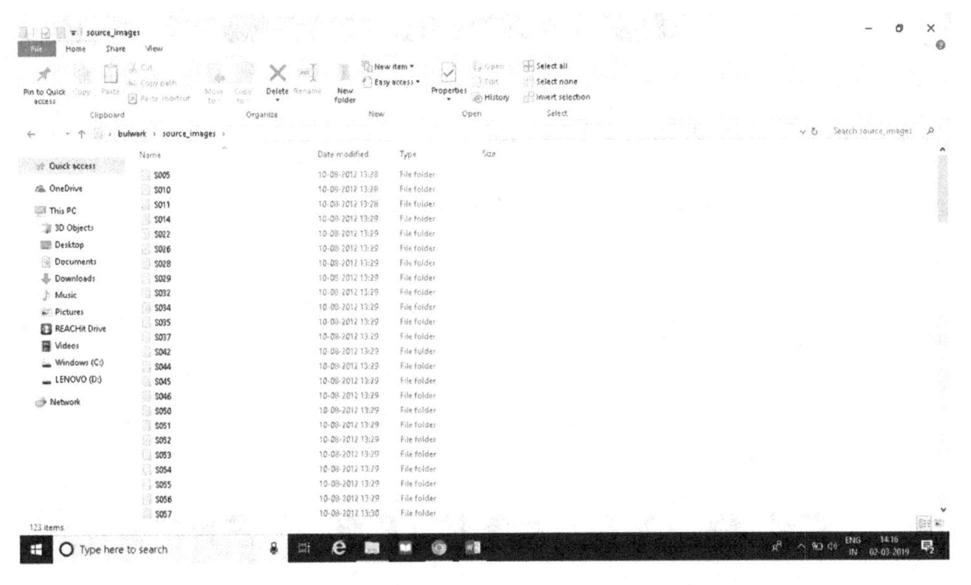

Figure 7. Each folder consists of similar kind of images

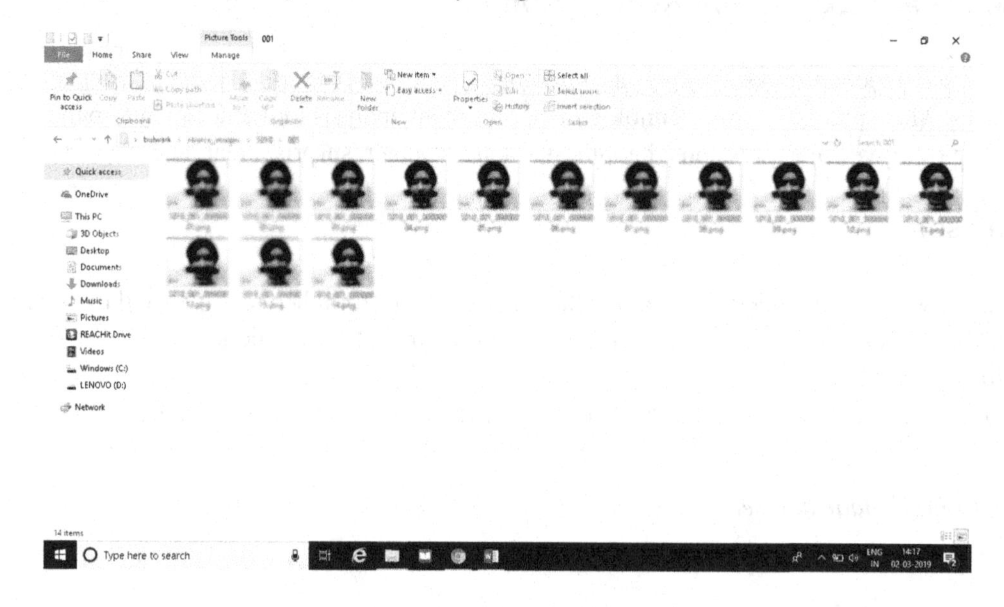

Sorting the Dataset For A Different Emotion

This raw dataset is first sorted and arranged into different folders to the corresponding expression (As per Figure 8).

Figure 8. Sorted dataset for 'disgust'

Training and Test Data

The training data is an initial set of data used to help a program understand how to apply technologies like neural networks to learn and produce sophisticated results. It may be complemented by subsequent sets of data called validation and testing sets.

Test data is data that has been specifically identified for use in tests, typically of a computer program. Some data may be used in a confirmatory way, typically to verify that a given set of input to a given function produces some expected result.

Figure 9. Dividing the sorted dataset into training and test data

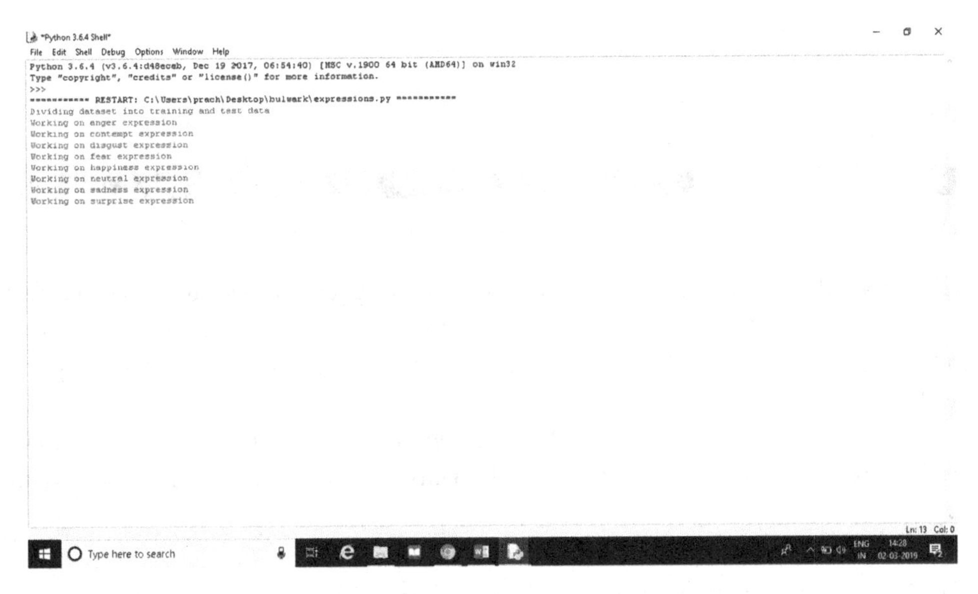

Then this dataset is divided into a ratio of 4:1 for training and testing respectively (As per Figure 9).

Training Support-Vector Machines(SVM) and Calculating Accuracies

In machine learning, SVM are supervised learning models with associated learning algorithms that analyze data used for classification and regression analysis.

Figure 10. The model gives an approximate accuracy of 82%

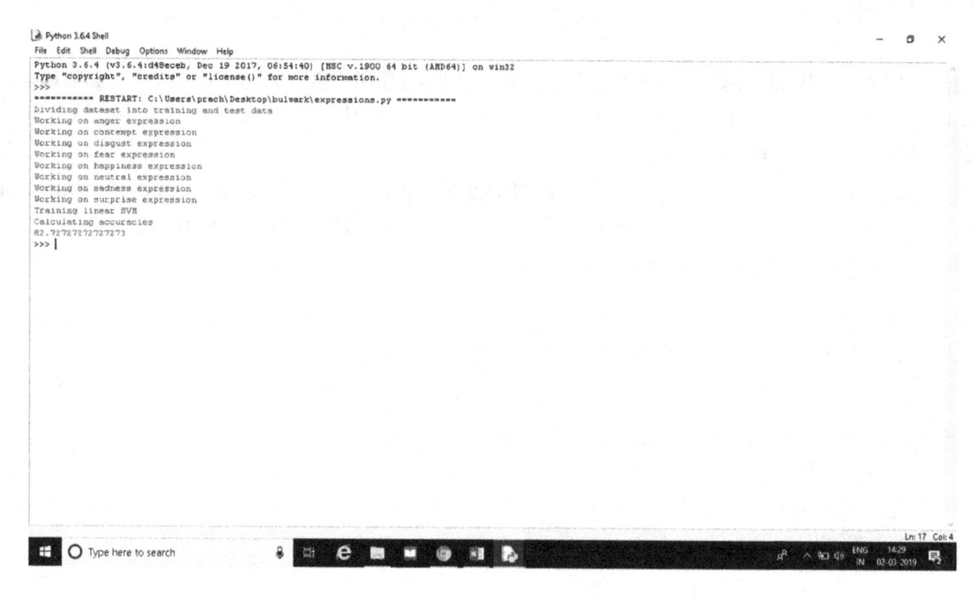

In the training set, the distance calculated from the centre of gravity of face to every coordinate of the face (found with the help of Dlib) is given. Here support vector classifier is used the parameters kernel and probability are set to linear and true respectively. The prediction score obtained is approximately 82% (As per Figure 10).

Some more datasets to added to improve the accuracy of this module: Dataset- faces94 from the University of Essex, U.K., Dataset- faces95 from the University of Essex, U.K(As per Figure 11 and Figure 12).

Figure 11. Dataset- faces94 from the University of Essex, U.K.

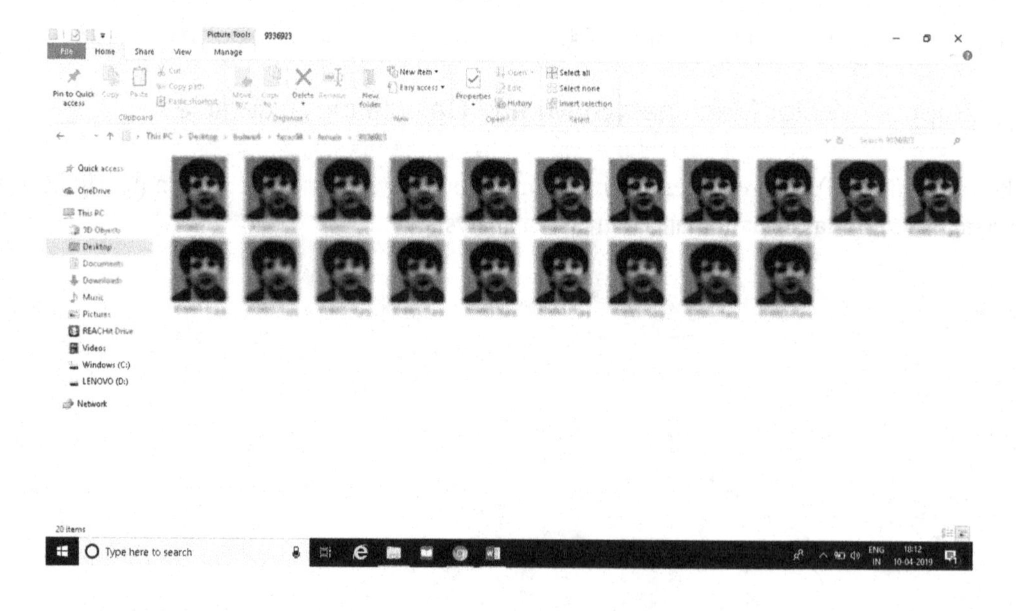

Figure 12. Dataset- faces95 from the University of Essex, U.K

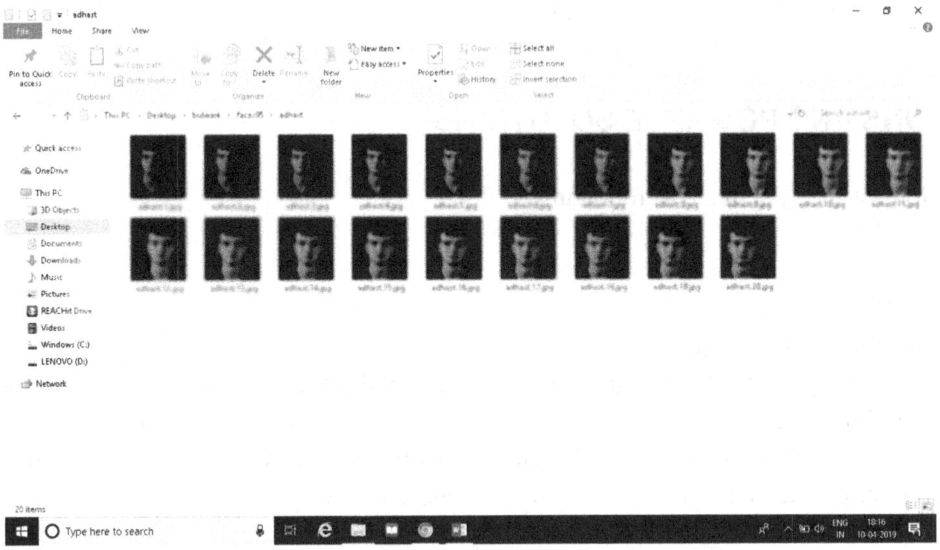

Result and discussion for every implemented module are mentioned below:

For module 1 which is intended to detect faces from a real-time video. The code was tested for a 10-second video which was 30 frames per second. For module 2 which is intended to recognize facial expression from the faces detected by module 1. Cohn Kanade's CK+ dataset is to be used to train the model for the prediction of facial expression (As per Figure 6). This raw dataset is first sorted and arranged into different folders to the corresponding expression (As per Figure 7). Then this dataset is divided into the ratio of 4:1 for training and testing respectively. In the training set, the distance calculated from the centre of gravity of face to every coordinate of the face (found with the help of Dlib) is given (As per Figure 8). Here support vector classifier is used the parameters kernel and probability are set to linear and true respectively. The prediction score obtained is approximately 82% (As per Figure 9).

CHAPTER CONTRIBUTION

We aim to develop a system which can detect suspicious activities, it will be developed by integrating the following modules:

1. Face detection
2. Face expression recognition
3. Object recognition
4. Speech recognition

Systems or software that have been developed so far are mostly based on individual modules that are mentioned above. But here we intend to develop an efficient system which can detect any kind of suspicious activity in the environment. The face detection module will detect the faces which are present in the scene to eliminate the background and make the process for further modules fast. The face expression recognition uses these faces to detect their expression and detects a possible combination of facial expression which would indicate suspicious activity.

LIMITATIONS AND POSSIBLE SOLUTIONS

The limitations of the system are as following:

Case 1

When there is more than 1 person

Suspicious activity, in most cases, takes place in crowded places but the facial expression recognition module will not be able to tell which expression belongs to which particular person. To solve this problem, we will need the object/face recognition system through which the system can distinguish between different people and can tell their facial expressions individually.

Case 2

Fake face expression.

There are situations when people may fake some facial expressions. This leads to wrong results. To solve this problem we may need the interference of the third person to verify the situation.

Here, the role of the third person will be played by police authority or security personnel.

APPLICATIONS

The device finds its applications in various situations: For the safety of solo travellers in a public conveyance in almost all public places (schools, colleges, etc.) (As per Figure 13).

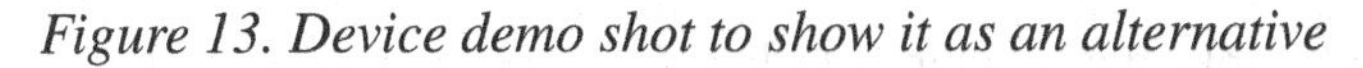

Figure 13. Device demo shot to show it as an alternative

Applications of AI technology and Cyber-Physical Systems (CPS) are increasing exponentially. However, framing resilient and correct smart contracts (SCs) for this smart application is a quite challenging task because of the complexity associated with them. SC is modernizing the traditional industrial, technical, and business processes. It is self-executable, self-verifiable, and embedded into the BC that eliminates the need for trusted third-party systems, which ultimately saves administration as well as service costs.

FUTURE RESEARCH DIRECTIONS

This system has a scintillating future scope that can overcome its limitations. Some of them are mentioned below:

A module of object recognition can be incorporated into this system. This would enhance the accuracy of the system and would give more reliable results. Through object recognition, we can detect objects which possess potential harm to society like a gun, knife, etc. This would also help to distinguish between two or more persons and assist the facial expression recognition module to assign face expression to the respective person.

CONCLUSION

The level of intelligence now embedded in our cars, homes, communication devices, consumer electronics, and other devices increases every day. Very soon, not only will humans interact with a rapidly growing array of smart products, but many of these products will interact autonomously with each other and other systems.

Moreover, factory production lines, process plants for energy and utilities, and smart cities are dependent on cyber-physical systems (CPS) to self-monitor; optimize; and even autonomously run infrastructure, transportation, and buildings. In the future, cyber-physical systems will rely less on human control and more on the intelligence embedded in the artificial intelligence (AI)-enabled core processors.

While manufacturers across all industrial sectors are ramping up to meet demand for this growing "smart product" market, they face major challenges developing and manufacturing these new and increasingly more complex products and systems. These cyber-physical systems require tight coordination and integration between the computational (virtual) and the physical (continuous) worlds. To meet these complexity and integration requirements, designers of cyber-physical intelligent systems are using advanced simulation platforms that cover model-based mechatronic systems engineering, embedded system design integration, and simulation models that validate product and system design in the physical world.

Crimes like kidnapping, murders, robbery, sexual assaults, etc. are increasing day by day in metro and semi-urban cities.

Figure 14. Map showing the comparative rate of violence against women in Indian states and Union Territories, 2012

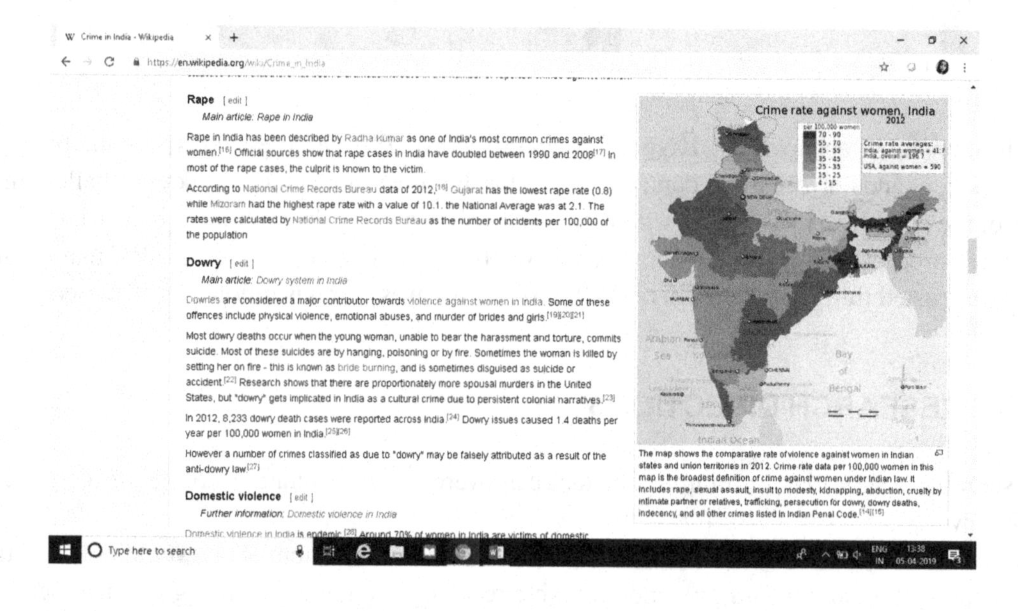

ACKNOWLEDGMENT

Our Sincere thanks to all direct and indirect supporters and wellwishers. Esteem gratitude to Management of Dev Sanskriti Vishwavidyalaya Hardwar and Shantikunj where the image processing experiments and study was conducted. We Are specially thanked for ABESEC, Ghaziabad management, and staff for their time contribution to be part of this study. We also convey gratitude to officers of the DayalBagh Educational Institute for their mentorship and timely valuable suggestions.

REFERENCES

Alaa, E. (2017). Comparative Study on Facial Expression Recognition using Gabor and Dual-Tree Complex Wavelet Transforms. *International Journal of Engineering & Applied Sciences, 9*(1).

Arman, S. (2017). Non-rigid registration based model-free 3D facial expression recognition Comput. *Vis. Image Underst., 162*, 146–165. doi:10.1016/j.cviu.2017.07.005

Bruunelli, R., & Poggio, T. (1993, October). Face Recognition: Features versus Templates. *IEEE Transactions on Pattern Analysis and Machine Intelligence, 15*(10), 1042–1052. doi:10.1109/34.254061

Bures, T., Weyns, D., Schmer, B., Tovar, E., Boden, E., Gabor, T., Gerostathopoulos, I., Gupta, P., Kang, E., Knauss, A., Patel, P., Rashid, A., Ruchkin, I., Sukkerd, R., & Tsigkanos, C. (2017). Software engineering for smart cyber-physical systems: Challenges and promising solutions. *Software Engineering Notes, 42*(2), 19–24. doi:10.1145/3089649.3089656

Cassell, J., Bickmore, T., Campbell, L., Vilhjalmsson, H., & Yan, H. (2000). Human Conversation As a System Framework: Designing Embodied Conversational Agents. In *Embodied conversational agents* (pp. 29–63). MIT Press. doi:10.7551/mitpress/2697.003.0004

Chowanda, A., Blanchfield, P., Flintham, M., & Valstar, M. (2016). Computational models of emotion, personality, and social relationships for interactions in games. *Proceedings of the 2016 International Conference on Autonomous Agents & Multiagent Systems,* 1343–1344.

Kaelbling, L. P., Littman, M. L., & Moore, A. W. (1996). Reinforcement learning: A survey. *Journal of Artificial Intelligence Research, 4*, 237–285. doi:10.1613/jair.301

Khatoun, R., & Zeadally, S. (2017). Cybersecurity and privacy solutions in smart cities. *IEEE Communications Magazine, 55*(3), 51–59. doi:10.1109/MCOM.2017.1600297CM

Krizhevsky, Sutskever, & Hinton. (2012). Imagenet Classification with Deep Convolutional Neural Networks. *Advances in Neural Information Processing Systems,* 1097–1105.

Liang, M., & Hu, X. (2015). Recurrent convolutional neural network for object recognition. *The IEEE Conference on Computer Vision and Pattern Recognition*. 3367-3375. 10.1109/CVPR.2015.7298958

Miles, B. (2018). The malicious use of artificial intelligence: Forecasting, prevention, and mitigation. Academic Press.

Nugrahaeni & Mutijarsa. (2017). Comparative analysis of machine learning KNN, SVM, and random forests algorithm for facial expression classification. *IEEE Xplore*.

Pradhan, A. (2012). Support vector machine-A survey. *International Journal of Emerging Technology and Advanced Engineering, 2*(8), 82–85.

Rastogi, Chaturvedi, Arora, Trivedi, & Singh. (2017). Role and efficacy of Positive Thinking on Stress Management and Creative Problem Solving for Adolescents. *International Journal of Computational Intelligence, Biotechnology and Biochemical Engineering, 2*(2).

Rastogi, Chaturvedi, Sharma, Bansal, & Agrawal. (2017). Understanding Human Behaviour and Psycho Somatic Disorders by Audio Visual EMG & GSR Biofeedback Analysis and Spiritual Methods. *International Journal of Computational Intelligence, Biotechnology and Biochemical Engineering, 2*(2).

Rastogi, Chaturvedi, Arora, Trivedi, & Mishra. (2018). Swarm Intelligent Optimized Method of Development of Noble Life in the perspective of Indian Scientific Philosophy and Psychology. *Journal of Image Processing andArtificial Intelligence, 4*(1). http://matjournals.in/index.php/JOIPAI/issue/view/463

Rastogi, Chaturvedi, Arora, Trivedi, & Chauhan. (2019). Framework for Use of Machine Intelligence on Clinical Psychology to Study the effects of Spiritual tools on Human Behavior and Psychic Challenges. *Journal of Image Processing and Artificial Intelligence, 4*(1).

Rastogi, R., Chaturvedi, D., Sharma, S., Bansal, A., & Agrawal, A. (2017). Audio-Visual EMG & GSR Biofeedback Analysis for Effect of Spiritual Techniques on Human Behaviour and Psychic Challenges. *Journal of Applied Information Science, 5*(2), 37-46. Retrieved from https://www.i-scholar.in/index.php/jais/article/view/167372

Rastogi, R., Chaturvedi, D. K., Arora, N., Trivedi, P., & Mishra, V. (2017). Swarm Intelligent Optimized Method of Development of Noble Life in the perspective of Indian Scientific Philosophy and Psychology. *Proceedings of NSC-2017 (National system conference) IEEE Sponsored conf. of Dayalbagh Educational Institute, Agra.*

Rastogi, R., Chaturvedi, D. K., Satya, S., Arora, N., Bansal, I., & Yadav, V. (2018). Intelligent Analysis for Detection of Complex Human Personality by Clinical Reliable Psychological Surveys on Various Indicators. *National Conference on 3rd Multi-Disciplinary National Conference Pre-Doctoral Research [MDNCPDR-2018] at Dayalbagh Educational Institute, Dayalbagh.*

Rastogi, R., Chaturvedi, D. K., Satya, S., Arora, N., Saini, H., Verma, H., Mehlyan, K., & Varshney, Y. (2018). Statistical Analysis of EMG and GSR Therapy on Visual Mode and SF-36 Scores for Chronic TTH. *Proceedings of International Conference on 5th IEEE Uttar Pradesh Section International Conference 2-4 Nov 2018 MMMUT Gorakhpur.* 10.1109/UPCON.2018.8596851

Rastogi, R., Chaturvedi, D. K., Satya, S., Arora, N., Singh, P., & Vyas, P. (2018). Statistical Analysis for Effect of Positive Thinking on Stress Management and Creative Problem Solving for Adolescents. *Proceedings of the 12th INDIACom; INDIACom-2018; IEEE Conference ID: 42835, Technically Sponsored by IEEE Delhi Section 2018, 5th International Conference on "Computing for Sustainable Global Development*, 245-251.

Rastogi, R., Chaturvedi, D. K., Satya, S., Arora, N., Yadav, V., & Chauhan, S. (2018). An Optimized Biofeedback Therapy for Chronic TTH between Electromyography and Galvanic Skin Resistance Biofeedback on Audio, Visual and Audio Visual Modes on Various Medical Symptoms. *National Conference on 3rd Multi-Disciplinary National Conference Pre-Doctoral Research [MDNCPDR-2018] at Dayalbagh Educational Institute.*

Ren, S., He, K., Girshick, R., & Sun, J. (2017). Faster R-CNN: Towards Real-Time Object Detection with Region Proposal Networks. *IEEE Transactions on Pattern Analysis and Machine Intelligence, 39*(6), 1137–1149. doi:10.1109/TPAMI.2016.2577031 PMID:27295650

Sharma & Rameshan. (2017). Dictionary Based Approach for Facial Expression Recognition from Static Images. *Int. Conf. Comput. Vision, Graph. Image Process.*, 39–49.

Singh, S., & Kaur, A. (2015, March). Face Recognition Technique using Local Binary Pattern Method. *International Journal of Advanced Research in Computer and Communication Engineering*, *4*(3).

Thomas, C. (2019, February 14). Artificial intelligence crime: An interdisciplinary analysis of foreseeable threats and solutions. *Science and Engineering Ethics*. Advance online publication. PMID:30767109

Toygar, O., & Acan, A. (2003). Face Recognition Using PCA, LDA, and ICA Approach on Colored Images. *Journal of Electrical and Electronics Engineering (Oradea)*, *3*(1), 735–743.

TrendsT. (2018). https://www.gartner.com/smarterwithgartner/gartner-top-10-strategic-technology-trends-for-2018

Tripathy, R., & Daschoudhary, R. N. (2014). Real-time face detection and Tracking using Haar Classifier on SoC. *International Journal of Electronics and Computer Science Engineering.*, *3*(2), 175–184.

Turing, A. M. (1950). Computing machinery and intelligence. *Mind*, *59*(236), 433–460. doi:10.1093/mind/LIX.236.433

Uřičář, M., Franc, V., & Hlaváč, V. (2012). The detector of facial landmarks learned by the structured output SVM. *VISAPP'12: Proceedings of the 7th International Conference on Computer Vision Theory and Applications*, 1, 547-556.

Viola, P., & Jones, M. (2001). Rapid object detection using a boosted cascade of simple features. *Computer Vision and Pattern Recognition.*, *1*, 511–518. doi:10.1109/CVPR.2001.990517

Viola, P., & Jones, M. J. (2004, May). Robust Real-Time Face Detection. *International Journal of Computer Vision*, *57*(2), 137–154. doi:10.1023/B:VISI.0000013087.49260.fb

Vishnu Priya, T. S., Vinitha Sanchez, G., & Raajan, N. R. (2018). Facial Recognition System Using Local Binary Patterns (LBP). *International Journal of Pure and Applied Mathematics*, *119*(15), 1895–1899.

Welinder, P., Branson, S., Mita, T., Wah, C., Schroff, F., Belongie, S., & Perona, P. (2010). *Caltech-UCSD Birds 200*. California Institute ofTechnology. CNS-TR-2010-001. 2010. https://towardsdatascience.com/face-recognition-how-lbph-works-90ec258c3d6b

Zhu, Chen, Zhao, Zhou, & Zhang. (2017). Emotion Recognition from Chinese Speech for Smart Affective Services Using a Combination of SVM and DBN. *New Advances in Identification Information & Knowledge on the Internet of Things*.

ADDITIONAL READING

Babiceanu, R. F., & Seker, R. (2017). Trustworthiness requirements for manufacturing cyber-physical systems. *Procedia Manufacturing*, *11*, 973–981. doi:10.1016/j.promfg.2017.07.202

Gunsekaran, A., & Ngai, E. W. T. (2014). Expert systems and Artiðcial Intelligence in the 21st-century logistics and supply chain management. *Expert Systems with Applications, 41*(1), 1–4. doi:10.1016/j.eswa.2013.09.006

Gupta, R., Tanwar, S., Al-Turjman, F., Italiya, P., Nauman, A., & Kim, S. W. (2020). Smart Contract Privacy Protection Using AI in Cyber-Physical Systems: Tools, Techniques and Challenges. *IEEE Access: Practical Innovations, Open Solutions, 8*, 24746–24772. doi:10.1109/ACCESS.2020.2970576

Klumpp, M. (2018). Innovation Potentials and Pathways Merging AI, CPS, and IoT. *Applied System Innovation., 1*(1), 5. doi:10.3390/asi1010005

Leitão, P., Karnouskos, S., Ribeiro, L., Lee, J., Strasser, T., & Colombo, A. (2016). Smart Agents in Industrial Cyber-Physical Systems. *Proceedings of the IEEE, 104*(5), 1086–1101. doi:10.1109/JPROC.2016.2521931

Mikusz, M. (2014). Towards an Understanding of Cyber-Physical Systems as Industrial Software-Product-ServiceSystems. *Procedia CIRP, 16*, 385–389. doi:10.1016/j.procir.2014.02.025

KEY TERMS AND DEFINITIONS

Face Analysis: A facial recognition system is a technology capable of identifying or verifying a person from a digital image or video frame from a video source. There are multiple methods in which facial recognition systems work, but in general, they work by comparing selected facial features from a given image with faces within a database.

Machine Learning: Machine learning (ML) is the scientific study of algorithms and statistical models that computer systems use to perform a specific task without using explicit instructions, relying on patterns and inference instead. It is seen as a subset of artificial intelligence.

Object Detection: It is a computer technology related to Computer Vision and image processing that deals with detecting instances of semantic objects of a certain class (such as humans, buildings, or cars) in digital images and videos. Well-researched domains of object detection include face detection and pedestrian detection.

Speech Recognition: Speech recognition is an interdisciplinary subfield of computational linguistics that develops methodologies and technologies that enables the recognition and translation of spoken language into text by computers. It is also known as automatic speech recognition (ASR), computer speech recognition, or speech to text (STT). It incorporates knowledge and research in the linf=guistics, computer Science and electrical engineering fields.

Suspicious: Having or showing a cautious distrust of someone or something or causing one to have the idea or impression that someone or something is questionable, dishonest, or dangerous or having the belief or impression that someone is involved in the illegal or dishonest activity.

SVM: In machine learning, support vector machines (SVMs, also support-vector networks) are supervised learning models with associated learning algorithms that analyze data used for classification and regression analysis.

Chapter 10
Nonlinear Filtering Methods in Conditions of Uncertainty

Rinat Galiautdinov
https://orcid.org/0000-0001-9557-5250
Independent Researcher, Italy

ABSTRACT

In the chapter, the author considers the possibility of applying modern IT technologies to implement information processing algorithms in UAV motion control system. Filtration of coordinates and motion parameters of objects under a priori uncertainty is carried out using nonlinear adaptive filters: Kalman and Bayesian filters. The author considers numerical methods for digital implementation of nonlinear filters based on the convolution of functions, the possibilities of neural networks and fuzzy logic for solving the problems of tracking UAV objects (or missiles), the math model of dynamics, the features of the practical implementation of state estimation algorithms in the frame of added additional degrees of freedom. The considered algorithms are oriented on solving the problems in real time using parallel and cloud computing.

INTRODUCTION

The growth and development of the IT sphere and the power of the computers allows to apply more advanced methods of information processing in the sphere of UAV and airplanes. The major algorithms for information processing in such the spheres are filtering and extrapolation of the parameters of the trajectories of objects upon their detection and tracking according to measurement data.

The greatest interest is represented by the implementation of numerical methods for nonlinear filtering of dynamic processes under conditions of uncertainty and insufficiency of a priori information since the quality of the state vector estimate directly affects the operation of a closed control system (Bain, A. & Crisan, D., 2009). To provide feedback in such a system, the control object is equipped with meters, with which you can determine the components of the state vector, while the measurements have limited accuracy. To improve the quality of state assessment, special algorithms are used that reduce the measurement noise, taking into account its parameters and the dynamics of the control object. In the case of

DOI: 10.4018/978-1-7998-7852-0.ch010

Copyright © 2023, IGI Global. Copying or distributing in print or electronic forms without written permission of IGI Global is prohibited.

nonlinear systems, for example, the methods of nonlinear Kalmanovskaya and Bayesian filtering, which are successfully used to solve problems of assessing the state of aircraft and other objects, are most popular.

Until recently, their practical value was small, since many algorithms were unrealizable due to the stringent requirements of real-time processing. Therefore, analytical approximations in the form of linear Kalman – Bucy filters, which make it possible to solve problems in real time, are most widely used in practice (Bergemann, K. & Reich, S., 2012).

The development of adaptive filtering methods for nonlinear dynamic processes described by a not fully known and time-varying mathematical model under the action of random perturbations with a changing family of probability densities is extremely important. One of the effective ways to solve the problem is the Bayesian algorithm. Since information on external influences is inaccurate for adaptive nonlinear filtering problems, such artificial intelligence methods as neural networks and fuzzy logic can serve as an excellent complement to traditional methods of statistical synthesis of adaptive nonlinear filtering systems. In this regard, the possibilities of using neural networks, fuzzy logic and knowledge bases when implementing numerical algorithms for adaptive nonlinear filtering to increase their efficiency are considered. Although an extended Kalman filter will be briefly considered here, we take the Bayesian approach to the synthesis of discrete nonlinear filtering algorithms, which allows us to completely solve the problem of both linear and nonlinear filtering in discrete time.

ADVANCED KALMAN FILTER

The extended Kalman filter uses the model of a stochastic continuous system:

$$x^{\cdot} = (x, t) + w \tag{1}$$

and discrete measurements:

$$z_k = h_k(x(t_k)) + y_k \tag{2}$$

where x is the state vector of the system, w is the noise of the system, y_k is the measurement noise.

The task of filtering is to find a measurement function z_k an estimate of the state vector of the x'_k system that minimizes the variance of the error $x'_{k\text{-}x}(t_k)$.

Let an estimate of the state vector x'_{k-1} be obtained at time t_{k-1}. Based on this estimate, the forecast of the state vector estimate $x'k$ '(a priori estimate) is built, then zk measurements and correction of the estimate a priori based on x'k'' measurement results (a posteriori estimate) are carried out (Wilson, R. C. & Finkel, L., 2009). An estimate a priori of the state vector x'k' is calculated by integrating the model equation (3):

$$\frac{dx'}{dt} = f(x', t)$$

with initial conditions $x'(t) = x'k\text{-}1$, $x'(0) = {}_{x0}$.

The a priori estimate for the covariance error matrix for the linearized equations in P'k increments *is* calculated as (4),(5), (6):

Pk'= φPk''-1φT +Q,

$$F = \frac{df(x,t)}{dx} \mid x - x'_{k-1}$$

φ = E + Fδt

With the initial conditions P'(t) = Pk''−1, P(0) = P0. *Where, Q is the co*variance m*at*rix of the noise of the system (Doucet, A. & Johansen, A., 2009). A posteriori estimate for the state vector and covariance matrix errors are constructed as follows:

x'k'' = x'k' + Kk(Zk − $_{Hkx}$'k') (7)

P_k'' = (I − $_{Kk}$Hk) Pk' (8)

Hk –linearized sensitivity matrix (9):

$$H_k = \frac{\partial h(x,t)}{\partial x} \mid x = x'_{k-1} \quad (9)$$

Kk –feedback correction matrix (10):

Kk = *Pk*' *H*Tk[*Hk Pk*' *H*Tk + *R*]-1, (10)

where *R* –covariance matrix of noise measurements.

Bayesian Approach to Solving Problems of Adaptive Nonlinear Filtering Under A Priori Uncertainty

For discrete observations, we write down nonlinear models of the processes of dynamics and measurements in the form of the following three difference vector equations (Doya, K., et al., 2007):

$$x_k = f_k(x_{k-1}, a_{k-1}) + g_k(x_{k-1})\xi_{k-1} \quad (11)$$

$$z_k = h_k(x_k, a_k) + u_k \quad (12)$$

$$a_k = \varphi_k(a_{k-1}) + \sigma_{k-1} \quad (13)$$

Where k is the moment in time;
Xk- is a stochastic n-dimensional vector;
ak - is the vector of accompanying parameters of dimension r;

fk(xk-1, ak-1) - in the general case, a *v*ector column function of dimension n non-linear with respect to its arguments;

gk(xk-1) - is a matrix function of dimension n * l;

îk-1 - is a sequence of independent l-dimensional random vectors of forming noise with density p(îk-1)

zk is the m-dimensional observation vector;

hk(xk,ak) - in the general case, a vector column function of dimension m non-linear with respect to its arguments;

uk - is a sequence of independent m-dimensional random measurement noise vectors with density p(uk);

ök(ak-1)- in the general case, a column vector function of dimension r non-linear with respect to its argument;

ók-1 - is a column of vector noise of the drift of parameters, which are purely random processes, has the dimension r and is described by the probability density p(ók-1);

the posterior probability density p(x0, a0) at the ze*ro* instant of time coincides with the a priori probability density ppr(x0, a0) of the *vectors x* and a.

The formulation of the adaptive estimation problem: it is necessary to obtain optimal estimates of the stochastic vector dynamic process xk from the available vector observations zk with a lack of *a* priori information about the vector of accompanying parameters ak and the characteristics of the noise of the measurement channel vk and the forming noise ók. Noises may depend on xk and ak. The vector ak, which is to be identified, takes into account a priori uncertainty in the specification of the model of the process being evaluated and perturbations (Galiautdinov, R., 2020).

Bayesian Methodology for Solving the Problem

Assume that the posterior probability density p(xk-1, ak-1 I z0k-1) for the time tk-1 is found. Here z0k-1 denotes a sequence of observations for time instants: t0, t1, ... tk-1. It is necessary to find the posterior probability density p(xk, ak I z0k-1) of the extrapolated values of xk and ak for the next time moment tk in the absence of the observation zk, the posterior probability density p(xk, ak I z0k) for the next time tk after obtaining the reference observations zk and a posterior probability density p(xk + 1, ak + 1 I $_z$0k) depending on p(xk, ak I z0k-1).

Accordingly, for the adaptive estimation problem, the following formulas were obtained for the recursive calculation of the conditional posterior probability density of the state vector of the system (14), (15), (16):

$$p(x_k, a_k \mid z_0^{\,k-1}) = \int_{-\infty}^{\infty} p\left[x_{pk} - f_{pk}\left(x_{pk-1}\right)\right] p(x_{k-1}, a_{k-1} \mid z_0^{\,k-1}) dx_{k-1} da_{k-1}$$

$$p(x_k, a_k \mid z_0^{\,k}) = c^{-1}(k)\, p\left[z_k - h_k\left(x_k, a_k\right)\right] \int_{-\infty}^{\infty} p\left[x_{pk} - f_{pk}\left(x_{pk-1}\right)\right] p(x_{k-1}, a_{k-1} \mid z_0^{\,k-1}) dx_{k-1} da_{k-1}$$

$$p(x_{k+1},a_{k+1}\mid z_0^{\,k})=c^{-1}(k)\int_{-\infty}^{\infty}p\left[x_{pk+1}-f_{pk+1}\left(x_{pk}\right)\right]p\left[z_k-h_k\left(x_k,a_k\right)\right]p(x_k,a_k\mid z_0^{\,k-1})dx_kda_k$$

Here we use the following notation for the extended state vector (17):

$$x_{pk}=\begin{bmatrix}x_k\\ a_k\end{bmatrix},f_{pk}=\begin{bmatrix}f_k\left(x_{k-1},a_{k-1}\right)\\ \varphi_k\left(a_{k-1}\right)\end{bmatrix},\xi_{pk}=\begin{bmatrix}\xi_k\\ \delta_k\end{bmatrix}$$

Knowing the posterior probability density allows you to find an estimate of the state vector of the system by any criterion, including the criterion of the minimum mean square error (Greaves-Tunnell, A., 2015). Formulas, in principle, completely solve the problem of both linear and nolinear filtering in discrete time, however, the solution of practical problems requires consideration of numerical algorithms in each specific case to obtain estimates and their corresponding covariance matrices (Huang, Y. & Rao, R. P. N., 2016). Since without considering specific practical examples, little can be said about the features of nonlinear filtering algorithms, in the future we will consider as an example the estimation of the motion parameters of maneuvering air objects (Kording, K. P. & Wolpert, D. M., 2004).

To solve the problem of tracking a maneuvering target, we take as a basis a variant of an adaptive algorithm for filtering the parameters of the trajectory of a maneuvering target based on the Bayesian approach (Sokoloski, S., 2015). A linear dynamic system is considered as a model of the target trajectory and linear measurements are used, and a linear Kalman filter is used to estimate the state vector (Moreno-Bote R., et al., 2011). Taking into account expressions (14) - (17), we generalize the result to a nonlinear dynamical system with nonlinear measurements and a nonlinear discrete optimal filter in order to obtain the maximum gain in accuracy in comparison with linear adaptive and non-adaptive models. As a model of indefinite perturbation, we choose a model of the semi-Markov process (Cappe, O., 2011).

The scalar perturbation a - is represented as a random process, the average value of which M[ak] changes stepwise, taking a number of fixed values (states) in the range from -amax to amax.

Transitions of the jump-like process from state i *to* state occur with probability ðij determined by a priori data. The residence time of the process in state i *until* the transition to state j *is* a random variable with an arbitrary distribution density p(*ti*). As a special case of the semi-Markov process, we can consider a pulsed random process, which is a sequence of rectangular video pulses with random amplitudes having a density p(*a*), with random durations and intervals between them having an arbitrary distribution density (Galiautdinov, R. et al., 2019A). The Bayesian approach to constructing an adaptive filtering algorithm will be considered initially for the case of continuous perturbation a.

The problem of optimal estimation of the vector parameter x'k for the quadratic loss function reduces to a weighted averaging of the estimates x'k (ak), which are the solution to the filtering problem for fixed values of ak (18):

$$x'_k=\int_{(x)}x_kp(x_k\mid z_0^{\,k})\mathrm{d}x_k=\int_{(x)}x_k\int_{(a)}p(x_k,a\mid z_0^{\,k})\mathrm{d}adx_k=\int_{(x)}x_k\int_{(a)}p(x_k,a\mid z_0^{\,k})p(a\mid z_0^{\,k})\mathrm{d}adx_k=\int_{(a)}x'(a)\,p(a\mid z_0^{\,k})\mathrm{d}a_k$$

where *(X)* is the space of possible values of the estimated parameter;

a- is the range of possible values of the perturbing parameter, *b*- is the $p(x_k \mid z_0^k)$ posterior probability density of the vector x_k according to the data of the $k + 1$-dimensional measurement sequence z_0^k,

c- is $p(x_k, a \mid z_0^k)$ - the posterior probability density of the vector x_k in the presence of the perturbing parameter a.

Estimates $x'_k(a)$ can be obtained in any way that minimizes the standard error of the mean square error, including using a recurrent linear Kalman filter (Galiautdinov, R. et al., 2019B).

The problem of optimal adaptive filtering will be solved if at each step the a posteriori probability density p of the perturbing parameter $p(a_k, z_0^k)$ is calculated. The calculation of this density from the sample of measurements z_0^k and its use for obtaining weighted estimates is the main feature of the adaptive filtering method, which we will use. In the case when the perturbing parameter takes only fixed values a_j $(j = -ì/2, ..., -1, 0, 1, ..., ì/2, ì \text{ is even})$, instead of expression (18), we can write (19):

$$x'_k = \sum_{j=-\mu/2}^{\mu/2} x'_k\left(a_{j,k}\right)P(a_{j,k} \mid z^k{}_0)$$

where $P(a_{j,k} \mid z_0^k)$ is the posterior probability of the event $-a_{j,k} = a_j$ according to the measurements of z_0^k.
To calculate the posterior probability $P(a_{j,k} \mid z_0^k)$ – we use the Bayes rule and consequently get (20):

$$P(a_{j,k} \mid z^k 0) = P_{k,j} = \frac{P(a_{j,k} \mid z^{k-1}{}_0)p(z_k \mid a_{j,k-1})}{\sum_{j=-\mu/2}^{\mu/2} P(a_{j,k} \mid z^{k-1}{}_0)p(z_k \mid a_{j,k-1})}$$

Here, according to the general scheme for applying the Bayes formula, the $k + 1$-dimensional sequence of measurements z_0^k is considered as event B.

In this expression, $p(x_k \mid z_0^{k-1})$ is the a priori probability of the parameter a_j at the k-step, obtained from the results of k-1 measurements and calculated by the formula (21):

$$P(a_{j,k} \mid z^{k-1}{}_0) = \sum_{j=-\mu/2}^{\mu/2} \pi_{ij} P(a_{i,k-1} \mid z^{k-1}{}_0)$$

Where $ð_{ij} = P(a_k = a_j \mid a_{k-1} = a_i)$ is the conditional probability of the perturbing process transitioning from state i to $(k-1)^{th}$ step in state j at the k-step;

$p(z_k \mid a_{j,k-1})$ is the conditional probability density of the measured value of the coordinate z_k, if the perturbing parameter at the previous step k-1 had the value a_j.

In view of (11), we obtain the following expression for the posterior probability $P_{k,j}$ (22):

$$P_{k,j} = \frac{\sum_{i=-\mu/2}^{\mu/2} \pi_{ij} P(a_{i,k-1} \mid z^{k-1}{}_0)p(z_k \mid a_{j,k-1})}{\sum_{j=-\mu/2}^{\mu/2}\sum_{i=-\mu/2}^{\mu/2} \pi_{ij} P(a_{i,k-1} \mid z^{k-1}{}_0)p(z_k \mid a_{j,k-1})}$$

The values of $P_{k,j}$ for each j are weighting coefficients when averaging the estimates of the filtered parameter.

In contrast to (13), to obtain the estimates $x'_k(a_k)$, we use the optimal discrete nonlinear filter (14) - (17).

In equations (14) - (17), the paratrometer a play the role of a “fixed state” of an indefinite disturbance; f_k, g_k, h_k are known.

We use formulas (14) - (17) to calculate estimates of the state of a maneuvering target.

It is believed that the posterior probability density $p[x_{k-1}(a_{j,k-1}) \mid z_0^{k-1}]$ for the time t_{k-1} is found. It is necessary to find the posterior probability density $p[x_k(a_{j,k}) \mid z_0^k]$ of the extrapolated values of x_k for the following time instants t_k in the absence of an observation count z_k, the posterior probability density $p[x_k(a_{j,k}) \mid z_0^k]$ for the next time t_k after obtaining the observation z_k.

The sought posterior probability density of the extrapolated values of x_k in the absence of an observation z_k in accordance with (4) has the following form (23):

$$p\left[x_k\left(a_{j,k}\right)\middle|z_0^{k-1}\right]=\int_{-\infty}^{\infty}p\left\{x_k\left[a_{j,k}\right]-f_k\left[x_{k-1}\left(a_{j,k-1}\right)\right]\right\}p\left[x_{k-1}\left(a_{j,k-1}\right)\middle|z_0^{k-1}\right]dx_{k-1}$$

The sought posterior probability density for the next time moment t_k after obtaining the observation count z_k in accordance with (5) has the form (24):

$$p\left[x_k\left(a_{j,k}\right)\middle|z_0^{k}\right]=c^{-1}(k)\,p\left\{z_k-h_k\left[x_k\left(a_{j,k}\right)\right]\right\}\int_{-\infty}^{\infty}p\left\{x_k\left[a_{j,k}\right]-f_k\left[x_{k-1}\left(a_{j,k-1}\right)\right]\right\}p\left[x_{k-1}\left(a_{j,k}\right)\middle|z_0^{k-1}\right]dx_{k-1}$$

The expression of the posterior density $p[x_{k+1}(a_{j,k+1})|z_0^k]$ depending on $p[x_k(a_{j,k})|z_0^{k-1}]$ according to (6) has the form (25):

$$p\left[x_{k+1}\left(a_{j,k+1}\right)\middle|z_0^{k}\right]=c^{-1}(k)\int_{-\infty}^{\infty}p\left\{x_{k+1}\left(a_{j,k+1}\right)-f_{k+1}\left[x_k\left(a_{j,k}\right)\right]\right\}p\left\{z_k-h_k\left[x_k\left(a_{j,k}\right)\right]\right\}p\left[x_k\left(a_{j,k}\right)\middle|z_0^{k-1}\right]dx_k$$

In formulas (24) - (25), c is the normalized constant.

To calculate the vector of estimates $x'_k(a_{j,k})$, we must use formula (10), as a result we obtain (26):

$$x'_k\left(a_{j,k}\right)=\int x_k\left(a_{j,k}\right)p[x_k\left(a_{j,k}\right)\mid z^k{}_0]dx_k$$

NUMERICAL METHODS FOR IMPLEMENTING ADAPTIVE NONLINEAR FILTERING ALGORITHMS

The main problems in the implementation of nonlinear filters is the integral (global) approximation of the conditional probability density and numerical integration. There are various numerical methods for representing probability density: replacing a continuous density function by a set of discrete points, approximation using Hermite polynomials, spline functions, Fourier series, and the use of poly-Gaussian distributions. We focus on numerical methods that are convenient for implementation on a computer. These include the convolution of functions and the grid method.

Using Convolution Functions and Fast Fourier Transform

An analysis of Bayesian recurrence formulas (4) - (6), (13) - (15) to calculate the posterior probability density shows that they are a convolution of two functions of the vector argument x. In accordance with the convolution theorem, the classical convolution of two functions $f(x)$ and $g(x)$ from the interval *(a, b)* is defined as (27):

$$\psi(y)=\int_a^b f(y-x)d(x)dx=\int_a^b g(y-x)f(x)dx$$

Function (27) is denoted by $f * g$.

For the numerical implementation of nonlinear filters based on expression (27), two discrete calculation schemes for nonlinear convolution can be proposed for discrete measurements to obtain optimal estimates. For convenience, we consider them for one-dimensional convolution and the discrete Fourier transform.

The First Scheme

The convolution can be quickly executed based on the Borel theorem and a pair of Fourier transforms. Let the Fourier transforms under consideration exist. Direct Fourier Transform (28):

$$c(\nu)=F\left|f(x)\right|\equiv\int_{-\infty}^{\infty} f(x)e^{-2\pi i\nu x}\mathrm{dx}=\sqrt{2\pi}C(2\pi\nu)=2\pi\Omega(\omega)$$

Note that the Fourier transform of the function f (x) is not only called c c(ν), but also C(ω) or Ω (ω): (29)

$$C(\omega)\equiv\frac{1}{\sqrt{2\pi}}\int_{-\infty}^{\infty} f(x)e^{-2i\omega x}\mathrm{dx}\equiv\frac{1}{\sqrt{2\pi}}c\left(\frac{\omega}{2\pi}\right)$$

$$\Omega(\omega)\equiv\int_{-\infty}^{\infty} f(x)e^{-i\omega x}\mathrm{dx}\equiv c\left(\frac{\omega}{2\pi}\right)$$

Where $\omega=2\pi\nu$ is the circular frequency, ν - is the cyclic frequency.

The function *f(x)*, having the Fourier transform c(ν), is the inverse Fourier transform $F^{-1}|\,c(\nu)\,|$ of the function c(ν) (30):

$$f(x)=\int_{-\infty}^{\infty}c(v)e^{2\pi ivx}\mathrm{d}v=\frac{1}{\sqrt{2\pi}}\int_{-\infty}^{\infty}C(\omega)e^{ivx}\mathrm{d}\omega=\frac{1}{2\pi}\int_{-\infty}^{\infty}\Omega(\omega)e^{ivx}\mathrm{d}\omega$$

Further, the Borel convolution theorem is used:

$$F[f_1(x)]\,F[f_2(x)]=F[f_1(x)*f_2(x)] \quad (31)$$

$$F[f_1(x)f_2(x)]=F[f_1(x)]*F[f_2(x)] \quad (32)$$

The dimension of the Fourier image is equal to the dimension *f(x)* times the dimension *x*. Integrals (28) - (30) exist (are converging to a finite value) only for the so-called "two-sided decaying processes". In numerical calculations, this contradiction is smoothed out, since in this case we are dealing only with processes of a limited duration, and the process itself in a given range must be set by its values in a limited number of points. In this case, the integration is replaced by summation, and instead of calculating the integral (29), they are limited to calculating the sum (33):

$$c\left[(k-1)\cdot\Delta v\right]=\Delta x\cdot\sum_{m=1}^{n}x\left[(m-1)\cdot\Delta x\right]\cdot e^{-i2\pi\cdot(k-1)\cdot(m-1)\cdot\Delta v\cdot\Delta x}$$

Here, in comparison with the integral (29), the following changes are made:

- the continuous integral is approximately replaced by a limited sum of the areas of the rectangles, one of the sides of which is equal to the discrete variable *Äx* with which the process values are presented, and the second to the instantaneous value of the process at the corresponding time;
- the continuous parameter *x* is replaced by its discrete values *(m-1)Äx*, where m is the point number from the beginning of the process;
- continuous values of the frequency í are replaced by its discrete values *(k-1)Äí*, where k is the number of the frequency value, and the frequency discrete is *Äí* = *1 / T*, where, T is the interval at which the process is specified;
- the differential *dx* is replaced by a limited increment of the parameter *Äx*. In solving practical problems, a finite number of samples N of the analog function is used, and in this case, a pair of Fourier transforms (19), (20) takes the form of the so-called discrete Fourier transform (DFT) [12] (24) (25):

$$c(k)=\sum_{m=0}^{n-1}f(m)\cdot e^{-i\left(\frac{2\pi}{n}\right)\cdot mk}$$

where $k=0,1,\ldots,n-1$

$$f(m)=\frac{1}{n}\sum_{k=0}^{n-1}c(k)\cdot e^{i\left(\frac{2\pi}{n}\right)\cdot(m-1)(k-1)}$$

where $m=0,1,\ldots,n-1$

Expression (24) defines the direct discrete Fourier transform, and expression (25) defines the inverse discrete Fourier transform. The discrete Fourier transform algorithm provides a large amount of computation. Elimination of redundant multiplication operations leads to the so-called fast Fourier transform (FFT).

So, in accordance with the Borel theorem, a convolution can be performed quickly, providing for the following sequence of actions:

1) using FFT, the calculation of the spectra *c(k)* and *G(k)* involved in the convolution of the functions *f(x)* and *g(x)*;
2) the calculation of the product of the spectra Z(k) = c(k)G(k);
3) using FFT, the inverse DFT is calculated from Z(k) = c(k)G(k), which is the desired result of the convolution y(k).

The second scheme. It is based on the convolution operation of two periodic sequences *f(k)* and *g(k)*, which is expressed by the formula (26)(27):

$$y(n)=\sum_{m=0}^{N}f(m)g(n-m),n=0,1,...,N-1$$

Or

$$y(n)=\sum_{m=0}^{N}f(n-m)g(m),n=0,1,...,N-1$$

Since the original functions *f(k)* and *g(k)* are not periodic, it is necessary to supplement the sequence of their samples with so many zero values so that when calculating the convolution, the values of the original functions would be taken only from the main period.

Let *f(k)* be an aperiodic sequence of length N_1, *g(k)* an aperiodic sequence of length N_2 samples. In this case, sequences of samples $f_1(k)$ and $g_1(k)$ are formed, each of the length $N_1 + N_2 - 1$ samples by including additional zero values (28) (29):

$$f_1(n)=\begin{cases}f(n), where\ n=0,1,\ldots,N_1-1\\0, where\ n=N_1,\ldots,N_1+N_2-1\end{cases}$$

$$g_1(n)=\begin{cases}g(n), where\ n=0,1,...,N_2-1\\0, where\ n=N_2,...,N_1+N_2-1\end{cases}$$

Moreover, the convolution of the sequences *f(k)* and *g(k)* is determined by *(N1 + N2-1)* - point convolution (30)(31):

$$y(n) = \sum_{m=0}^{N_1+N_2-1} f_1(m) g_1(n-m), where\ n = 0,1,\ldots,N_1+N_2-2$$

Or

$$y(n) = \sum_{m=0}^{N_1+N_2-1} f_1(n-m) g_1(m), where\ n = 0,1,\ldots,N_1+N_2-2$$

Computing such a convolution using FFT will require an $N_1 + N_2 - 2$ -point DFT. Despite the fact that the first method of convolution calculation involves triple DFT calculation, it is more economical than direct convolution calculation using formulas (26) and (27). Similarly, to expressions for one-dimensional convolution and FFT, expressions for multidimensional transformations can be written.

UAV Controlled Dynamics with Rotary Rotors

The UAV state vector with rotary rotors includes the position of the center of mass, speed, the quaternion of the orientation of the UAV body, and angular velocity (32):

$x = (r\ \upsilon\ q\ \grave{\text{U}})$T

The movement of the apparatus is described by equations (33) (34) (35) (36):

$$r' = v$$

$$v' = q \circ \left(\sum_{i=1}^{4} qBR_i \circ (-1)^{i+1} k\hat{I}_i \left|\hat{I}_i\right| \circ q'BR_i \right) \circ q'$$

$$q' = \frac{1}{2} q \circ \Omega$$

$$\Omega' = J_B^{-1} \left[-\Omega \times J_B \Omega + \sum_{i=1}^{4} r_i^B \times (-1)^{i+1} k\hat{I}_i \left|\hat{I}_i\right| e_{zi}^I - \sum_{i=1}^{4} qBR_i \circ b\hat{I}_i \left|\hat{I}_i\right| e_{ri}^{Ri} \circ q'BR_i + \sum_{i=1}^{4} qBR_i \circ \left(J_{Ri}\hat{I}_i^{Ri} + \hat{I}_i^{Ri} + JR_i\omega_i^{Ri} \right) \circ q'BR_i \right]$$

where

M - is the total mass of the apparatus,
g – is the acceleration of gravity,

e^I_{zi} – unit vector along the axis of symmetry of the *i*th rotor,
k and *b* – is aerodynamic propeller coefficients,
ω'i – is the rotation speed of the *i*-th propeller,
JB — is the inertia tensor of the body in the main axes of the body,
*r*Bi – is the radius-vector, drawn from the center of mass of the *i*-th UAV towards the rotor,
*e*Bri – is the unit vector of *r*Bi,
qB^R_i – *is* a quaternion that determines the orientation of the *i*-th rotor with respect to the body of the apparatus,
JRi – is the inertia tensor of the *i*-th rotor with a propeller written in its own main axes,
$\omega i\,Ri$ – is the total angular velocity of the *i*-th rotor with a propeller recorded in its own main axes.

To ensure independent control by position and orientation, regulators of the form (37) are used:

$$r''_d(t) = r''^0(t) + K_{r1}\left(r'^0(t) - r'(t)\right) + K_{r2}\delta r$$

$$\Omega'_d(t) = \Omega'^0(t) + K_{\Omega 1}\left(\Omega^0(t) - \Omega(t)\right) + K_{\Omega 2}\delta q$$

where δq – is a vector part of quaternion mismatch,
δr – is a mismatch of the current position,
r''0, '0, Ù''0, Ù'0 –they are target acceleration, velocity, angular acceleration and angular velocity,
*Kr*1, *Kr*2, *K*Ù1, *K*Ù2 – is a diagonal matrix of coefficients.

Maintaining the values of the revolutions ωi and the orientation *q*BR*i* of the engines satisfying the outputs of the regulators (10) according to the model (12) ensures the achievement of the target motion parameters.

Measurement Noise Parameters

Table 1. Measurement noise parameters

σ_{rx}	1 m
σ_{rz}	2 m
σ_v	0.5 m/s
σ_α	0.5°
σ_β	0.5°
σ_γ	1.5°
σ_Ω	0.6°/s

In this paper, we do not consider the solution of the problem of the inverse dynamics of the UAV and the synthesis of the control loop, but it is believed that the method for determining the control parameters is known.

EXPERIMENTAL PART

To determine the motion path, the UAV model (9) is numerically integrated, while the control parameters $\omega' i$, $qBRi$ are some $_{\text{fu}}$nctions $_{\text{of}}$ the current state and the target path. The integration of the model is carried out using the MATLAB package of the 4th order Runge – Kutt method with a step of δ = 10-2 s. The measurement v^{ec}tor is composed in a manner similar to the state vector (32) (38):
$z = (r\ \upsilon\ q\ \grave{\text{U}})$

The measurements are modeled by adding to the results of integration of the equations of motion (9) white Gaussian noise, the parameters of which are selected in such a way as to correspond to the parameters of existing devices.

The standard deviations of the noise measurements of horizontal position components, vertical position components, velocity components, pitch angles, roll, yaw and angular velocity components are given in the Table 1:
As a model of aircraft dynamics in nonlinear filtering algorithms, a simplified model of UAV motion is used, which does not take into account the inertia of rotary rotors with propellers. Then the expression for angular acceleration (33) can be written as (39):

$$\Omega' = J_B^{-1}\left[-\Omega \times J_B\Omega + \sum_{i=1}^{4} r_i^B \times (-1)^{i+1} k\hat{I}_i |\hat{I}_i| e_{zi}^I - \sum_{i=1}^{4} qBR_i \circ b\hat{I}_i |\hat{I}_i| e_{ri}^{Ri} \circ q' BR_i\right]$$

This simplification is often resorted to in practice, since when designing UAVs with rotary rotors, even approximate identification of the main parameters of the dynamics of the executive organs of the control system is a laborious process and requires special measurements. Thus, the expressions associated with the dynamics of the executive bodies of the control system form the noise vector w of the system from the equation of the stochastic continuous system (1).

As the parameter determining the filter performance, the standard deviation of the components of the state estimation vector from the results of integration of the equations of motion was selected. To exclude the influence of the target trajectory on the experimental results, the standard deviation is averaged over 100 spans over the characteristic trajectories.

The duration of each flight mission is 90s, during this time the UAV moves along a curved path in space, and its body unfolds according to the target orientation parameters at the current moment. The maximum permissible speed of the UAV is limited to 5 m/s, and the angular speed is 3m/s.

All three filters use the same measurements and work on the same set of paths.

The performance of each of the filtering algorithms is studied for various observation intervals that are multiples of the system integration step, that is, measurements and an assessment of the state are performed at intervals $TN = N\delta$.

The m*atrices Q* covariance of the system noise and R covariance of the measurement noise used in all three algorithms were chosen taking into account the knowledge about the parameters of the measurement noise and are adjusted so that the algorithms show high performance for the measurement interval $T1 = \delta$ (40):

$Q = 10\text{-}6\text{diag}(1, 1, 1, 1, 1, 1, 1, 1, 1, 1, 50, 50, 50),$

$R = \text{diag}(1, 1, 1, 0.5, 0.5, 0.5, 0.003, 0.003, 0.003, 0.006, 0.003, 0.003, 0.003).$

Figure 1. Demo of flight

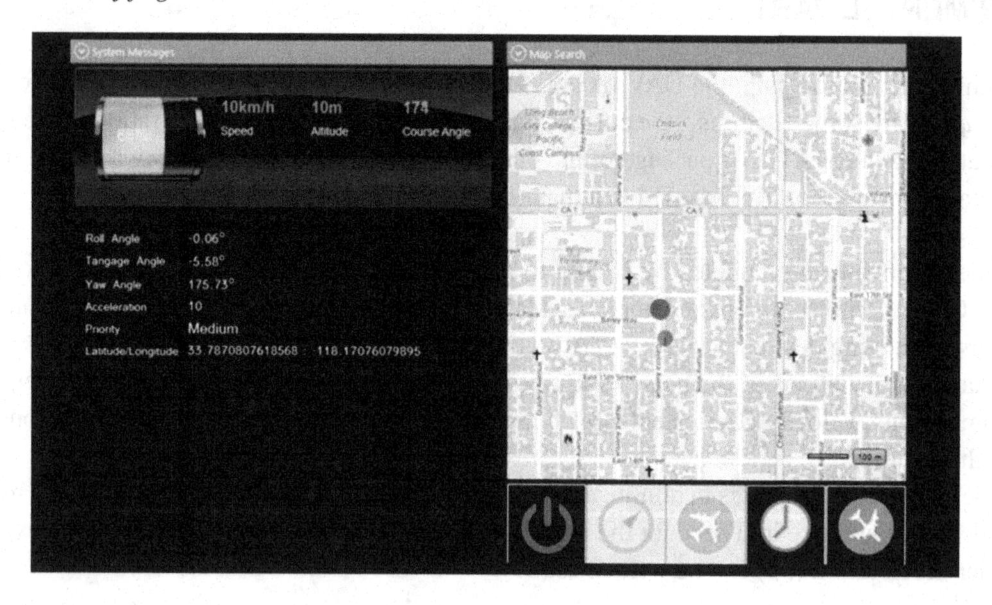

The experiments showed that the quality of the estimation of the state vector correlates with the interval of operation of the filters — a decrease in the measurement frequency leads to a deterioration of the estimation parameters. The extended Kalman filter shows the lowest performance at large measurement intervals: even for $N > 2$, state estimation errors in some cases are beyond the acceptable range, and for $N > 5$, the position estimate becomes incorrect.

The performance of the Kalman sigma-point and cubature filters also decreases with increasing N, but not so noticeably, while the failures in the operation of these algorithms are much smaller. A comparative analysis of the UKF and CKF results showed that the CKF algorithm makes a more accurate state assessment in most of the cases under consideration and is more resistant to increasing the measurement interval.

The best result was shown by an algorithm based on Bayesian filters.

As a practical implementation the described above approach was applied in the AI of the Drone Management System developed by Rinat Galiautdinov, where the additional details are described at http://airhighway.online.

CONCLUSION

Here the author considered the possibility of applying modern IT technologies to implement information processing algorithms in UAV motion control system. Filtration of coordinates and motion parameters of objects under a priori uncertainty is carried out using nonlinear adaptive filters: Kalman and Bayesian filters. The author considers numerical methods for digital implementation of nonlinear filters based on the convolution of functions, the possibilities of neural networks and fuzzy logic for solving the problems of tracking UAV objects (or missiles), the math model of dynamics, the features of the practical implementation of state estimation algorithms in the frame of added additional degrees of freedom. The considered algorithms are oriented on solving the problems in real time using parallel and cloud computing.

A comparative analysis of the performance of the extended Kalman filter and the Bayesian filter for various measurement intervals was carried out (Galiautdinov, R., 2019). As a quality criterion, the standard deviation of the state estimate averaged over 100 experiments from the trajectory of the model for each of the components of the state vector is used. It is shown that extended Kalman filter has significant sensitivity to measurement intervals, its increase leads to a rapid increase in the error of state estimation and, subsequently, to completely incorrect estimation of some components of the state vector.

The extended Kalman filter will be briefly considered here, mostly focusing on the Bayesian approach to the synthesis of discrete nonlinear filtering algorithms, which allows us to completely solve the problem of both linear and nonlinear filtering in discrete time.

The Bayesian approach was applied, and this allowed to obtain a general solution to the nonlinear filtering problem in discrete time. Based on the Bayesian methodology, formulas are obtained for the recursive calculation of the conditional posterior probability density of the extended state vector, which includes the state vector of the dynamic system and the vector of accompanying parameters, which takes into account the uncertainty in the statistical characteristics of the dynamic process and real perturbations, the inaccuracy in setting the mathematical parameters models. For the numerical implementation of algorithms using the grid method and convolution of functions, it is proposed to use Kalman filter estimates or estimates generated by a neural network as a reference solution. The problem of estimating the motion parameters of maneuvering objects based on neural networks and the use of knowledge bases and fuzzy logic to increase the efficiency of filtering algorithms are considered. With the practical implementation of parallel and pipelined computing, neurocomputers, it becomes possible to increase the accuracy and reliability of evaluations in real-time implementation of optimal nonlinear filtering algorithms. They provide developers of devices and systems with greater accuracy in estimating the parameters of dynamic processes. The numerical methods of Bayesian adaptive filtering algorithms considered in the paper allow almost unlimited parallelization and pipelining of computational processes. For the numerical implementation of nonlinear filtering methods, effective methods of parallel programming can be applied here. In addition, to increase the speed of information processing during programming, it is advisable to use dynamic data structures such as lists, trees, table functions, hashing, high-speed sorting algorithms and an arithmetic machine.

REFERENCES

Bain, A., & Crisan, D. (2009). Fundamentals *of Stochastic Filtering. Springer.* doi:10.1007/978-0-387-76896-0

Beck, J. M., & Pouget, A. (2007). Exact inferences in a neural implementation of a hidden Markov model. Neural C*omputation, 19(5), 1344*–1361. doi:10.1162/neco.2007.19.5.1344 PMID:17381269

Bergemann, K., & Reich, S. (2012). An ensemble Kalman-Bucy filter for continuous data assimilation. Meteo*rologische Zeitschrift (Berlin), 21(*3), 213–219. doi:10.1127/0941-2948/2012/0307

Berkes, P., Orban, G., Lengyel, M., & Fiser, J. (2011). Spontaneous Cortical Activity Reveals Hallmarks of an Optimal Internal Model of the Environment. S*cience, 331(6*013), 83–87. doi:10.1126cience.1195870 PMID:21212356

Beskos, A., Crisan, D., & Jasra, A. (2014). On the Stability of Sequential Monte Carlo Methods in High Dimensions. *PLoS Computational Biology, 46.*

Beskos, A., Crisan, D., Jasra, A., Kamatani, K., & Zhou, Y. (2014). *A Stable Particle Filter in High-Dimensions.* ArXiv:1412.350.

Cappe, O. (2011). Online EM Algorithm for Hidden Markov Models. *Journal of Computational and Graphical Statistics*, *20*(3), 1–20. doi:10.1198/jcgs.2011.09109

Churchland, A. K., Kiani, R., Chaudhuri, R., Wang, X.-J., Pouget, A., & Shadlen, M. N. (2011). Variance as a signature of neural computations during decision making. *Neuron*, *69*(4), 818–831. doi:10.1016/j.neuron.2010.12.037 PMID:21338889

Churchland, M. M., Yu, B. M., Cunningham, J. P., Sugrue, L. P., Cohen, M. R., Corrado, G. S., Newsome, W. T., Clark, A. M., Hosseini, P., Scott, B. B., Bradley, D. C., Smith, M. A., Kohn, A., Movshon, J. A., Armstrong, K. M., Moore, T., Chang, S. W., Snyder, L. H., Lisberger, S. G., ... Shenoy, K. V. (2010). Stimulus onset quenches neural variability: A widespread cortical phenomenon. *Nature Neuroscience*, *13*(3), 369–378. doi:10.1038/nn.2501 PMID:20173745

Crisan, D., & Xiong, J. (2010). Approximate McKean–Vlasov representations for a class of SPDEs. *Stochastics*, *82*(1), 53–68. doi:10.1080/17442500902723575

Deneve, S., Duhamel, J.-R., & Pouget, A. (2007). Optimal sensorimotor integration in recurrent cortical networks: A neural implementation of Kalman filters. *The Journal of Neuroscience: The Official Journal of the Society for Neuroscience*, *27*(21), 5744–5756. doi:10.1523/JNEUROSCI.3985-06.2007 PMID:17522318

Doucet, A., Godsill, S., & Andrieu, C. (2000). On sequential Monte Carlo sampling methods for Bayesian filtering. *Statistics and Computing*, *10*(3), 197–208. doi:10.1023/A:1008935410038

Doucet, A., & Johansen, A. (2009). A tutorial on particle filtering and smoothing: Fifteen years later. Handbook of Nonlinear Filtering, 4–6.

Doya, K. (2007). *Bayesian Brain: Probabilistic Approaches to Neural Coding*. The MIT Press.

Ernst, M. O., & Banks, M. S. (2002). Humans integrate visual and haptic information in a statistically optimal fashion. *Nature*, *415*(6870), 429–433. doi:10.1038/415429a PMID:11807554

Fiser, A., Mahringer, D., Oyibo, H. K., Petersen, A. V., Leinweber, M., & Keller, G. B. (2016). Experience-dependent spatial expectations in insect visual cortex. *Nature Neuroscience*, *19*(12), 1658–1664. doi:10.1038/nn.4385 PMID:27618309

Fiser, J., Berkes, P., Orban, G., & Lengyel, M. (2010). Statistically optimal perception and learning: From behavior to neural representations. *Trends in Cognitive Sciences*, *14*(3), 119–130. doi:10.1016/j.tics.2010.01.003 PMID:20153683

Galiautdinov, R. (2019). Advanced method of planning the trajectory of swarm's drones possessing Artificial Intelligence. *International Journal of Scientific Research.* Advance online publication. doi:10.21275/ART20203362

Galiautdinov, R. (2020). Brain machine interface: The accurate interpretation of neurotransmitters' signals targeting the muscles. *International Journal of Applied Research in Bioinformatics, 10*(1), 26–36. Advance online publication. doi:10.4018/IJARB.2020010102

Galiautdinov, R., & Mkrttchian, V. (2019A). Math model of neuron and nervous system research, based on AI constructor creating virtual neural circuits: Theoretical and Methodological Aspects. In V. Mkrttchian, E. Aleshina, & L. Gamidullaeva (Eds.), *Avatar-Based Control, Estimation, Communications, and Development of Neuron Multi-Functional Technology Platforms* (pp. 320–344). IGI Global. doi:10.4018/978-1-7998-1581-5.ch015

Galiautdinov, R., & Mkrttchian, V. (2019B). Brain machine interface – for Avatar Control & Estimation in Educational purposes Based on Neural AI plugs: Theoretical and Methodological Aspects. In V. Mkrttchian, E. Aleshina, & L. Gamidullaeva (Eds.), *Avatar-Based Control, Estimation, Communications, and Development of Neuron Multi-Functional Technology Platforms* (pp. 345–360). IGI Global. doi:10.4018/978-1-7998-1581-5.ch016

Greaves-Tunnell, A. (2015). *An optimization perspective on approximate neural filtering* [Master thesis].

Hennequin, G., Aitchison, L., & Lengyel, M. (2014). Fast Sampling-Based Inference in Balanced Neuronal Networks. *Advances in Neural Information Processing Systems.*

Huang, Y. & Rao, R., (2014). Neurons as Monte Carlo Samplers: Bayesian Inference and Learning in Spiking Networks. *Neural Information Processing Systems*, 1–9.

Huang, Y., & Rao, R. P. N. (2016). Bayesian Inference and Online Learning in Poisson Neuronal Networks. *Neural Computation, 28*(8), 1503–1526. doi:10.1162/NECO_a_00851 PMID:27348304

Kantas, N., Doucet, A., Singh, S. S., Maciejowski, J., & Chopin, N. (2015). On Particle Methods for Parameter Estimation in State-Space Models. *Statistical Science, 30*(3), 328–351. doi:10.1214/14-STS511

Kappel, D., Habenschuss, S., Legenstein, R., & Maass, W. (2015). Network Plasticity as Bayesian Inference. *PLoS Computational Biology, 11*(11), 1–31. doi:10.1371/journal.pcbi.1004485 PMID:26545099

Knill, D. C., & Pouget, A. (2004). The Bayesian brain: The role of uncertainty in neural coding and computation. *Trends in Neurosciences, 27*(12), 712–719. doi:10.1016/j.tins.2004.10.007 PMID:15541511

Kording, K., Tenenbaum, J. B., & Shadmehr, R. (2007). The dynamics of memory as a consequence of optimal adaptation to a changing body. *Nature Neuroscience, 10*(6), 779–786. doi:10.1038/nn1901 PMID:17496891

Kording, K. P., & Wolpert, D. M. (2004). Bayesian integration in sensorimotor learning. *Nature, 427*(6971), 244–247. doi:10.1038/nature02169 PMID:14724638

Legenstein, R., & Maass, W. (2014). Ensembles of Spiking Neurons with Noise Support Optimal Probabilistic Inference in a Dynamically Changing Environment. *PLoS Computational Biology, 10*(10), e1003859. doi:10.1371/journal.pcbi.1003859 PMID:25340749

Ma, W. J., Beck, J. M., Latham, P. E., & Pouget, A. (2006). Bayesian inference with probabilistic population codes. *Nature Neuroscience*, *9*(11), 1432–1438. doi:10.1038/nn1790 PMID:17057707

MacKay, D. J. (2005). *Information Theory, Inference and Learning Algorithms*. Cambridge University Press.

Makin, J. G., Dichter, B. K., & Sabes, P. N. (2015). Learning to Estimate Dynamical State with Probabilistic Population Codes. *PLoS Computational Biology*, *11*(11), 1–28. doi:10.1371/journal.pcbi.1004554 PMID:26540152

Mkrttchian, V., Gamidullaeva, L., & Galiautdinov, R. (2019). Design of Nano-scale Electrodes and Development of Avatar-Based Control System for Energy-Efficient Power Engineering: Application of an Internet of Things and People (IOTAP) Research Center. *International Journal of Applied Nanotechnology Research*, *4*(1), 41–48. Advance online publication. doi:10.4018/IJANR.2019010104

Mongillo, G., & Deneve, S. (2008). Online learning with hidden markov models. *Neural Computation*, *20*(7), 1706–1716. doi:10.1162/neco.2008.10-06-351 PMID:18254694

Moreno-Bote, R., Knill, D. C., & Pouget, A. (2011). Bayesian sampling in visual perception. *Proceedings of the National Academy of Sciences of the United States of America*, *108*(30), 12491–12496. doi:10.1073/pnas.1101430108 PMID:21742982

Movellan, J. R., Mineiro, P., & Williams, R. J. (2002). A Monte Carlo EM approach for partially observable diffusion processes: Theory and applications to neural networks. *Neural Computation*, *14*(7), 1507–1544. doi:10.1162/08997660260028593 PMID:12079544

Orban, G., Berkes, P., Fiser, J., & Lengyel, M. (2016). Neural Variability and Sampling-Based Probabilistic Representations in the Visual Cortex. *Neuron*, *92*(2), 530–543. doi:10.1016/j.neuron.2016.09.038 PMID:27764674

Pecevski, D., Buesing, L., & Maass, W. (2011). Probabilistic inference in general graphical models through sampling in stochastic networks of spiking neurons. *PLoS Computational Biology*, *7*(12), 7. doi:10.1371/journal.pcbi.1002294 PMID:22219717

Poyiadjis, G., Doucet, A., & Singh, S. S. (2011). Particle approximations of the score and observed information matrix in state space models with application to parameter estimation. *Biometrika*, *98*(1), 65–80. doi:10.1093/biomet/asq062

Rebeschini, P., & Van Handel, R. (2015). Can local particle filters beat the curse of dimensionality? *Annals of Applied Probability*, *25*(5), 2809–2866. doi:10.1214/14-AAP1061

Sokoloski, S. (2015). *Implementing a Bayes Filter in a Neural Circuit: The Case of Unknown Stimulus Dynamics*. ArXiv, ArXiv:1512.07839.

Surace, S. C., & Pfister, J.-P. (2016). *Online Maximum Likelihood Estimation of the Parameters of Partially Observed Diffusion Processes*. ArXiv:1611.00170.

Welling, M., & Teh, Y. (2011), Bayesian Learning via Stochastic Gradient Langevin Dynamics. *Proceedings of the 28th International Conference on Machine Learning*.

Wilson, R. C., & Finkel, L. (2009). A neural implementation of the Kalman filter. *Advances in Neural Information Processing Systems*, 22.

Yang, T., Laugesen, R. S., Mehta, P. G., & Meyn, S. P. (2016). Multivariable feedback particle filter. *Automatica*, *71*, 10–23. doi:10.1016/j.automatica.2016.04.019

Yang, T., Mehta, P. G., & Meyn, S. P. (2013). Feedback particle filter. *IEEE Transactions on Automatic Control*, *58*(10), 2465–2480. doi:10.1109/TAC.2013.2258825

Compilation of References

A History Of Flexible Manufacturing Systems. (2017). Retrieved January 20, 2018, from https://www.ukessays.com/essays/information-technology/a-history-of-flexible-manufacturing-systems-information-technology-essay.php

ABB. (n.d.a). *GoFa CRB 15000*. Retrieved from https://assets.ctfassets.net/gt89rl895hgs/1MBowsjHDvAEykEwKKBWwl/e2d1a0446a0b0ab046c5b8a510730121/GoFa_CRB15000-datasheet.pdf

ABB. (n.d.b). *SWIFTI CRB 1100*. Retrieved from https://assets.ctfassets.net/gt89rl895hgs/7lxW2lwo38EpADo8WYr0ja/0f36e51c736ccd9800b2efa59bcb17a3/SWIFTI_CRB1100-datasheet.pdf

Abdulhameed, O., Al-Ahmari, A., Ameen, W., & Mian, S. H. (2019). Additive manufacturing: Challenges, trends, and applications. *Advances in Mechanical Engineering*, *11*(2), 1–27. doi:10.1177/1687814018822880

Accenture. (n.d.). Closing the Data-value Gap: How to Become Data Driven and Pivot to the New. *White Paper, Accenture*. https://www.accenture.com/_acnmedia/pdf-108/accenture-closing-data-value-gap-fixed.pdf

Ackerman, H., & King, J. (2019). *Operationalizing the Data Lake – Building and Extracting Value from a Data Lake with a Cloud Native Data Platform*. O'Reilly Media, Incorporated.

Afroz, S., Shimanto, Z. H., Jahan, R. S., & Parvez, M. Z. (2019). Exploring the cognitive learning process by measuring cognitive load and emotional states. *Biomedical Engineering: Applications, Basis and Communications*, *31*(04), 1950032.

Ahmed, Ferzund, Rehman, Usman Ali, Sarwar, & Mehmood. (2017). Modern Data Formats for Big Bioinformatics Data Analytics. *Int'l Journal of Advanced Computer Sc. & Applications*, *8*(4).

Ahmed, S., Ferzund, J., Rehman, A., Usman Ali, A., Sarwar, M., & Mehmood, A. (2017). Modern Data Formats for Big Bioinformatics Data Analytics. *Int'l Journal of Advanced Computer Sc. & Applications (IJACSA)*, *8*(4).

Aktan, E. (2018). Büyük Veri: Uygulama Alanları [Big Data: Application Areas]. *Analitiği ve Güvenlik Boyutu.*, *1*(1), 1–22.

Akyol, S., & Uçar, A. (2019). Rp-Lidar ve Mobil Robot Kullanılarak Eş Zamanlı Konum Belirleme ve Haritalama [Simultaneous Positioning and Mapping Using Rp-Lidar and Mobile Robot]. *Fırat Üniversitesi Mühendislik Bilimleri Dergisi*, *31*(1), 137–143.

Alaa, E. (2017). Comparative Study on Facial Expression Recognition using Gabor and Dual-Tree Complex Wavelet Transforms. *International Journal of Engineering & Applied Sciences*, *9*(1).

Alenazi, M., Niu, N., Wang, W., & Savolainen, J. (2018). Using obstacle analysis to support SysML-based model testing for cyber physical systems. *Proceedings - 2018 8th International Model-Driven Requirements Engineering Workshop, MoDRE 2018*, (pp. 46–55). 10.1109/MoDRE.2018.00012

Alting, D. L., & Jogensen, J. (1993). The Life Cycle Concept as a Basis for Sustainable Industrial Production. *CIRP Annals*, *42*(1), 163–167. doi:10.1016/S0007-8506(07)62417-2

Amin, H., Darwish, A., & Hassanien, A. E. (2021). Classification of COVID19 X-ray Images Based on Transfer Learning InceptionV3 Deep Learning Model. In A. E. Hassanien & A. Darwish (Eds.), *Digital Transformation and Emerging Technologies for Fighting COVID-19 Pandemic: Innovative Approaches. Studies in Systems, Decision and Control* (Vol. 322). Springer. doi:10.1007/978-3-030-63307-3_7

Anderson, E. W., Potter, K. C., Matzen, L. E., Shepherd, J. F., Preston, G. A., & Silva, C. T. (2011). A user study of visualization effectiveness using EEG and cognitive load. *Computer Graphics Forum*, *30*(3), 791–800. doi:10.1111/j.1467-8659.2011.01928.x

Anderson, G. (2016). *The Economic Impact of Technology Infrastructure for Advanced Robotics*. National Institute of Standards and Technology. doi:10.6028/NIST.EAB.2

Angeles, R. (2005). RFID technologies: Supply-chain applications and implementation issues. *Information Systems Management*, *22*, 51–65.

Antonenko, P., Paas, F., Grabner, R., & van Gog, T. (2010). Using electroencephalography to measure cognitive load. *Educational Psychology Review*, *22*(4), 425–438. doi:10.100710648-010-9130-y

Apache Arrow. (2019). *Feather File Format*. Apache Arrow. https://arrow.apache.org/docs/python/feather.html#:~:text=There%20are%20two%20file%20format,available%20in%20Apache%20Ar
row%200.17

Apache Arrow. (n.d.). *Feather File Format*. Apache Arrow White Paper. Available: https://arrow.apache.org/docs/python/feather.html

Arman, S. (2017). Non-rigid registration based model-free 3D facial expression recognition Comput. *Vis. Image Underst.*, *162*, 146–165. doi:10.1016/j.cviu.2017.07.005

Armbrust, M., Das, T., Sun, L., Yavuz, B., Zhu, S., Murthy, M., Torres, J., van Hovell, H., Ionescu, A., Łuszczak, A., Świtakowski, M., Szafrański, M., Li, X., Ueshin, T., Mokhtar, M., Boncz, P., Ghodsi, A., Paranjpye, S., Senster, P., ... Zaharia, M. (2020). Delta Lake: High Performance ACID Table Storage over Cloud Object Stores. *PVLDB*, *13*(12), 3411–3424. doi:10.14778/3415478.3415560

Arnold, R., Fidan, I., & Tantawi, K. (2018). Transforming Industry towards Smart Manufacturing in the United States. In *NSF-ATE 2018 Conference*. Washington, DC: National Science Foundation.

Attaran, M. (2017). Additive Manufacturing: The Most Promising Technology to Alter the Supply Chain and Logistics. *Journal of Service Science and Management*, *10*(03), 189–205. doi:10.4236/jssm.2017.103017

Ayati, N., Saiyarsarai, P., & Nikfar, S. (2020). Short and long term impacts of COVID-19 on the pharmaceutical sector. *Daru: Journal of Faculty of Pharmacy, Tehran University of Medical Sciences*, *28*(2), 799–805. Advance online publication. doi:10.100740199-020-00358-5 PMID:32617864

Aydın, C. (2018). Makine Öğrenmesi Algoritmaları Kullanılarak İtfaiye İstasyonu İhtiyacının Sınıflandırılması [Classification of Fire Station Needs Using Machine Learning Algorithms]. *European Journal of Science and Technology*, (14), 169–175. doi:10.31590/ejosat.458613

Babiceanu, R., & Chen, F. F. (2006). Development and applications of holonic manufacturing systems: A survey. *Journal of Intelligent Manufacturing*, *17*(1), 111–131. doi:10.100710845-005-5516-y

Baesens, A. (2014). *Analytics in a big data world: The essential guide to data science and its applications*. John Wiley & Sons.

Baheti, R., & Gill, H. (2011). Cyber-Physical Systems. *The Impact of Control Technology, 12*(1), 161–166. Available: http://ieeecss.org/sites/ieeecss/files/2019-07/IoCT-Part3-02 CyberphysicalSystems.pdf

Bain, A., & Crisan, D. (2009). *Fundamentals of Stochastic Filtering*. Springer. doi:10.1007/978-0-387-76896-0

Ballou, R. H. (2007). *Business Logistics/supply Chain Management, 5/E (With Cd)*. Pearson Education India.

Baltrusaitis, T., Ahuja, C., & Morency, L. (2017). Multimodal Machine Learning: A Survey and Taxonomy. arXiv:1705.09406v2 [cs.LG]

Barhhan, P. (2003). *International Trade Growth and Development: Essays*. Wiley-Blackwell.

Bateman, J. (2018). *Why China is spending billions to develop an army of robots to turbocharge its economy*. CNBC. Retrieved from https://www.cnbc.com/2018/06/22/chinas-developing-an-army-of-robots-to-reboot-its-economy.html

Beck, J. M., & Pouget, A. (2007). Exact inferences in a neural implementation of a hidden Markov model. *Neural Computation*, *19*(5), 1344–1361. doi:10.1162/neco.2007.19.5.1344 PMID:17381269

Belov, Tatarintsev, & Nikulchev. (2021). Choosing a Data Storage Format in the Apache Hadoop System. *Symmetry, 13*(195). 13020195 doi:10.3390/sym

Belov, V., Tatarintsev, A., & Nikulchev, E. (2021). Choosing a Data Storage Format in the Apache Hadoop System. *Symmetry, 13*.

Bergemann, K., & Reich, S. (2012). An ensemble Kalman-Bucy filter for continuous data assimilation. *Meteorologische Zeitschrift (Berlin)*, *21*(3), 213–219. doi:10.1127/0941-2948/2012/0307

Berggren, M., Nilsson, D., & Robinson, N. D. (2007). Organic materials for printed electronics. *Nature Materials*, *6*(1), 3–5. doi:10.1038/nmat1817 PMID:17199114

Berkes, P., Orban, G., Lengyel, M., & Fiser, J. (2011). Spontaneous Cortical Activity Reveals Hallmarks of an Optimal Internal Model of the Environment. *Science*, *331*(6013), 83–87. doi:10.1126cience.1195870 PMID:21212356

Bert, J., Schellong, D., Hagenmaier, M., Hornstein, D., Wegscheider, A. K., & Palme, T. (2020). *How COVID-19 Will Shape Urban Mobility*. Boston Consulting Group.

Beskos, A., Crisan, D., Jasra, A., Kamatani, K., & Zhou, Y. (2014). *A Stable Particle Filter in High-Dimensions*. ArXiv:1412.350.

Beskos, A., Crisan, D., & Jasra, A. (2014). On the Stability of Sequential Monte Carlo Methods in High Dimensions. *PLoS Computational Biology*, *46*.

Bhatia, R. (2021). Big Data File Format. *White Paper, Clairvoyant*. https://www.clairvoyant.ai/blog/big-data-file-formats

Biswas, A., Chakraborty, S., Rifat, A. N. M. Y., Chowdhury, N. F., & Uddin, J. (2020). Comparative Analysis of Dimension Reduction Techniques Over Classification Algorithms for Speech Emotion Recognition. In M. H. Miraz, P. S. Excell, A. Ware, S. Soomro, & M. Ali (Eds.), *Emerging Technologies in Computing. iCETiC 2020. Lecture Notes of the Institute for Computer Sciences, Social Informatics and Telecommunications Engineering* (Vol. 332). Springer. doi:10.1007/978-3-030-60036-5_12

Bodkhe, U., Bhattacharya, P., Tanwar, S., Tyagi, S., Kumar, N., & Obaidat, M. S. (2019). BloHosT: Blockchain Enabled Smart Tourism and Hospitality Management. *2019 International Conference on Computer, Information and Telecommunication Systems (CITS)*. IEEE. 10.1109/CITS.2019.8862001

Bo-hu, L., Lin, Z., Shi-long, W., Fei, T., Jun-wei, C., Xiao-dan, J., . . . Xu-dong, C. (2010). Cloud manufacturing:a new service-oriented networked manufacturing model. Computer Integrated Manufacturing Systems, 16.

Boursianis, Papadopoulou, Diamantoulakis, Liopa-Tsakalidi, Barouchas, Salahas, . . . Goudos. (2020). Internet of Things (IoT) and Agricultural Unmanned Aerial Vehicles (UAVs) in smart farming: A comprehensive review. *Internet of Things*.

Bowman, P., Ng, J., Harrison, M., Lopez, S., & Illic, A. (2009). Sensor based condition monitoring. *Building Radio frequency IDentification for the Global Environment, (Bridge)*. Euro RFID project.

Brandi, G., & Matteo, T. (2021). Predicting Multidimensional Data via Tensor Learning. arXiv:2002.04328v3 [stat.ML]

Bruunelli, R., & Poggio, T. (1993, October). Face Recognition: Features versus Templates. *IEEE Transactions on Pattern Analysis and Machine Intelligence*, *15*(10), 1042–1052. doi:10.1109/34.254061

Bulut, F. (2016). Sınıflandırıcı Topluluklarının Dengesiz Veri Kümeleri Üzerindeki Performans Analizleri [Performance Analyzes of Classifier Communities on Unbalanced Datasets]. *Bilişim Teknolojileri Dergisi*, *9*(2), 153–159. doi:10.17671/btd.81137

Bures, T., Weyns, D., Schmer, B., Tovar, E., Boden, E., Gabor, T., Gerostathopoulos, I., Gupta, P., Kang, E., Knauss, A., Patel, P., Rashid, A., Ruchkin, I., Sukkerd, R., & Tsigkanos, C. (2017). Software engineering for smart cyber-physical systems: Challenges and promising solutions. *Software Engineering Notes*, *42*(2), 19–24. doi:10.1145/3089649.3089656

Business Insider. (2015). China Solidifies Its Position as the World's Largest Manufacturer. Manufacturers Alliance for Productivity and Innovation (MAPI).

Bynum, A. (2022). *Parsing Improperly Formatted JSON Objects in the Databricks Lakehouse*. White Paper, Databricks. Available: https://www.databricks.com/blog/2022/09/07/parsing-improperly-formatted-json-objects-databricks-lakehouse.html

Calderisi, M., Galatolo, G., Ceppa, I., Motta, T., & Vergentini, F. (2019). Improve image classification tasks using simple convolutional architectures with processed metadata injection. *Proceedings - IEEE 2nd International Conference on Artificial Intelligence and Knowledge Engineering, AIKE 2019*, (pp. 223–230). IEEE. 10.1109/AIKE.2019.00046

Cappe, O. (2011). Online EM Algorithm for Hidden Markov Models. *Journal of Computational and Graphical Statistics*, *20*(3), 1–20. doi:10.1198/jcgs.2011.09109

Cardona, M., Cortez, F., Palacios, A., & Cerros, K. (2020). Mobile Robots Application Against Covid-19 Pandemic. 2020 IEEE ANDESCON.

Case Western Reserve University. (n.d.). *Open Source Data Set*. Case Western Reserve University. https://csegroups.case.edu/bearingdatacenter/pages/welcome-case-western-reserve-university-bearing-data-center-website

Cassell, J., Bickmore, T., Campbell, L., Vilhjalmsson, H., & Yan, H. (2000). Human Conversation As á System Framework: Designing Embodied Conversational Agents. In *Embodied conversational agents* (pp. 29–63). MIT Press. doi:10.7551/mitpress/2697.003.0004

Cerdan, R., Candel, C., & Leppink, J. (2018). Cognitive load and learning in the study of multiple documents. *Frontiers in Education.*, *3*, 59. doi:10.3389/feduc.2018.00059

Chandra, S., Sharma, G., Verma, K.L., Mittal, A., & Jha, D. (2015). EEG based cognitive workload classification during NASA MATB-II multitasking. *International Journal of Cognitive Research in Science, Engineering and Education, 3*(1).

Chehaibi, M. (2017). Parquet Data Format Used in Thing Worx Analytics. *PTC Community*. https://community.ptc.com/t5/IoT-Tech-Tips/Parquet-Data-Format-used-in-ThingWorx-Analytics/td-p/535228

Chen, J., Adebomi, O. E., Olusayo, O. S., & Kulesza, W. (2010). The Evaluation of the Gaussian Mixture Probability Hypothesis Density approach for Multi-target Tracking. *2010 IEEE International Conference on Imaging Systems and Techniques*, 182-185. 10.1109/IST.2010.5548541

Chen, X., & Guhl, J. (2018). Industrial robot control with object recognition based on deep learning. *Procedia CIRP*, *76*, 149–154. doi:10.1016/j.procir.2018.01.021

Choi, S., Kim, B. H., & Noh, S. D. (2015). A diagnosis and evaluation method for strategic planning and systematic design of a virtual factory in smart manufacturing systems. *International Journal of Precision Engineering and Manufacturing, 16*, 1107–1115.

Chowanda, A., Blanchfield, P., Flintham, M., & Valstar, M. (2016). Computational models of emotion, personality, and social relationships for interactions in games. *Proceedings of the 2016 International Conference on Autonomous Agents & Multiagent Systems,* 1343–1344.

Churchland, A. K., Kiani, R., Chaudhuri, R., Wang, X.-J., Pouget, A., & Shadlen, M. N. (2011). Variance as a signature of neural computations during decision making. *Neuron*, *69*(4), 818–831. doi:10.1016/j.neuron.2010.12.037 PMID:21338889

Churchland, M. M., Yu, B. M., Cunningham, J. P., Sugrue, L. P., Cohen, M. R., Corrado, G. S., Newsome, W. T., Clark, A. M., Hosseini, P., Scott, B. B., Bradley, D. C., Smith, M. A., Kohn, A., Movshon, J. A., Armstrong, K. M., Moore, T., Chang, S. W., Snyder, L. H., Lisberger, S. G., ... Shenoy, K. V. (2010). Stimulus onset quenches neural variability: A widespread cortical phenomenon. *Nature Neuroscience*, *13*(3), 369–378. doi:10.1038/nn.2501 PMID:20173745

Clark, L. (2013). Tesco Uses Supply Chain Analytics to Save £100 m a Year. *Computer Weekly*. https://www.computerweekly.com/news/2240182951/Tesco-uses-supply-chain-analytics-to-save-10-m-a-year

Coleman, L. (2018). *Inside Trends And Forecast For The $3.9T AI Industry.* Forbes.

Coleman, L. (2018). *Inside Trends And Forecast For The $3.9T AI Industry.* Forbes. (2020). *COVID-19 Impact on the Industrial Robotics Market by Type (Articulated, SCARA, Parallel, Cartesian Robots), Industry (Automotive; Electrical and Electronics; Food & Beverages; Pharmaceuticals and Cosmetics), and Region – Global Forecast to 2025.* Research and Markets.

Collie, B., Wachtmeister, A., Waas, A., Kirn, R., Krebs, K., & Quresh, H. (2020). *Covid-19's Impact on the Automotive Industry.* Boston Consulting Group.

Cooke, J. A. (2013). Three trends to watch in 2013, Perspective. *Supply chain Quarterly, 1*, 11.

Coronavirus disease (COVID-19) pandemic. (2021). Retrieved September 6, 2021, from World Health Organization: https://covid19.who.int/

Coronis, A., & Tantawi, K. (2019). Advances in Energy-Efficient Manufacturing using Industrial Robotics. In *NSF-ATE 2019 Conference.* Washington, DC: National Science Foundation.

Cortez, P., Rio, M., Rocha, M., & Sousa, P. (2012). Multi-scale Internet traffic forecasting using neural networks and time series methods. *Expert Systems: International Journal of Knowledge Engineering and Neural Networks, 29*(2), 143–155.

Coşkun, C., & Baykal, A. (2011). *Veri Madenciliğinde Sınıflandırma Algoritmalarının Bir Örnek Üzerinde Karşılaştırılması. Akademik Bilişim 2011 [Comparison of Classification Algorithms in Data Mining on a Sample. Academic Informatics 2011].* Akademik Bilişim.

COVID-19 Impact on the Industrial Robotics Market by Type (Articulated, SCARA, Parallel, Cartesian Robots), Industry (Automotive; Electrical and Electronics; Food & Beverages; Pharmaceuticals and Cosmetics), and Region – Global Forecast to 2025. (2020). Research and Markets.

Crisan, D., & Xiong, J. (2010). Approximate McKean–Vlasov representations for a class of SPDEs. *Stochastics, 82*(1), 53–68. doi:10.1080/17442500902723575

Data Flair. (2018). What are the File Format in Hadoop. *Data Flair.* https://data-flair.training/forums/topic/what-are-the-file-format-in-hadoop/

Databerg. (2019). Why is High-Quality Data Governance a Key Tool in Industry 4.0? *White Paper, Databerg.*https://blog.datumize.com/why-is-high-quality-data-governance-a-key-tool-in-industry-4.0

Davenport, T. H., & O'Dwyer, J (2011). Tap into the Power of Analytics. *Supply Chain Quarterly*, 28-31.

Davenport, T. H. (2006). *Competing on analytics, Harvard Business Review.* Harvard Business Press.

Davenport, T. H., Barth, P., & Bean, R. (2012). How Big Data is different. *MIT Sloan Management Review*, (Fall), 22–24.

Davenport, T. H., & Harris, J. G. (2007). *Competing on analytics – the new science of wining.* Harvard Business School Publishing Corporation.

Davenport, T. H., Harris, J. G., & Morison, R. (2010). *Analytics at work – smart decisions, better results.* Harvard Business Press.

Davenport, T. H., & Prusak, L. (2000). *Working knowledge: how organizations manage what they know.* Harvard Business Press.

Davis, J., & Swink, D. (n.d.). *Smart Manufacturing as a Real-Time Networked Enterprise and a Market-Driven Innovation Platform.* Smart Manufacturing Leadership Coalition.

Davis, J., Swink, D., Tran, J., Wetzel, J., Profozich, G., McKewen, E., & Thys, R. (2015). *Smart Manufacturing The Next Revolution in Manufacturing.* California Manufacturing Technology Consulting.

Davis-Sramek, B., Germain, R., & Iyer, K. (2010). Supply Chain Technology: The Role of Environment in Predicting Performance. *Journal of the Academy of Marketing Science, 38*(1), 42–55. doi:10.100711747-009-0137-1

Decker, C., Berchtold, M. L., Chaves, W. F., Beigl, M., Roehr, D., & Riedel, T. (2008). Cost benefit model for smart items in the supply chain. In The Internet of Things, 155–172. Springer.

Deloitte & MHI (2016). The 2016 MHI Annual Industry Report – Accelerating change: How innovation is driving digital, always-on. *Supply Chains.*

Deloitte & MHI. (2014). *The 2014 MHI Annual Industry Report – Innovation the driven supply chain.* MHI.

Demaitre, E. (2018). *RBR50 2018 Names the Leading Robotics Companies of the Year.* Robotics Business Review.

Demirkan, H., & Delen, D. (2012). Levering the Capabilities of Service-oriented Decision Support Systems: Putting Analytics and Big Data in Cloud. *Decision Support Systems, 55*(1), 412–421. doi:10.1016/j.dss.2012.05.048

Deneve, S., Duhamel, J.-R., & Pouget, A. (2007). Optimal sensorimotor integration in recurrent cortical networks: A neural implementation of Kalman filters. *The Journal of Neuroscience: The Official Journal of the Society for Neuroscience, 27*(21), 5744–5756. doi:10.1523/JNEUROSCI.3985-06.2007 PMID:17522318

Denso-Wave. (2017). *Denso Cobotta user manual.* Denso. Retrieved from http://eidtech.dyndns-at-work.com/support/Cobotta_Manual/007 260.html

Desai, A. (2014). *Economy League's 2014 World Class Summit: Tracking Philadelphia's Progress on Growth and Opportunity.* Global Philadelphia Association. Retrieved from https://globalphiladelphia.org/news/eceonmy-leagues-2014-wor ld-class-summit-tracking-philadelphia%E2%80%99s-progress-gro wth-and-opportunity

Dietrich, B., Plachy, E. C., & Norton, M. F. (2014). *Analytics across the enterprise: How IBM realize business value from big data and analytics.* IBM Press Books.

Digitale Wirtschaft und Gesellschaft. (n.d.). Federal Ministry of Education and Research-Germany. Retrieved August 17, 2017, from https://www.bmbf.de/de/zukunftsprojekt-industrie-4-0-848.htm l

Doğan, A., Sönmez, B., & Cankül, D. (2020). Yiyecek-İçecek İşletmelerinde İnovasyon Ve Artırılmış Gerçeklik Uygulamaları [Innovation And Augmented Reality Applications in Food And Beverage Companıes]. *Journal of Business Research - Turk, 10*(3), 576–591. doi:10.20491/isarder.2018.488

Doucet, A., & Johansen, A. (2009). A tutorial on particle filtering and smoothing: Fifteen years later. Handbook of Nonlinear Filtering, 4–6.

Doucet, A., Godsill, S., & Andrieu, C. (2000). On sequential Monte Carlo sampling methods for Bayesian filtering. *Statistics and Computing, 10*(3), 197–208. doi:10.1023/A:1008935410038

Dowling, J. (2019). *Guide to File Formats for Machine Learning: Columnar, Training, Inferencing, and the Feature Store.* Towards Data Science. Available: https://towardsdatascience.com/guide-to-file-formats-for-mac hine-learning-columnar-training-inferencing-and-the-feature- store-2e0c3d18d4f9

Doya, K. (2007). *Bayesian Brain: Probabilistic Approaches to Neural Coding.* The MIT Press.

Dremio. (n.d.). "What is Apache Parquet, online, Available: https://www.dremio.com/resources/guides/intro-apache-parquet /

Duan, J., He, Y., Du, B., Ghandour, R. M. R., Wu, W., & Zhang, H. (2019). Intelligent Localization of Transformer Internal Degradations Combining Deep Convolutional Neural Networks and Image Segmentation. *IEEE Access: Practical Innovations, Open Solutions, 7*, 62705–62720. doi:10.1109/ACCESS.2019.2916461

Duff, I., & Lewis, J. (n.d.). The Rutherford-Boeing Sparse Matrix Collection. White Paper, Rutherford Appleton Laboratory.

Duniam, G., Kitaeff, S., Wicenec, A., German, G., & Shen, A. (2021). Source finding with SoFiA-2 and very large source files. White Paper, University of Western Australia.

Dye, S. (2019). Feather Files: Faster Than the Speed of Light. *Medium*. https://medium.com/@steven.p.dye/feather-files-faster-than-the-speed-of-light-d4666ce24387

Energy, U. S. (2016). *Industrial Assessment Centers (IACs)*. Retrieved June 12, 2017, from https://energy.gov/eere/amo/industrial-assessment-centers-iacs

Epply, T., & Nagengast, J. (n.d.). *The Lean Manufacturing Handbook* (2nd ed.). Continental Design and Engineering.

Ericsson, K. A., & Kintsch, W. (1995). Long-term working memory. *Psychological Review*, *102*(2), 211–245. doi:10.1037/0033-295X.102.2.211 PMID:7740089

Ernst, M. O., & Banks, M. S. (2002). Humans integrate visual and haptic information in a statistically optimal fashion. *Nature*, *415*(6870), 429–433. doi:10.1038/415429a PMID:11807554

Erol, S., Schumacher, A., & Sihn, W. (2016). Strategic guidance towards Industry 4.0 – a three-stage process model. *International Conference on Competitive Manufacturing*.

Eşref, Y. (2019). *Türkçe Dizi Etiketleme İçin Sinir Ağ Modelleri [Neural Network Models for Sequence Labeling in Turkish]*. Hacettep Üniveristesi.

Estolatan, E., Geuna, A., Guerzoni, M., & Nuccio, M. (2018). *Mapping the Evolution of the Robotics Industry: A Cross Country Comparison*. University of Toronto.

Estolatan, E., Geuna, A., Guerzoni, M., & Nuccio, M. (2018). *Mapping the evolution of the robotics industry: a cross country comparison*. University of Toronto. Innovation Policy White Paper Series.

Everton, K., Hirsch, M., Stravroulakis, P., Leach, R. K., & Clare, A. T. (2016). Review of in-situ process monitoring and in-situ metrology for metal additive manufacturing. *Materials & Design*, *95*, 431–445. doi:10.1016/j.matdes.2016.01.099

Ezeelive Technologies. (n.d.). *JSON – Its Advantages and Disadvantages*. Available: https://ezeelive.com/json-advantages-disadvantages/

Faggella, D. (2018, May 29). *Global Competition Rises for AI Industrial Robotics*. Retrieved September 24, 2018, from https://www.techemergence.com/global-competition-rises-ai-industrial-robotics/

Falsey, A. R., McCann, R. M., Hall, W., Criddle, M. M., Formica, M. A., Wycoff, D., & Kolassa, J. E. (1997). The “Common Cold” in Frail Older Persons: Impact of Rhinovirus and Coronavirus in a Senior Daycare Center. *Journal of the American Geriatrics Society*, *45*(6), 706–711. doi:10.1111/j.1532-5415.1997.tb01474.x PMID:9180664

Fang, C., Liu, X., Pardalos, P. M., & Pei, J. (2016). Optimization for a three-stage production system in the Internet of Things: Procurement, production and product recovery, and acquisition. *International Journal of Advanced Manufacturing Technology*, *83*, 689–710.

Fang, J., & Ma, A. (2020). IoT application modules placement and dynamic task processing in edge-cloud computing. *IEEE Internet of Things Journal*, *8*(6), 12771–12781.

Fang, J., Qu, T., Li, Z. G., Xu, G., & Huang, G. Q. (2013). Agent-based gateway operating system for RFID-enabled ubiquitous manufacturing enterprise. *Robotics and Computer-integrated Manufacturing, 29*, 222–231.

FANUC America Corporation. (n.d.). *Basic Robot Operations*. FANUC America Corporation.

Ferrari, P., Rinaldi, S., Sisinni, E., Colombo, F., Ghelfi, F., Maffei, D., & Malara, M. (2019). Performance evaluation of full-cloud and edge-cloud architectures for Industrial IoT anomaly detection based on deep learning. *IEEE International Workshop on Metrology for Industry 4.0 and IoT, MetroInd 4.0 and IoT 2019 - Proceedings*, (pp. 420–425). 10.1109/METROI4.2019.8792860

Firth, S. (2022). *CSV on the Web: Creating Descriptive Metadata Files*. Available: https://www.stevenfirth.com/csv-on-the-web-creating-descriptive-metadata-files/

Fiser, A., Mahringer, D., Oyibo, H. K., Petersen, A. V., Leinweber, M., & Keller, G. B. (2016). Experience-dependent spatial expectations in insect visual cortex. *Nature Neuroscience, 19*(12), 1658–1664. doi:10.1038/nn.4385 PMID:27618309

Fiser, J., Berkes, P., Orban, G., & Lengyel, M. (2010). Statistically optimal perception and learning: From behavior to neural representations. *Trends in Cognitive Sciences, 14*(3), 119–130. doi:10.1016/j.tics.2010.01.003 PMID:20153683

Fogg, E. (2020). 5 Steps to Bring Your Legacy System Online with IIoT. *White Paper, Machine Metrics*. https://www.machinemetrics.com/blog/legacy-systems-online-iiot

Forslund, H., & Jonsson, P. (2007). The Impact of Forecast Information Quality on Supply Chain Performance. *International Journal of Operations & Production Management, 27*(1), 90–107. doi:10.1108/01443570710714556

Fournier, L. R., Wilson, G. F., & Swain, C. R. (1999). Electrophysiological, behavioral, and subjective indexes of workload when performing multiple tasks: Manipulations of task difficulty and training. *International Journal of Psychophysiology, 31*(2), 129–145. doi:10.1016/S0167-8760(98)00049-X PMID:9987059

Fox, J. C., Moylan, S. P., & Lane, B. M. (2016). Effect of process parameters on the surface roughness of overhanging structures in laser powder bed fusion additive manufacturing. *Procedia CIRP, 45*, 131–134. doi:10.1016/j.procir.2016.02.347

France, G. o. (2013). *The New Face of Industry in France*. Author.

Francis, S. (2018). *ABB claims to have connected 7,000 of its industrial robots to its IIoT platform*. Retrieved from https://www.i-scoop.eu/industry-4-0-society-5-0/

Fraser, K. L., Ayres, P., & Sweller, J. (2015). Cognitive load theory for the design of medical simulations. *Simulation in Healthcare, 10*(5), 295–307. doi:10.1097/SIH.0000000000000097 PMID:26154251

FROSTT. (n.d.). *Formidable Repository of Open Sparse Tensor and Tools*. Available: http://frostt.io/tensors/file-formats.html

Gabriel, E. (n.d.). *Big Data Analytics Data Formats – HDF5 and Parquet files*. COSC 6339, University of Houston. Available: http://cs.uh.edu/~gabriel/courses/cosc6339_f18/BDA_16_DataFormats.pdf

Galiautdinov, R. (2019). Advanced method of planning the trajectory of swarm's drones possessing Artificial Intelligence. *International Journal of Scientific Research*. Advance online publication. doi:10.21275/ART20203362

Galiautdinov, R. (2020). Brain machine interface: The accurate interpretation of neurotransmitters' signals targeting the muscles. *International Journal of Applied Research in Bioinformatics, 10*(1), 26–36. Advance online publication. doi:10.4018/IJARB.2020010102

Galiautdinov, R., & Mkrttchian, V. (2019A). Math model of neuron and nervous system research, based on AI constructor creating virtual neural circuits: Theoretical and Methodological Aspects. In V. Mkrttchian, E. Aleshina, & L. Gamidullaeva (Eds.), *Avatar-Based Control, Estimation, Communications, and Development of Neuron Multi-Functional Technology Platforms* (pp. 320–344). IGI Global. doi:10.4018/978-1-7998-1581-5.ch015

Galiautdinov, R., & Mkrttchian, V. (2019B). Brain machine interface – for Avatar Control & Estimation in Educational purposes Based on Neural AI plugs: Theoretical and Methodological Aspects. In V. Mkrttchian, E. Aleshina, & L. Gamidullaeva (Eds.), *Avatar-Based Control, Estimation, Communications, and Development of Neuron Multi-Functional Technology Platforms* (pp. 345–360). IGI Global. doi:10.4018/978-1-7998-1581-5.ch016

Gallagher, R. M. (2019, August 30). *New Safety Standards for Collaborative Robots*. Retrieved from https://www.engineering.com/AdvancedManufacturing/ArticleID/19403/New-Safety-Standards-for-Collaborative-Robots.aspx

GeeksforGeeks. (n.d.). *Numpy Arrays in Python*. Available: https://www.geeksforgeeks.org/numpy-array-in-python/

Gereffi, G. (1999). International trade and industrial upgrading in the apparel commodity chain. *Journal of International Economics*, *48*(1), 37–70. doi:10.1016/S0022-1996(98)00075-0

Gerstenberger, M. (2019). Industrial Mobile Robots Safety Standard Update. *National Robot Safety Conference*.

Gevins, A., Smith, M. E., McEvoy, L., & Yu, D. (1997). High-resolution EEG mapping of cortical activation related to working memory: Effects of task difficulty, type of processing, and practice. *Cerebral Cortex*, *7*(4), 374–385. doi:10.1093/cercor/7.4.374 PMID:9177767

Giffi, C., Dollar, B., Drew, M., McNelly, J., Carrick, G., & Gangula, B. (2015). *The Skills Gap in U.S. Manufacturing 2015 and Beyond*. Deloitte Development LLC.

Giret, A. (2008). Holonic Manufacturing Systems. In *A Multi-Agent Methodology for Holonic Manufacturing* (pp. 7-20). Springer. Retrieved from http://impactphl.org/wp-content/uploads/2016/07/ELGP-BFT-Impact-Report.pdf

Gopinath, V., Kerstin, J., Derelov, M., Gustafsson, A., & Axelsson, S. (2021, February). Safe Collaborative Assembly on a Continuously Moving Line with Large Industrial Robots. *Robotics and Computer-integrated Manufacturing*, *67*, 67. doi:10.1016/j.rcim.2020.102048

Graves, A., & Schmidhuber, J. (2005). Framewise Phoneme Classification with Bidirectional LSTM and other Neural Network Architectures. *Neural Networks*, *18*(5–6), 602–610. doi:10.1016/j.neunet.2005.06.042 PMID:16112549

Greaves-Tunnell, A. (2015). *An optimization perspective on approximate neural filtering* [Master thesis].

Gubbi, J., Buyya, R., Murusic, S., & Palaniswami, M. (2013). Internet of Things (IoT): A vision, architectural elements, and future directions. *Future Generation Computer Systems*, *29*(7), 1645–1660.

Handfield, R. (2006). *Supply Market Intelligence: A Managerial Handbook for Building Sourcing Strategies*. Taylor & Francis. doi:10.4324/9780203339527

Handfield, R., & Nichols, E. Jr. (2004). Key Issues in Global Supply Base Management. *Industrial Marketing Management*, *33*(1), 29–35. doi:10.1016/j.indmarman.2003.08.007

Hao, M., Li, H., Luo, X., Xu, G., Yang, H., & Liu, S. (2019). Efficient and Privacy-enhanced Federated Learning for Industrial Artificial Intelligence. *IEEE Transactions on Industrial Informatics*, *3203*, 1–11. doi:10.1109/TII.2019.2945367

Harrington, J. (1979). *Computer Integrated Manufacturing*. Krieger Pub Co.

Hasan, J., Kim, J.-M., & Manjurul Islam, M. M. (2019). Acoustic spectral imaging and transfer learning for reliable bearing fault diagnosis under variable speed conditions. *Measurement, 138,* 620-631. doi:10.1016/j.measurement.2019.02.075

Hasan, M. J., & Kim, J.-M. (2018). Bearing Fault Diagnosis under Variable Rotational Speeds Using Stockwell Transform-Based Vibration Imaging and Transfer Learning. *Applied Sciences (Basel, Switzerland), 8*(12), 2357. doi:10.3390/app8122357

Hennequin, G., Aitchison, L., & Lengyel, M. (2014). Fast Sampling-Based Inference in Balanced Neuronal Networks. *Advances in Neural Information Processing Systems.*

Hernández, D. E., & Trujillo, L. (2018). Detecting Epilepsy in EEG Signals Using Time, Frequency and Time-Frequency Domain Features. Computer Science and Engineering—Theory and Applications.

Huang, Y. & Rao, R., (2014). Neurons as Monte Carlo Samplers: Bayesian Inference and Learning in Spiking Networks. *Neural Information Processing Systems*, 1–9.

Huang, Y., & Rao, R. P. N. (2016). Bayesian Inference and Online Learning in Poisson Neuronal Networks. *Neural Computation, 28*(8), 1503–1526. doi:10.1162/NECO_a_00851 PMID:27348304

Hulsebos, M. (2019). Sherlock: A Deep Learning Approach to Semantic Data Type Detection. arXiv:1905.10688v1 [cs.LG

IATA. (2020). *Air Passenger Market Analysis.* International Air Transport Association (IATA).

IBM. (2021). XML Schemas Overview. *White Paper, IBM.* https://www.ibm.com/docs/en/control-desk/7.6.1?topic=schemas-xml-overview

IBM. (n.d.a). *Node Side Network Configuration.* IBM Cloud Pak for Data System, Documentation. Available: https://www.ibm.com/docs/en/cloud-paks/cloudpak-data-system/1.0?topic=configuration-node-side-network

IBM. (n.d.b). *What Are IBM Cloud Paks?* IBM Cloud Pak for Data System, Documentation. Available: https://www.ibm.com/docs/en/cloud-paks

Idreos, S., Alagiannis, I., Johnson, R., & Ailamaki, A. (2011). Here are my Data Files. Here are my Queries: Where are my Results? *5th Biennial Conf. on Innovative Data System Research (CIDR 2011)*; Asilomar, California, USA.

IFR Press Conference. (2018). Robots double worldwide by 2020. International Federation of Robotics.

International Federation of Robotics. (2020). *IFR Executive Summary Report for 2019.* IFR.

International Federation of Robotics. (2021). *Executive Summary World Robotics 2021 Industrial Robots.* IFR.

International Organization for Standardization. (2012). *ISO 8373:2012 Robots and robotic devices — Vocabulary.* ISO.

Internet of Things Global Standards Initiative. (n.d.). Retrieved 6 16, 2017, from https://www.itu.int/en/ITU-T/gsi/iot/Pages/default.aspx

Japan, G. o. (n.d.). *Realizing Society 5.0.* Author.

Jiang, J., Li, Z., Tian, Y., & Al-Nabhan, N. (2020). A review of techniques and methods for IoT applications in collaborative cloud-fog environment. *Security and Communication Networks*, 1–15.

Ji, Y., Wang, Q., Li, X., & Liu, J. (2019). A Survey on Tensor Techniques and Applications in Machine Learning. *IEEE Access : Practical Innovations, Open Solutions, 7*, 162950–162990. doi:10.1109/ACCESS.2019.2949814

Joda, T., Gallucci, G. O., Wismeijer, D., & Zitzmann, N. U. (2019). Augmented and virtual reality in dental medicine: A systematic review. *Computers in Biology and Medicine*, *108*, 93–100. doi:10.1016/j.compbiomed.2019.03.012 PMID:31003184

John, S. M., & Oyekanlu, E. (2010). Impact of Packet Losses on the Quality of Video Streaming. *Technical Report*. Blekinge Institute of Technology.

Jung, K., Morris, K., Lyons, K. W., Leong, S., & Cho, H. (2015). Mapping Strategic Goals and Operational Performance Metrics for Smart Manufacturing Systems. *Procedia Computer Science*, *44*, 184–193. doi:10.1016/j.procs.2015.03.051

Jurcak, V., Tsuzuki, D., & Dan, I. (2007). 10/20, 10/10, and 10/5 systems revisited: Their validity as relative head-surface-based positioning systems. *NeuroImage*, *34*(4), 1600–1611. doi:10.1016/j.neuroimage.2006.09.024 PMID:17207640

Kaelbling, L. P., Littman, M. L., & Moore, A. W. (1996). Reinforcement learning: A survey. *Journal of Artificial Intelligence Research*, *4*, 237–285. doi:10.1613/jair.301

Kahn, J. S., & McIntosh, K. (2005). History and Recent Advances in Coronavirus Discovery. *The Pediatric Infectious Disease Journal*, *24*(11), S223–S227. doi:10.1097/01.inf.0000188166.17324.60 PMID:16378050

Kalimeri, K., & Saitis, C. (2016). Exploring multimodal biosignal features for stress detection during indoor mobility. *Proceedings of the 18th ACM international conference on multimodal interaction*, 53-60. 10.1145/2993148.2993159

Kamiya, T., Kolesnikov, A., Murphy, B., Watson, K., & Widell, N. (2021). Characteristics of IIoT Information Models. *White Paper, Industrial Internet Consortium*. https://www.iiconsortium.org/pdf/Characteristics-of-IIoT-Information-Models.pdf

Kancharla, C. R., (2021). Augmented Reality Based Machine Monitoring for Legacy Machines: a retrofitting use case. *2021 XXX International Scientific Conference Electronics (ET)*, (pp. 1-5). IEEE. 10.1109/ET52713.2021.9579936

Kang, H. S., Lee, J. Y., Choi, S., Kim, H., Park, J. H., Son, J. Y., . . . Noh, S. D. (2016). Smart Manufacturing: Past Research, Present Findings, and Future Directions. *International Journal of Precision Engineering and Manufacturing-Green Technologies, 3*, 111-128.

Kantas, N., Doucet, A., Singh, S. S., Maciejowski, J., & Chopin, N. (2015). On Particle Methods for Parameter Estimation in State-Space Models. *Statistical Science*, *30*(3), 328–351. doi:10.1214/14-STS511

Kappel, D., Habenschuss, S., Legenstein, R., & Maass, W. (2015). Network Plasticity as Bayesian Inference. *PLoS Computational Biology*, *11*(11), 1–31. doi:10.1371/journal.pcbi.1004485 PMID:26545099

Karagiorgou, S., Vafeiadis, G., Ntalaperas, D., Lykousas, N., Vergeti, D., & Alexandrou, D. (2019). Unveiling Trends and Predictions in Digital Factories. *15th Annual International Conference on Distributed Computing in Sensor Systems, DCOSS 2019*, (pp. 326–332). IEEE. 10.1109/DCOSS.2019.00073

Karakose, E., & Gencoglu, M. T. (2013). An analysis approach for condition monitoring and fault diagnosis in pantograph-catenary system. *IEEE International Conference on In Systems, Man, and Cybernetics (SMC)*, (pp. 1963-1968). IEEE. 10.1109/SMC.2013.337

Kellner, T. (2017). GE, CFM Expect $15 Billion. In *New Business In Paris; New LEAP Engines Are Giving A Lift To The Aviation Industry*. General Electric.

Kennell, B. (2015). *Smart Manufacturing: A Path to Profitable Growth*. Retrieved from https://www.huffingtonpost.com/brian-kennell/smart-manufacturing-a-pat_b_7314828.html

Khandelwal, D. (2020). An Introduction to Big Data Formats. *Big Data and Cloud Practice*. https://bd-practice.medium.com/an-introduction-to-big-data-formats-450c8db3d29a

Khan, S. A., & Kim, J. M. (2016). Automated bearing fault diagnosis using 2d analysis of vibration acceleration signals under variable speed conditions. *Shock and Vibration, 2016*, 8729572. doi:10.1155/2016/8729572

Khatoun, R., & Zeadally, S. (2017). Cybersecurity and privacy solutions in smart cities. *IEEE Communications Magazine, 55*(3), 51–59. doi:10.1109/MCOM.2017.1600297CM

Klimesch, W. (1999). EEG alpha and theta oscillations reflect cognitive and memory performance: A review and analysis. *Brain Research. Brain Research Reviews, 29*(2-3), 169–195. doi:10.1016/S0165-0173(98)00056-3 PMID:10209231

Knill, D. C., & Pouget, A. (2004). The Bayesian brain: The role of uncertainty in neural coding and computation. *Trends in Neurosciences, 27*(12), 712–719. doi:10.1016/j.tins.2004.10.007 PMID:15541511

Knowledgebase. (2005). *Difference Between CPU Time and Wall Time*. Available: https://service.futurequest.net/index.php?/Knowledgebase/Article/View/407

Kohli, R., & Grover, V. (2008). Business Value of IT: An Essay on Expanding Research Directions to Keep up with the times. *Journal of the Association for Information Systems, 9*(1), 23–39. doi:10.17705/1jais.00147

Koidan, K. (2019). *Array vs. List in Python – What's the Difference?* LearnPython. Available: https://learnpython.com/blog/python-array-vs-list/

Kong, Q., Siauw, T., & Bayen, A. (2021). *Python Programming and Numerical Methods – A Guide for Engineers and Scientists*. Academy Press.

Kording, K. P., & Wolpert, D. M. (2004). Bayesian integration in sensorimotor learning. *Nature, 427*(6971), 244–247. doi:10.1038/nature02169 PMID:14724638

Kording, K., Tenenbaum, J. B., & Shadmehr, R. (2007). The dynamics of memory as a consequence of optimal adaptation to a changing body. *Nature Neuroscience, 10*(6), 779–786. doi:10.1038/nn1901 PMID:17496891

Kotz, F., Arnold, K., Bauer, W., Schild, D., Keller, N., Sachsenheimer, K., Nargang, T. M., Richter, C., Helmer, D., & Rapp, B. E. (2017). Three-dimensional printing of transparent fused silica glass. *Nature, 544*(7650), 337–339. doi:10.1038/nature22061 PMID:28425999

Kramer, A.F. (1991). Physiological metrics of mental workload: A review of recent progress. *Multiple-task performance*, 279-328.

Krigolson, O.E., Hassall, C.D., Satel, J., & Klein, R.M. (2015). The impact of cognitive load on reward evaluation. *Brain Research, 1627*, 225-32.

Krizhevsky, Sutskever, & Hinton. (2012). Imagenet Classification with Deep Convolutional Neural Networks. *Advances in Neural Information Processing Systems,* 1097–1105.

Kumar, A. (2022). *Tensor Explained with Python Numpy Examples*. Available: https://vitalflux.com/tensor-explained-with-python-numpy-examples/

Kumar, J., Goomer, R., Singh, A. K., & Recurrent, L. S. T. M. (2018). Neural Network (LSTM-RNN) Based Workload Forecasting Model For Cloud Datacenters. *Procedia Computer Science, 125*, 676-682. Science Direct. doi:10.1016/j.procs.2017.12.087

Kumar, N., & Kumar, J. (2016). Measurement of cognitive load in HCI systems using EEG power spectrum: An experimental study. *Procedia Computer Science, 84*, 70–78. doi:10.1016/j.procs.2016.04.068

Kusiak, A. (2017). Smart manufacturing must embrace big data. *Nature, 544*(7648), 23–25. doi:10.1038/544023a PMID:28383012

Laing, T. (2017). *UK Industrial Strategy: Navigating a changing world.* University of Cambridge Institute for Sustainability Leadership.

Lai, R., Lin, W., & Wu, Y. (2018). Review of Research on the Key Technologies, Application Fields and Development Trends of Intelligent Robots. *International Conference on Intelligent Robotics and Applications, 10985*, 449-458. 10.1007/978-3-319-97589-4_38

Lambrecht, J., Kästner, L., Guhl, J., & Krüger, J. (2021). Towards commissioning, resilience and added value of Augmented Reality in robotics: Overcoming technical obstacles to industrial applicability. *Robotics and Computer-integrated Manufacturing, 71*, 71. doi:10.1016/j.rcim.2021.102178

Lavalle, S., Lesser, E., Shockey, R. H., & Crosthwait, N. M. (2011). Big Data, Analytics and the Path from Insight to Value. *MIT Sloan Management Review, 52*(2), 21–32.

Leathers, M. L. (2016). How to Prepare Your Workforce for Smart Manufacturing. *Industry Week.*

Lecun, Y., Bengio, Y., & Hinton, G. (2015). Deep learning. *Nature, 521*(7553), 436–444. doi:10.1038/nature14539 PMID:26017442

Lee, D., & Heintz, B. (2019). Productionizing Machine Learning with Delta Lakes. *Databricks.* https://databricks.com/blog/2019/08/14/productionizing-machine-learning-with-delta-lake.html

Legenstein, R., & Maass, W. (2014). Ensembles of Spiking Neurons with Noise Support Optimal Probabilistic Inference in a Dynamically Changing Environment. *PLoS Computational Biology, 10*(10), e1003859. doi:10.1371/journal.pcbi.1003859 PMID:25340749

Levy, E. (2022). What is Parquet File Format and Why You Should Use It. *White Paper, Upsolver.* https://www.upsolver.com/blog/apache-parquet-why-use

Levy, E. (2022). *What is the Parquet File Format? Use Cases and Benefits.* Upsolver. Available: https://www.upsolver.com/blog/apache-parquet-why-use

Liang, M., & Hu, X. (2015). Recurrent convolutional neural network for object recognition. *The IEEE Conference on Computer Vision and Pattern Recognition.* 3367-3375. 10.1109/CVPR.2015.7298958

Li, B., Yang, C., & Huang, S. (2014). Study on supply chain disruption management under service level dependent demand. *Journal of Networking, 9*, 1432–1439.

Liberatore, M., & Luo, W. (2010). The Analytics Movement. *Interface: a Journal for and About Social Movements, 40*(4), 313–324. doi:10.1287/inte.1100.0502

Library of Congress. (n.d.). *TSV, Tab Separated Values*. Available: https://www.loc.gov/preservation/digital/formats/fdd/fdd000533.shtml

Li, D. (2016). Perspective for smart factory in petrochemical industry. *Computers & Chemical Engineering, 91*, 136–148.

Liebhart, R., & Hohmann, L. (2016). *Korea: Evolution of manufacturing industry.* Maschinen Markt International.

Lima, F., Massote, A. A., & Maia, R. F. (2019). IoT Energy Retrofit and the Connection of Legacy Machines Inside the Industry 4.0 Concept. *IECON 2019 - 45th Annual Conference of the IEEE Industrial Electronics Society*, (pp. 5499-5504). IEEE. 10.1109/IECON.2019.8927799

Litvin, Y. (2019). *Petastorm: A Light-Weight Approach to Building ML Pipelines @Uber*. Uber White Paper. Available: https://qcon.ai/system/files/presentation-slides/yevgeni_-_petastorm_16th_apr_2019_.pdf

Liu, H., Xie, T., Ran, J., & Gao, S. (2017). An Efficient Algorithm for Server Thermal Fault Diagnosis Based on Infrared Image. *Journal of Physics Conference Series, 910*, 1240-1256. . doi:10.1088/1742-6596/910/1/012031

Liu, R., Wang, Z., Sparks, T., Liou, F., & Newkirk, J. (2016). Aerospace Applications of Laser Additive Manufacturing. In Laser Additive Manufacturing: Materials, Design, Technologies, and Applications (pp. 351-353). Woodhead Publishing-Elsevier.

Liu, C., & Tomizuka, M. (2017). *Towards Intelligent Industrial Co-robots- Democratization of Robots in Factories.* Berkeley Artificial Intelligence Research.

Liu, F., Shi, Y., & Li, P. (2017). Analysis of the Relation between Artificial Intelligence and the Internet from the Perspective of Brain Science. *Procedia Computer Science, 122*, 377–383. doi:10.1016/j.procs.2017.11.383

Liu, J., Liu, Y., Cheng, J., & Feng, F. (2013). Extraction of Gear Fault Feature Based on the Envelope and Time-Frequency Image of S Transformation. *Chemical Engineering Transactions, 33*, 55–60. doi:10.1016/j.ces.2013.01.060

Liukkonen, M., & Tsai, T. N. (2016). Toward decentralized intelligence in manufacturing: Recent trends in automatic identification of things. *International Journal of Advanced Manufacturing Technology, 87*, 2509–2531.

Li, W., Li, H., Member, S., Wu, Q., Chen, X., & Ngan, K. N. (2019). *Simultaneously Detecting and Counting Dense Vehicles From Drone Images., 66*(12), 9651–9662.

Lu, M. C. (2020). Future pandemics can be prevented, but that'll rely on unprecedented global cooperation. *The Washington Post.*

Lu, Y., Morris, K., & Frechette, S. (2016). Current Standards Landscape for Smart Manufacturing Systems. *National Institute of Standards and Technology NISTIR 8107.*

Luis, A., Casarez, P., Cuadrado-Gallego, J., & Patricio, M. (2021). PSON: A Serialization Format for IoT Sensor. *Sensors (Basel), 21*(13), 4559. doi:10.339021134559 PMID:34283115

MacKay, D. J. (2005). *Information Theory, Inference and Learning Algorithms.* Cambridge University Press.

Maggipinto, M., Terzi, M., Masiero, C., Beghi, A., & Susto, G. A. (2018). A Computer Vision-Inspired Deep Learning Architecture for Virtual Metrology Modeling with 2-Dimensional Data. *IEEE Transactions on Semiconductor Manufacturing, 31*(3), 376–384. doi:10.1109/TSM.2018.2849206

Mahdavinejad, M. S., Rezvan, M., Barekatain, M., Adibi, P., Barnaghi, P., & Sheth, A. P. (2018, August 1). Machine learning for internet of things data analysis: A survey. *Digital Communications and Networks*, *4*(3), 161–175. doi:10.1016/j.dcan.2017.10.002

Makin, J. G., Dichter, B. K., & Sabes, P. N. (2015). Learning to Estimate Dynamical State with Probabilistic Population Codes. *PLoS Computational Biology*, *11*(11), 1–28. doi:10.1371/journal.pcbi.1004554 PMID:26540152

Malik, N., & A, J. (2016, January). US expects energy savings through smart manufacturing. *MRS Bulletin, 41*(1), 10-11.

Malik, A. A., Masood, T., & Kousar, R. (2020). Repurposing factories with robotics in the face of COVID-19. *Science Robotics*, *5*(43), eabc2782. doi:10.1126cirobotics.abc2782 PMID:33022618

Maly, I., Sedlacek, D., & Leitao, P. (2016). Augmented Reality Experiments with Industrial Robot in Industry 4.0 Environment. *IEEE International Conference on Industrial Informatics*, 176-181. 10.1109/INDIN.2016.7819154

Martinez, B., Cano, C., & Vilajosana, X. (2018). *A Square Peg in a Round Hole: The Complex Path for Wireless in the Manufacturing Industry.* Cornell University.

Marvel, J. A., & Norcross, R. (2017, April). Implementing speed and separation monitoring in collaborative robot workcells. *Robotics and Computer-integrated Manufacturing*, *44*, 144–155. doi:10.1016/j.rcim.2016.08.001 PMID:27885312

Mastromatteo, D. (2017). The Python Pickle Module: How to Persist Objects in Python. *Real Python.* https://realpython.com/python-pickle-module/

Matthias, B., Oberer-Treitz, S., Staab, H., Schuller, E., & Peldschus, S. (2010). *Injury Risk Quantification for Industrial Robots in Collaborative Operation with Humans*. ROBOTIK.

Ma, W. J., Beck, J. M., Latham, P. E., & Pouget, A. (2006). Bayesian inference with probabilistic population codes. *Nature Neuroscience*, *9*(11), 1432–1438. doi:10.1038/nn1790 PMID:17057707

Mayika, J. M., Chui, B., Brown, J. Bughin, R., Dobbs, Roxburgh, C., & Byers, A. (2011). Big Data: The Next Frontier for Innovation, Competition, and Productivity. *McKinsey Report.*

McAfee, A., & Brynjolfsson, E. (2012). Big data: The management revolution. *Harvard Business Review*, *90*(10), 61–68. PMID:23074865

McKee, D. W., Clement, S. J., Almutairi, J., & Xu, J. (2018). Survey of advances and challenges in intelligent autonomy for distributed cyber-physical systems. *CAAI Transactions on Intelligence Technology*, *3*(2), 75–82. doi:10.1049/trit.2018.0010

McKinney, W., & Richardson, N. (2020). Feather V2 with Compression Support in Apache Arrow 0.17.0, *White Paper, URSA Labs.* https://ursalabs.org/blog/2020-feather-v2/

Megagon Labs. (n.d.). *Learning to Detect Semantic Types from Large Table Corpora*. White Paper. Available: https://megagonlabs.medium.com/learning-to-detect-semantic-types-from-large-table-corpora-fe22fcd97060

MessagePack. (n.d.). *It's Like JSON but Fast and Small.* MessagePack White Paper. Available: https://msgpack.org/index.html

Michael Franzino, A. G. (2020). *The $8.5 Trillion Talent Shortage.* Korn Ferry. Retrieved from https://www.kornferry.com/insights/this-week-in-leadership/talent-crunch-future-of-work

Miles, B. (2018). The malicious use of artificial intelligence: Forecasting, prevention, and mitigation. Academic Press.

Miller, G. A. (1956). The magical number seven, plus or minus two: Some limits on our capacity for processing information. *Psychological Review*, *63*(2), 81–97. doi:10.1037/h0043158 PMID:13310704

Mishra, S. (2021). Demystifying Delta Lake. *Analytics Vidya*. https://medium.com/analytics-vidhya/demystifying-delta-lake-d15869fd3470

Miškuf, M., & Zolotová, I. (2016). Comparison between multi-class classifiers and deep learning with focus on industry 4.0. *2016 Cybernetics and Informatics, K and I 2016 - Proceedings of the 28th International Conference*. IEEE. 10.1109/CYBERI.2016.7438633

Mithas, S., Ramasubbu, N., & Sambamurthy, V. (2011). How Information Management Capability Influences Firm Performance. *Management Information Systems Quarterly*, *35*(1), 237–256. doi:10.2307/23043496

Miyake, A., & Shah, P. (1999). *Models of working memory. Mechanisms of active maintenance and executive control*. Cambridge University Press. doi:10.1017/CBO9781139174909

Mkrttchian, V., Gamidullaeva, L., & Galiautdinov, R. (2019). Design of Nano-scale Electrodes and Development of Avatar-Based Control System for Energy-Efficient Power Engineering: Application of an Internet of Things and People (IOTAP) Research Center. *International Journal of Applied Nanotechnology Research*, *4*(1), 41–48. Advance online publication. doi:10.4018/IJANR.2019010104

Mohamed, E., Tantawi, K. H., Pemberton, A., Pickard, N., Dyer, M., Hickman, E., . . . Nasab, A. (2020). Real Time Gesture-Controlled Mobile Robot using a Myo Armband. In *Proceedings of the 2nd African International Conference on Industrial Engineering and Operations Management* (pp. 2432-2437). Harare, Zimbabwe: IEOM Society International.

Mongillo, G., & Deneve, S. (2008). Online learning with hidden markov models. *Neural Computation*, *20*(7), 1706–1716. doi:10.1162/neco.2008.10-06-351 PMID:18254694

Morency, L., & Baltrusaitis, T. (n.d.). *Tutorial on Multimodal Machine Learning*. White Paper, Language Technologies Institute, Carnegie Mellon University. Available: https://www.cs.cmu.edu/~morency/MMML-Tutorial-ACL2017.pdf

Moreno-Bote, R., Knill, D. C., & Pouget, A. (2011). Bayesian sampling in visual perception. *Proceedings of the National Academy of Sciences of the United States of America*, *108*(30), 12491–12496. doi:10.1073/pnas.1101430108 PMID:21742982

Movellan, J. R., Mineiro, P., & Williams, R. J. (2002). A Monte Carlo EM approach for partially observable diffusion processes: Theory and applications to neural networks. *Neural Computation*, *14*(7), 1507–1544. doi:10.1162/08997660260028593 PMID:12079544

MPHY0021. (2022). *Scientific File Formats*. University College London. http://github-pages.ucl.ac.uk/rsd-engineeringcourse/ch01data/070hdf5.html#:~:text=HDF5%20is%20the%20current%20version,hierarchy%2C%20similar%20to%20a%20filesystem

Musa, Y., Tantawi, O., Bush, V., Johson, B., Dixon, N., Kirk, W., & Tantawi, K. (2019). Low-Cost Remote Supervisory Control System for an Industrial Process using Profibus and Profinet. 2019 IEEE SoutheastCon, 1-4.

Nagowah, S. D., Ben Sta, H., & Gobin-Rahimbux, B. A. (2018). An Overview of Semantic Interoperability Ontologies and Frameworks for IoT. *2018 Sixth International Conference on Enterprise Systems (ES)*, 82-89. 10.1109/ES.2018.00020

Naidu, V. (2022). Performance of Using Appropriate File Formats in Big Data Hadoop Ecosystem. *Int'l Research Journal of Engnr & Tech*, *9*(1).

Ng, I. C., & Wakenshaw, S. Y. (2017). The Internet-of-Things: Review and research directions. *International Journal of Research in Marketing*, *34*(1), 3–21.

NIST. (n.d.). *Framework for Cyber-Physical Systems: Volume 1, Overview*. NIST Special Publication 1500-201. Available: https://nvlpubs.nist.gov/nistpubs/SpecialPublications/NIST.SP.1500-201.pdf

Nizamis, A., Vergori, P., Ioannidis, D., & Tzovaras, D. (2018). Semantic Framework and Deep Learning Toolkit Collaboration for the Enhancement of the Decision Making in Agent-Based Marketplaces. *Proceedings - 2018 5th International Conference on Mathematics and Computers in Sciences and Industry, MCSI 2018*, (pp. 135–140). IEEE. 10.1109/MCSI.2018.00039

Nugrahaeni & Mutijarsa. (2017). Comparative analysis of machine learning KNN, SVM, and random forests algorithm for facial expression classification. *IEEE Xplore*.

O'dwyer, J., & Renner, R. (2011). The Promise of Advanced Supply Chain Analytics, Supply Chain. *Management Review*, *15*, 32–37.

Okuda, H., Haraguchi, R., Domae, Y., & Shiratsuchi, K. (2016). Novel Intelligent Technologies for Industrial Robot in Manufacturing - Architectures and Applications. *Proceedings of ISR 2016: 47st International Symposium on Robotics*.

Olival, K. J., Hosseini, P. R., Zambrana-Torrelio, C., Ross, N., Bogich, T. L., & Daszak, P. (2017). Host and viral traits predict zoonotic spillover from mammals. *Nature*, *546*(7660), 646–650. doi:10.1038/nature22975 PMID:28636590

Oliva, R., & Watson, N. (2011). Cross-Functional Alignment in Supply Chain Planning: A Case Study of Sales & Operations Planning. *Journal of Operations Management*, *29*(5), 434–448. doi:10.1016/j.jom.2010.11.012

Oliveiros, S. (2016). *How to Choose a Data Format*. KDnuggets White Paper. Available: https://www.kdnuggets.com/2016/11/how-to-choose-data-format.html

Oostenveld, R., & Praamstra, P. (2001). The five percent electrode system for high-resolution EEG and ERP measurements. *Clinical Neurophysiology*, *112*(4), 713–719. doi:10.1016/S1388-2457(00)00527-7 PMID:11275545

Orban, G., Berkes, P., Fiser, J., & Lengyel, M. (2016). Neural Variability and Sampling-Based Probabilistic Representations in the Visual Cortex. *Neuron*, *92*(2), 530–543. doi:10.1016/j.neuron.2016.09.038 PMID:27764674

Oyekanlu, E. (2018a). Fault-Tolerant Real-Time Collaborative Network Edge Analytics for Industrial IoT and Cyber Physical Systems with Communication Network Diversity. *2018 IEEE 4th International Conference on Collaboration and Internet Computing (CIC)*, (pp. 336-345). IEEE. 10.1109/CIC.2018.00052

Oyekanlu, E. (2018b). Osmotic Collaborative Computing for Machine Learning and Cybersecurity Applications in Industrial IoT Networks and Cyber Physical Systems with Gaussian Mixture Models. *2018 IEEE 4th International Conference on Collaboration and Internet Computing (CIC)*, (pp. 326-335). IEEE. 10.1109/CIC.2018.00051

Oyekanlu, E., Nelatury, C., Fatade, A. O., Alaba, O., & Abass, O. (2017). Edge computing for industrial IoT and the smart grid: Channel capacity for M2M communication over the power line. *2017 IEEE 3rd International Conference on Electro-Technology for National Development (NIGERCON)*, 1-11, 10.1109/NIGERCON.2017.8281938

Oyekanlu, E., Onidare, S., & Oladele, P. (2018). Towards statistical machine learning for edge analytics in large scale networks: Real-time Gaussian function generation with generic DSP. *2018 First International Colloquium on Smart Grid Metrology (SmaGriMet)*, 1-6. 10.23919/SMAGRIMET.2018.8369850

Oyekanlu, E., Scoles, K., & Oladele, P. O. (2018). Advanced Signal Processing for Communication Networks and Industrial IoT Machines Using Low-Cost Fixed-Point Digital Signal Processor. *2018 10th International Conference on Advanced Infocomm Technology (ICAIT)*, 93-101. 10.1109/ICAIT.2018.8686577

Oyekanlu, E. (2017). Predictive Edge Computing for Time Series of Industrial IoT and Large Scale Critical Infrastructure based on Open-source software Analytic of Big Data, *2017 IEEE International Conference on Big Data (Big Data)*, (pp. 1663-1669). IEEE. 10.1109/BigData.2017.8258103

Oyekanlu, E. (2018c). Distributed Osmotic Computing Approach to Implementation of Explainable Predictive Deep Learning at Industrial IoT Network Edges with Real-Time Adaptive Wavelet Graphs. *2018 IEEE First International Conference on Artificial Intelligence and Knowledge Engineering (AIKE)*, (pp. 179-188). IEEE. 10.1109/AIKE.2018.00042

Oyekanlu, E. (2018c). *Powerline Communication for the Smart Grid and Internet of Things - Powerline Narrowband Frequency Channel Characterization Based on the TMS320C2000 C28x Digital Signal Processor*. Drexel University ProQuest Dissertations Publishing.

Oyekanlu, E. A., Smith, A. C., Thomas, W. P., Mulroy, G., Hitesh, D., Ramsey, M., Kuhn, D. J., Mcghinnis, J. D., Buonavita, S. C., Looper, N. A., Ng, M., Ng'oma, A., Liu, W., Mcbride, P. G., Shultz, M. G., Cerasi, C., & Sun, D. (2020). A Review of Recent Advances in Automated Guided Vehicle Technologies: Integration Challenges and Research Areas for 5G-Based Smart Manufacturing Applications. *IEEE Access : Practical Innovations, Open Solutions*, *8*, 202312–202353. doi:10.1109/ACCESS.2020.3035729

Oyekanlu, E., Mulroy, G., & Kuhn, D. (2022). Data Engineering for Factory of the Future: from Factory Floor to the Cloud – Part I: Performance Evaluation of State-of-the-Art Data Formats for Time Series Applications. In *Applied AI and Multimedia Technologies for Smart Manufacturing and CPS Applications. IGI Global.*

Oyekanlu, E., & Scoles, K. (2018a). Towards Low-Cost, Real-Time, Distributed Signal and Data Processing for Artificial Intelligence Applications at Edges of Large Industrial and Internet Networks. *2018 IEEE First International Conference on Artificial Intelligence and Knowledge Engineering (AIKE)*, (pp. 166-167). IEEE. 10.1109/AIKE.2018.00037

Oyekanlu, E., & Scoles, K. (2018b). Real-Time Distributed Computing at Network Edges for Large Scale Industrial IoT Networks. *2018 IEEE World Congress on Services (SERVICES)*, (pp. 63-64). IEEE. 10.1109/SERVICES.2018.00045

Paas, F., & Sweller, J. (2012). An evolutionary upgrade of cognitive load theory: Using the human motor system and collaboration to support the learning of complex cognitive tasks. *Educational Psychology Review*, *24*(1), 27–45. doi:10.100710648-011-9179-2

Paas, F., Tuovinen, J. E., Tabbers, H., & van Gerven, P. W. (2003). Cognitive load measurement as a means to advance cognitive load theory. *Educational Psychologist*, *38*(1), 63–71. doi:10.1207/S15326985EP3801_8

Paksoy, T., Karaoğlan, I., Gökçen, H., Pardalos, P. M., & Torğul, B. (2016). Experimental research on closed loop supply chain management with internet of things, Journal of Economy. *Bibliograph.*, *3*, 1–20.

Pal, K. (2017). Supply Chain Coordination Based on Web Services, in H K. Chan, N. Subramanian, & M. D. Abdulrahman (Eds), Supply Chain Management in the Big Data Era, 137-171. IGI Global Publishing, Hershey PA, USA.

Pal, K. (2018). A Big Data Framework for Decision Making in Supply Chain, in P K Gupta, T. Oren, & M. Singh (Edited). Predictive Intelligence Using Big Data and the Internet of Things, 51-76. IGI Global Publication.

Pal, K. (2019). Quality Assurance Issues for Big Data Applications in Supply Chain Management, in P K Gupta, T. Oren, & M. Singh (Eds.), Predictive Intelligence Using Big Data and the Internet of Things, 51-76. IGI Global Publication.

Pal, K. (2020). Information sharing for manufacturing supply chain management based on blockchain technology, I. Williams (Edited), in Cross-Industry Use of Blockchain Technology and Opportunities for the Future, 1-17. IGI Global Publication.

Pal, K. (2021). Applications of Secured Blockchain Technology. In S. K. Pani, B. Patnaik, S. Lun, & X Liu (Edited), Manufacturing Industry, in Blockchain and AI Technology in the Industrial Internet of Things. IGI Global Publication.

Pal, K. (2022). A Decentralized Privacy-Preserving Healthcare Blockchain for IoT, Challenges, and Solutions. In M. D. Borah, P. Zhang, & G. C. Deka (Edited), Prospects of Blockchain Technology for Accelerating Scientific Advancement in Healthcare, 158-188. IGI Global Publication.

Pal, K. (2023). Security Issues and Solutions for Resource-Constrained IoT Applications Using Lightweight Cryptography, in S. Verma, V. Vyas, & K. Kaushik (eds.) Cybersecurity Issues, Challenges, and Solutions in the Business World, 158-188. IGI Global Publication.

Pal, K. (2016). *Supply Chain Coordination Based on Web Service, Supply Chain Management in the Big Data Era, Hing Kai Chan, Nachiappan Subramanian, and Muhammad Dan-Asabe Abdulrahman (edited Book Chapter).* IGI Publication.

Pal, K. (2021). Privacy, Security and Policies: A Review of Problems and Solutions with Blockchain-Based Internet of Things Applications in Manufacturing Industry. *Procedia Computer Science*, *191*, 176–183.

Pal, K., & Yasar, A. (2020). Internet of Things and blockchain technology in apparel manufacturing supply chain data management. *Procedia Computer Science*, *170*, 450–457. doi:10.1016/j.procs.2020.03.088

Pal, K., & Yasar, A. U. H. (2020). Internet of Things and Blockchain Technology in Apparel Manufacturing Supply Chain Data Management. *Procedia Computer Science*, *170*, 450–457.

Park, S., & Choi, Y. (2020). Applications of Unmanned Aerial Vehicles in Mining from Exploration to Reclamation: A Review. *Minerals (Basel)*, *10*(8), 663. doi:10.3390/min10080663

Parvez, M. Z., & Paul, M. (2014). Epileptic seizure detection by analyzing EEG signals using different transformation techniques. *Neurocomputing*, *145*, 190–200. doi:10.1016/j.neucom.2014.05.044

Parvez, M. Z., & Paul, M. (2016). Seizure prediction using undulated global and local features. *IEEE Transactions on Biomedical Engineering*, *64*(1), 208–217. doi:10.1109/TBME.2016.2553131 PMID:27093309

Pathak, O. (2018). *Understanding Python Pickling with Example.* GeeksforGeeks Publication. Available: https://www.geeksforgeeks.org/understanding-python-pickling-example/

Pecevski, D., Buesing, L., & Maass, W. (2011). Probabilistic inference in general graphical models through sampling in stochastic networks of spiking neurons. *PLoS Computational Biology*, *7*(12), 7. doi:10.1371/journal.pcbi.1002294 PMID:22219717

Pekmezci, M. (2012). *Kısıtlanmış Boltzman Makinesi ile Zaman Serilerinin Tahmini, Y [Estimation of Time Series with Restricted Boltzman Machine, Y].* Lisans Tezi, Bilgisayar Mühendisliği Anabilim Dallı. Maltepe Üniversitesi.

Pellegrinelli, S., Orlandini, A., Pedrocchi, N., Umbrico, A., & Tolio, T. (2017). Motion planning and scheduling for human and industrial-robot collaboration. *CIRP Annals*, *66*(1), 1–4. doi:10.1016/j.cirp.2017.04.095

Peternel, G. (2021). *The Fundamentals of Data Warehouse + Data Lake = Lake House.* Towards Data Science. Available: https://towardsdatascience.com/the-fundamentals-of-data-warehouse-data-lake-lake-house-ff640851c832

Pillai, S., Punnoose, N. J., Vadakkepat, P., Loh, A. P., & Lee, K. J. (2018). An Ensemble of fuzzy Class-Biased Networks for Product Quality Estimation. *23rd IEEE International Conference on Emerging Technologies and Factory Automation, ETFA IEEE International Conference on Emerging Technologies and Factory Automation, ETFA, 2018-Septe*, (pp. 615–622). IEEE. 10.1109/ETFA.2018.8502492

Pivarski, J., Osborne, I., Das, P., Biswas, A., & Elmer, P. (2020). Awkward Array: JSON-like Data, Numpy-like Idioms. *Proc. of the 19th Python in Science Conference*. 10.25080/Majora-342d178e-00b

Plase, D., Niedrite, L., & Taranovs, R. (2017). A Comparison of HDFS Compact Data Formats: Avro Versus Parquet. *Elektronika ir Elektrotechnika*, *9*(3), 267–276.

Poluha, R. G. (2007). *Application of the SCOR Model in Supply Chain Management*. Youngstown.

Pool Party. (n.d.). *Introducing Semantic AI Ingredients for a sustainable Enterprise AI Strategy*. White Paper. Available: http://www.baonenterprises.com/uploads/1/0/1/6/101618342/semantic-ai-white-paper_en6.7.18.pdf

Porter, M. E. (1985). *Competitive Advantage: Creating and Sustaining Superior Performance*. The Free Press.

Power, D. J. (2007). A Brief History of Decision Support Systems. *DSS Resource*. http://DSSResource.COM/history/dsshistory.html

Poyiadjis, G., Doucet, A., & Singh, S. S. (2011). Particle approximations of the score and observed information matrix in state space models with application to parameter estimation. *Biometrika*, *98*(1), 65–80. doi:10.1093/biomet/asq062

Pradhan, A. (2012). Support vector machine-A survey. *International Journal of Emerging Technology and Advanced Engineering*, *2*(8), 82–85.

Prahalad, C. K., & Mashelkar, R. A. (2010). Innovation's Holy Grail. *Harvard Business Review*, (July-August), 2010.

Progress. (2017). *Benefits of JSON*. Available: https://docs.progress.com/en-US/bundle/openedge-abl-use-json-117/page/Benefits-of-JSON.html

Pykes, K. (2021). *Tensors and Arrays – What's the Difference?* Towards Data Science. Available: https://towardsdatascience.com/tensors-and-arrays-2611d48676d5

Python, Pickle. (2019). *Python Object Serialization*. Python. https://docs.python.org/3/library/pickle.html#:~:text=serialization%20and%20deserialization.-,Data%20stream%20format,to%20reconstruct%20pickled%20Python%20objects

Python. (n.d.). *CSV File Reading and Writing*. White Paper. Available: https://docs.python.org/3/library/csv.html

PyYAML 6.0. (n.d.). *Pip Install YAML – Project Description*. Available: https://pypi.org/project/PyYAML/

Rabanser, S., Shchur, O., & Gunnemann, S. (2017). Introduction to Tensor Decomposition and Their Applications in Machine Learning. arXiv:1711.10781v1 [stat.ML]

Ramm, J. (2021). Feather Documentation. *White Paper, Build Media*. https://buildmedia.readthedocs.org/media/pdf/plume/stable/plume.pdf

Ramm, J. (2021). *Feather Documentation*. White Paper, Release 0.1.0. Available: https://buildmedia.readthedocs.org/media/pdf/plume/stable/plume.pdf

Rashidibajgan, S., Hupperich, T., Doss, R., & Pan, L. (2020). Opportunistic Tracking in Cyber-Physical Systems. *2020 IEEE 19th International Conference on Trust, Security and Privacy in Computing and Communications (TrustCom)*, 1672-1679. 10.1109/TrustCom50675.2020.00230

Rastogi, Chaturvedi, Arora, Trivedi, & Chauhan. (2019). Framework for Use of Machine Intelligence on Clinical Psychology to Study the effects of Spiritual tools on Human Behavior and Psychic Challenges. *Journal of Image Processing and Artificial Intelligence, 4*(1).

Rastogi, Chaturvedi, Arora, Trivedi, & Mishra. (2018). Swarm Intelligent Optimized Method of Development of Noble Life in the perspective of Indian Scientific Philosophy and Psychology. *Journal of Image Processing and Artificial Intelligence, 4*(1). http://matjournals.in/index.php/JOIPAI/issue/view/463

Rastogi, Chaturvedi, Arora, Trivedi, & Singh. (2017). Role and efficacy of Positive Thinking on Stress Management and Creative Problem Solving for Adolescents. *International Journal of Computational Intelligence, Biotechnology and Biochemical Engineering, 2*(2).

Rastogi, Chaturvedi, Sharma, Bansal, & Agrawal. (2017). Understanding Human Behaviour and Psycho Somatic Disorders by Audio Visual EMG & GSR Biofeedback Analysis and Spiritual Methods. *International Journal of Computational Intelligence, Biotechnology and Biochemical Engineering, 2*(2).

Rastogi, R., Chaturvedi, D. K., Satya, S., Arora, N., Bansal, I., & Yadav, V. (2018). Intelligent Analysis for Detection of Complex Human Personality by Clinical Reliable Psychological Surveys on Various Indicators. *National Conference on 3rd Multi-Disciplinary National Conference Pre-Doctoral Research [MDNCPDR-2018] at Dayalbagh Educational Institute, Dayalbagh.*

Rastogi, R., Chaturvedi, D. K., Satya, S., Arora, N., Saini, H., Verma, H., Mehlyan, K., & Varshney, Y. (2018). Statistical Analysis of EMG and GSR Therapy on Visual Mode and SF-36 Scores for Chronic TTH. *Proceedings of International Conference on 5th IEEE Uttar Pradesh Section International Conference 2-4 Nov 2018 MMMUT Gorakhpur*. 10.1109/UPCON.2018.8596851

Rastogi, R., Chaturvedi, D. K., Satya, S., Arora, N., Singh, P., & Vyas, P. (2018). Statistical Analysis for Effect of Positive Thinking on Stress Management and Creative Problem Solving for Adolescents. *Proceedings of the 12th INDIACom; INDIACom-2018; IEEE Conference ID: 42835, Technically Sponsored by IEEE Delhi Section 2018, 5th International Conference on "Computing for Sustainable Global Development*, 245-251.

Rastogi, R., Chaturvedi, D. K., Satya, S., Arora, N., Yadav, V., & Chauhan, S. (2018). An Optimized Biofeedback Therapy for Chronic TTH between Electromyography and Galvanic Skin Resistance Biofeedback on Audio, Visual and Audio Visual Modes on Various Medical Symptoms. *National Conference on 3rd Multi-Disciplinary National Conference Pre-Doctoral Research [MDNCPDR-2018] at Dayalbagh Educational Institute.*

Rastogi, R., Chaturvedi, D., Sharma, S., Bansal, A., & Agrawal, A. (2017). Audio-Visual EMG & GSR Biofeedback Analysis for Effect of Spiritual Techniques on Human Behaviour and Psychic Challenges. *Journal of Applied Information Science, 5*(2), 37-46. Retrieved from https://www.i-scholar.in/index.php/jais/article/view/167372

Rastogi, R., Chaturvedi, D. K., Arora, N., Trivedi, P., & Mishra, V. (2017). Swarm Intelligent Optimized Method of Development of Noble Life in the perspective of Indian Scientific Philosophy and Psychology. *Proceedings of NSC-2017 (National system conference) IEEE Sponsored conf. of Dayalbagh Educational Institute, Agra.*

Rebeschini, P., & Van Handel, R. (2015). Can local particle filters beat the curse of dimensionality? *Annals of Applied Probability, 25*(5), 2809–2866. doi:10.1214/14-AAP1061

Ren, S., He, K., Girshick, R., & Sun, J. (2017). Faster R-CNN: Towards Real-Time Object Detection with Region Proposal Networks. *IEEE Transactions on Pattern Analysis and Machine Intelligence, 39*(6), 1137–1149. doi:10.1109/TPAMI.2016.2577031 PMID:27295650

Research, F. &. (2016). *Global Additive Manufacturing Market, Forecast to 2025.* Frost & Sullivan MB74-10.

Retter, A., Underdown, D., & Walpole, R. (n.d.). *CSV Schema Language 1.2 - A Language for Defining and Validating CSV Data.* W3C White Paper. Available: https://digital-preservation.github.io/csv-schema/csv-schema-1.2.html

Reuters. (2018, June 25). *Facing US blowback, Beijing softens its 'Made in China 2025' message.* Retrieved September 24, 2018, from https://www.cnbc.com/2018/06/25/facing-us-blowback-beijing-softens-its-made-in-china-2025-message.html

Richter, J., Streitferdt, D., & Rozova, E. (2017). On the Development of Intelligent Optical Inspections. *7th Annual Computing and Communication Workshop and Conference, CCWC 2017.* IEEE. 10.1109/CCWC.2017.7868455

Robotics, I. F. (2018). Industrial robot sales increase worldwide by 31 percent. International Federation of Robotics.

Robotics, I. F. (2017). *How robots conquer industry worldwide.* Author.

Robotics, I. F. (2017). *How robots conquer industry worldwide.* IFR Press Conference.

Roy, R. N., Bonnet, S., Charbonnier, S., & Campagne, A. (2013). Mental fatigue and working memory load estimation: interaction and implications for EEG-based passive BCI. *2013 35th Annual International Conference of the IEEE Engineering in Medicine and Biology Society (EMBC),* 6607-6610.

Ruffaldi, E., Brizzi, F., Tecchia, F., & Bacinelli, S. (2016). Third Point of View Augmented Reality for Robot Intentions Visualization. *International Conference on Augmented Reality, Virtual Reality and Computer Graphics, 9768.* 10.1007/978-3-319-40621-3_35

Ruiz, M., Mujica, L. E., Alférez, S., Acho, L., Tutivén, C., Vidal, Y., Rodellar, J., & Pozo, F. (2018). *Wind turbine fault detection and classification by means of image texture analysis* (Vol. 107). Mechanical Systems and Signal Processing. doi:10.1016/j.ymssp.2017.12.035

Rüßmann, M., Lorenz, M., Gerbert, P., Waldner, M., Justus, J., Engel, P., & Harnisch, M. (2015). *Industry 4.0: The Future of Productivity and Growth in Manufacturing Industries.* Boston Consulting Group.

Ryu, K., & Myung, R. (2005). Evaluation of mental workload with a combined measure based on physiological indices during a dual task of tracking and mental arithmetic. *International Journal of Industrial Ergonomics, 35*(11), 991–1009. doi:10.1016/j.ergon.2005.04.005

Saidur, R. (2010). A review on electrical motors energy use and energy savings. *Renewable & Sustainable Energy Reviews, 14*(3), 877–898. doi:10.1016/j.rser.2009.10.018

Saikia, A. R., Bora, K., Mahanta, L. B., & Das, A. K. (2019). *Comparative assessment of CNN architectures for classification of breast FNAC images* (Vol. 57). Tissue and Cell. , doi:10.1016/j.tice.2019.02.001

Sanchez, A. (n.d.). MessagePack, Racket White Paper, online. Available: https://docs.racket-lang.org/msgpack/index.html#:~:text=MessagePack%20is%20an%20efficient%20binary,addition%20to%20the%20strings%20themselves

Santur, Y., Karaköse, M., & Akın, E. (2016). Learning Based Experimental Approach For Condition Monitoring Using Laser Cameras In Railway Tracks. *International Journal of Applied Mathematics,Electronics and Computers*, *4*(Special Issue-1), 1–5. doi:10.18100/ijamec.270656

Saravanan, M., Satheesh Kumar, P., & Sharma, A. (2019). IoT enabled indoor autonomous mobile robot using CNN and Q-learning. *2019 IEEE International Conference on Industry 4.0, Artificial Intelligence, and Communications Technology, IAICT 2019*, (pp. 7–13). IEEE. 10.1109/ICIAICT.2019.8784847

SAS. (2012). *Supply Chain Analytics: Beyond ERP and SCM*. SAS.

Sathi, A. (2012). Big data analytics: Disruptive technologies for changing the game. MC Press Online.

Sertkaya, M. E. (2018). *Derin Öğrenme Tekniklerinin Biyomedikal İmgeler Üzerine Uygulamaları [Applications of Deep Learning Techniques on Biomedical Images]*. Fırat Üniversitesi.

Shaheen, J. (2015, April 22). *Shaheen Introduces Bill to Emhance Innovation, Energy Efficiency and Economic Competitiveness of Nation's Manufacturers*. Retrieved June 2017, from https://www.shaheen.senate.gov/news/press/shaheen-introduces-bill-to-enhance-innovation-energy-efficiency-and-economic-competitiveness-of-nations-manufacturers

Shapiro, J. (2010). Advanced Analytics for Sales & Operations Planning. *Analytics Magazine*, (May-June), 20–26.

Sharma & Rameshan. (2017). Dictionary Based Approach for Facial Expression Recognition from Static Images. *Int. Conf. Comput. Vision, Graph. Image Process.*, 39–49.

Sharma, A. (2022). What is YAML? A Beginner's Guide. *White Paper, Circleci*. https://circleci.com/blog/what-is-yaml-a-beginner-s-guide/#:~:text=YAML%20is%20a%20digestible%20data,that%20JSON%20can%20and%20more;

Sharma, S. (2021). *Ansible for Beginners – Overview, Architecture and Use Cases*. White Paper, K21 Academy. Available: https://k21academy.com/ansible/ansible-for-beginners/

Shin, T. H., Chin, S., Yoon, S. W., & Kwon, S. W. (2011). A service-oriented integrated information framework for RFID/WSN-based intelligent construction supply chain management. *Automation in Construction*, *20*, 706–715.

Siddiqui, S. A., Mercier, D., Munir, M., Dengel, A., & Ahmed, S. (2019). TSViz: Demystification of Deep Learning Models for Time-Series Analysis. *IEEE Access: Practical Innovations, Open Solutions*, *7*, 67027–67040. doi:10.1109/ACCESS.2019.2912823

Siegel, E. (2013). *Predictive analytics: The power to predict who will click, buy, lie or die*. John Wiley & Sons Inc.

Silva, B., Sousa, J., & Alenya, G. (2021). Data Acquisition and Monitoring System for Legacy Injection Machines. *2021 IEEE International Conference on Computational Intelligence and Virtual Environments for Measurement Systems and Applications (CIVEMSA)*, (pp. 1-6). IEEE. 10.1109/CIVEMSA52099.2021.9493675

Singh, C. (2018). Advantages and Disadvantages of XML. *BeginnersBook*. https://beginnersbook.com/2018/10/advantages-and-disadvantages-of-xml/

Singh, S., & Kaur, A. (2015, March). Face Recognition Technique using Local Binary Pattern Method. *International Journal of Advanced Research in Computer and Communication Engineering*, *4*(3).

Skulmowski, A., & Rey, G. D. (2017). Measuring cognitive load in embodied learning settings. *Frontiers in Psychology*, *8*, 1191. doi:10.3389/fpsyg.2017.01191 PMID:28824473

Smart Manufacturing Operations Planning and Control Program. (n.d.). National Institute of Standards and Technology. Retrieved 6 16, 2017, from https://www.nist.gov/sites/default/files/documents/2017/05/09/FY2014_SMOPAC_ProgramPlan.pdf

Snyder, J. (2019). Data Cleansing: An Omission from Data Analytics Coursework. *Information Systems Education Journal (ISEDJ)*, *17*(6).https://files.eric.ed.gov/fulltext/EJ1224578.pdf

Sokoloski, S. (2015). *Implementing a Bayes Filter in a Neural Circuit: The Case of Unknown Stimulus Dynamics*. ArXiv, ArXiv:1512.07839.

Soualhi, A., Clerc, G., & Razik, H. (2013). Detection and diagnosis of faults in induction motor using an improved artificial ant clustering technique. *IEEE Transactions on Industrial Electronics*, *60*(9), 4053–4062. doi:10.1109/TIE.2012.2230598

Souza, R., Nascimento, E., Miranda, U., Silva, W., & Lepikson, H. (2021). Deep learning for diagnosis and classification of faults in industrial rotating machinery. *Computers & Industrial Engineering*, *153*. doi:10.1016/j.cie.2020.107060

Stack Overflow. (2019). Why do my hdf5 files seem so unnecessarily large? *Stack Overflow*. https://stackoverflow.com/questions/65119241/why-do-my-hdf5-files-seem-so-unnecessarily-large

Stadler, S., Kain, K., Giuliani, M., Mirnig, N., Stollnberger, G., & Tscheligi, M. (2016). Augmented reality for industrial robot programmers: Workload analysis for task-based, augmented reality-supported robot control. *25th IEEE International Symposium on Robot and Human Interactive Communication*. 10.1109/ROMAN.2016.7745108

Stanford University. (n.d.). *Parsing JSON with jq*. Available: http://www.compciv.org/recipes/cli/jq-for-parsing-json/

Staubli, G. (2017). *Real Time Big Data Analytics: Parquet (and Spark) + Bonus*. Linkedin. https://www.linkedin.com/pulse/real-time-big-data-analytics-parquet-spark-bonus-garren-staubli/

Stiglitz, J. (2017). *Globalization and its Discontents Revisited: Anti-Globalization in the Era of Trump*. Penguin.

Stock, J. R. (2013). Supply chain management: A look back, a look ahead. *Supply Chain Quarterly*, *2*, 22–26.

Stock, T., & Seliger, G. (2016). Opportunities of sustainable manufacturing in industry 4.0. *Procedia*, *40*, 536–541.

Strickland, J. D. (2016). Applications of Additive Manufacturing in the Marine Industry. *Proceedings of PRADS2016*.

Subakti, H., & Jiang, J. R. (2018). Indoor Augmented Reality Using Deep Learning for Industry 4.0 Smart Factories. *42nd IEEE International Conference on Computer Software & Application*, *2*, (pp. 63–68). IEEE. 10.1109/COMPSAC.2018.10204

Supply Chain Council. (2010). *Homepage*. ASCM. http://supply-chain.org/f/down-load/726710733/SCOR10.pdf

Surace, S. C., & Pfister, J.-P. (2016). *Online Maximum Likelihood Estimation of the Parameters of Partially Observed Diffusion Processes*. ArXiv:1611.00170.

Sustainable Manufacturing. (n.d.). United States Environmental Protection Agency. Retrieved June 17, 2017, from https://www.epa.gov/sustainability/sustainable-manufacturing

Süzen, A. A., & Kayaalp, K. (2018a). Derin Öğrenme ve Türkiye'deki Uygulamaları [Deep Learning and its Applications in Turkey] (1st ed.). Research Gate. https://www.researchgate.net/publication/327666072_Derin_Ogrenme_ve_Turkiye'deki Uygulamalari

Süzen, A. A., & Kayaalp, K. (2018b). Derin Öğrenme Yöntemleri İle Sıcaklık Tahmini: Isparta İli Örneği [Temperature Estimation with Deep Learning Methods: Example of Isparta Province]. *International Academic Research Congress INES 2018*, (December).

Süzen, A. A., & Kayaalp, K. (2019). Endüstri 4.0 ve Adli Bilişim [Industry 4.0 and Forensic Informatics]. *Bilişim ve Teknoloji Araştırmaları, 2019*, 23–31.

Svilvar, M., Charkraborty, A. & Kanioura, A. (2013). Big data analytics in marketing. *OR/MS Today.*

Sweller, J. (1998). Cognitive Load During Problem Solving: Effects on Learning. *Cognitive Science, 12*(2), 257–285. doi:10.120715516709cog1202_4

Sweller, J., van Merrienboer, J. J., & Paas, F. G. (1998). Cognitive Architecture and Instructional Design. *Educational Psychology Review, 10*(3), 251–296. doi:10.1023/A:1022193728205

Sztipanovits. (2013). Strategic R&D Opportunities for 21st Century Cyber-Physical Systems – Connecting Computer and Information Systems with the Physical World. NIST White Paper. Available: https://www.nist.gov/system/files/documents/el/12-Cyber-Physical-Systems020113_final.pdf

TACCO. (n.d.). *Defining Tensors.* TACCO White Paper. Available: http://tensor-compiler.org/docs/tensors

Tan, J., & Koo, S. (2014). A survey of technologies in internet of things, in IEEE. *Computers & Society*, 269–274.

Tantawi, K., Fidan, I., & Tantawy, A. (2019). Status of Smart Manufacturing in the United States. 2019 IEEE SoutheastCon.

Tantawi, K. (2020). Literature Review: Rethinking BioMEMS in the aftermath of CoVid-19. *Biomedical Journal of Scientific & Technical Research, 31*(1), 23944–23946. doi:10.26717/BJSTR.2020.31.005053

Tantawi, K., Ashcroft, J., Cossette, M., Kepner, G., & Friedman, J. (2022). Investigation of the Post-Pandemic STEM Education (STEM 3.0). *Journal of Advanced Technological Education, 1*(1).

Tantawi, K., Fidan, I., Chitiyo, G., & Cossette, M. (2021). Offering Hands-on Manufacturing Workshops Through Distance Learning. *ASEE Annual Conference.*

Tantawi, K., Sokolov, A., & Tantawi, O. (2019). Advances in Industrial Robotics: From Industry 3.0 Automation to Industry 4.0 Collaboration. *4th Technology Innovation Management and Engineering Science International Conference (TIMES-iCON).* 10.1109/TIMES-iCON47539.2019.9024658

Technology, N. I. (2012). Measurement Science Roadmap for Metal-Based Additive Manufacturing. Academic Press.

Tekin, Z., & Karakuş, K. (2018). Gelenekselden Ak ı ll ı Üretime Spor Endüstrisi 4 . 0-. *İnsan Ve Toplum Bilimleri Araştırmaları Dergisi,*[Sports Industry 4.0 from Conventional to Smart Production] *7*(3), 2103–2117.

TensorFlow. (n.d.). *TFRecords and tf.train.Example.* Available: https://www.tensorflow.org/tutorials/load_data/tfrecord

Terry, S., Fidan, I., Zhang, Y., & Tantawi, K. (2019). *Smart Manufacturing for Energy Conservation and Savings.* NSF-ATE Conference.

Terry, S., Lu, H., Fidan, I., Zhang, Y., Tantawi, K., Guo, T., & Asiabanpour, B. (2020). The Influence of Smart Manufacturing Towards Energy Conservation: A Review. *Technologies, 8*(2), 31. doi:10.3390/technologies8020031

The Economist. (2010). Data, Data Everywhere. *The Economist.* https://www.economist.com/node/15557443

The Quietly Emerging Artificial Intelligence Arms Race. (2019). *China Youth Daily*. Retrieved from https://m.chinanews.com/wap/detail/zw/gn/2019/10-17/8981224.shtml

The Robot Revolution: The New Age of Manufacturing | Moving Upstream S1-E9. (2018). *Wall Street Journal.*

The skills gap in U.S. manufacturing 2015 and beyond. (2015). The Manufacturing Institute.

Thilak, V., Devadasan, S., & Sivaram, N. (2015). A Literature Review on the Progression of Agile Manufacturing Paradigm and Its Scope of Application in Pump Industry. *TheScientificWorldJournal.*

Thomas, C. (2019, February 14). Artificial intelligence crime: An interdisciplinary analysis of foreseeable threats and solutions. *Science and Engineering Ethics*. Advance online publication. PMID:30767109

Timescale. (2020). Time-Series Compression Algorithms, Explained. *Timescale*. https://www.timescale.com/blog/time-series-compression-algorithms-explained/

Tokcan, N., Gryak, J., Najarian, K., & Deriksen, H. (2020). Algebraic Methods for Tensor Data. arXiv:2005.12988v1 [math.RT]

Toygar, O., & Acan, A. (2003). Face Recognition Using PCA, LDA, and ICA Approach on Colored Images. *Journal of Electrical and Electronics Engineering (Oradea)*, *3*(1), 735–743.

TrendsT. (2018). https://www.gartner.com/smarterwithgartner/gartner-top-10-strategic-technology-trends-for-2018

Tripathy, R., & Daschoudhary, R. N. (2014). Real-time face detection and Tracking using Haar Classifier on SoC. *International Journal of Electronics and Computer Science Engineering.*, *3*(2), 175–184.

Trkman, P., McCormack, K., de Oliveira, M. P. V., & Ladeira, M. B. (2010). The Impact of Business Analytics on Supply Chain Performance. *Decision Support Systems*, *49*(3), 318–327. doi:10.1016/j.dss.2010.03.007

Tsai, S. Y., & Chang, J. Y. J. (2018). Parametric Study and Design of Deep Learning on Leveling System for Smart Manufacturing. *2018 IEEE International Conference on Smart Manufacturing, Industrial and Logistics Engineering, SMILE 2018, 2018-Janua*, (pp. 48–52). IEEE. 10.1109/SMILE.2018.8353980

Tsutsui, T., & Matsuzawa, T. (2019). Virtual metrology model robustness against chamber condition variation using deep learning. *IEEE Transactions on Semiconductor Manufacturing*, *32*(4), 428–433. doi:10.1109/TSM.2019.2931328

Turban, E., & Sepehri, M. (1986). Applications of Decision Support and Expert Systems in Flexible Manufacturing Systems. *Journal of Operations Management*, *6*(34), 433–448. doi:10.1016/0272-6963(86)90015-X

Turban, E., Sharda, R., Delen, D., & King, D. (2011). *Business Intelligence: A Managerial Approach* (2nd ed.). Prentice-Hall.

Turing, A. M. (1950). Computing machinery and intelligence. *Mind*, *59*(236), 433–460. doi:10.1093/mind/LIX.236.433

Turner-Trauring, I. (2021). *Where's Your Bottleneck? CPU Time vs Wallclock Time*. Available: https://pythonspeed.com/articles/blocking-cpu-or-io/

Tutorialspoint. (n.d.). *JSON Quick Guide*. Available: https://www.tutorialspoint.com/json/pdf/json_quick_guide.pdf

Ty, J., Chen, P. H., Te, J., Nanfei, D. W. (2016). *The Fault Feature Extraction of Rolling Bearing Based on EMD and Difference Spectrum of Singular Value*. doi:. doi:10.1155/2016/5957179

Tyson, M. (n.d.). *What is JSON? The Universal Data Format*. White Paper, Infoworld. Available: https://www.infoworld.com/article/3222851/what-is-json-a-better-format-for-data-exchange.html

Uber. (n.d.). *Introducing Petastorm: Uber ATG's Data Access Library for Deep Learning*. Uber White Paper. Available: https://www.uber.com/blog/petastorm/

UIC-712 R. (2002). *Rail defects, International Union of Railways (UIC)*. Fransa.

Uřičář, M., Franc, V., & Hlaváč, V. (2012). The detector of facial landmarks learned by the structured output SVM. *VISAPP'12: Proceedings of the 7th International Conference on Computer Vision Theory and Applications*, 1, 547-556.

Vafeiadis, T., Nizamis, A., Pavlopoulos, V., Giugliano, L., Rousopoulou, V., Ioannidis, D., & Tzovaras, D. (2019). Data Analytics Platform for the Optimization of Waste Management Procedures. *15th Annual International Conference on Distributed Computing in Sensor Systems, DCOSS 2019*, (pp. 333–338). IEEE. 10.1109/DCOSS.2019.00074

Van Merrienboer, J. J., & Sweller, J. (2005). Cognitive load theory and complex learning: Recent developments and future directions. *Educational Psychology Review*, *17*(2), 147–177. doi:10.100710648-005-3951-0

Verdouw, C. N., Wolfert, J., Beulens, A., & Rialland, A. (2016). Virtualization of food supply chains with the internet of things. *Journal of Food Engineering*, *176*, 128–136.

Veza, I., Mladineo, M., & Gjeldum, N. (2015). Managing innovative production network of smart factories. *IFAC-PapersOnLine*, *48*, 555–560.

Viola, P., & Jones, M. (2001). Rapid object detection using a boosted cascade of simple features. *Computer Vision and Pattern Recognition.*, *1*, 511–518. doi:10.1109/CVPR.2001.990517

Viola, P., & Jones, M. J. (2004, May). Robust Real-Time Face Detection. *International Journal of Computer Vision*, *57*(2), 137–154. doi:10.1023/B:VISI.0000013087.49260.fb

Vishnu Priya, T. S., Vinitha Sanchez, G., & Raajan, N. R. (2018). Facial Recognition System Using Local Binary Patterns (LBP). *International Journal of Pure and Applied Mathematics*, *119*(15), 1895–1899.

Viswanathan, N., & Sadlovska, V. (2010). *Supply Chain Intelligence: Adopt Role-based Operational Business Intelligence and Improve Visibility*. Aberdeen Group.

Vosburg, J., & Kumar, A. (2011). Managing Dirty Data in Organizations Using ERP: Lessons from a Case Study. *Industrial Management & Data Systems*, *101*(1), 21–31. doi:10.1108/02635570110365970

W3C. (2012). *SPARQL 1.1 Query Results CSV and TSV Formats*. White Paper. Available: https://www.w3.org/2009/sparql/docs/csv-tsv-results/results-csv-tsv.html

Wang, L., Laszewski, G. V., Young, K. M., & Tao, J. (2010). Cloud Computing: A Perspective Study. *New Generation Computing*, *28*(2), 137–146.

Wang, T., Zhang, Y., & Zang, D. (2016) Real-time visibility and traceability framework for discrete manufacturing shop-floor. In *Proceedings of the 22nd International Conference on Industrial Engineering and Engineering Management*,(pp. 763–772). IEEE.

Wasser, L. (2020). Hierarchical Data Formats – What is HDF5? *Neon Science*. https://www.neonscience.org/resources/learning-hub/tutorials/about-hdf5#:~:text=About%20Hierarchical%20Data%20Formats%20%2D%20HDF5,-The%20Hierarchical%20Data&text=HDF5%20uses%20a%20%22file%20directory,metadata%20making%20it%20self%2Ddescribing

Watson, M., Lewis, S., Cacioppi, P., & Jayaraman, J. (2013). Supply chain network design – applying optimization and analytics to the global supply chain. *FT Press*.

Watson, H. J. (2009). Tutorial: Business Intelligence – Past, Present, and Future. *Communications of the Association for Information Systems*, *39*(25). doi:10.17705/1CAIS.02539

Webster, G., Creemers, R., Triolo, P., & Kania, E. (2017). *Full Translation: China's 'New Generation Artificial Intelligence Development Plan*. New America.

Webster, S. A. (2016). *FANUC Launches New IoT System for Smart Manufacturing Era*. Society of Manufacturing Engineering.

Welinder, P., Branson, S., Mita, T., Wah, C., Schroff, F., Belongie, S., & Perona, P. (2010). *Caltech-UCSD Birds 200*. California Institute ofTechnology. CNS-TR-2010-001. 2010. https://towardsdatascience.com/face-recognition-how-lbph-works-90ec258c3d6b

Welling, M., & Teh, Y. (2011), Bayesian Learning via Stochastic Gradient Langevin Dynamics. *Proceedings of the 28th International Conference on Machine Learning*.

Wilson, R. C., & Finkel, L. (2009). A neural implementation of the Kalman filter. *Advances in Neural Information Processing Systems*, 22.

Xavier, L. (2021). Evaluation and Performance of Reading from Big Data Formats, [Bachelor's Thesis, Federal Univ. of Rio Grande do Sul, Brazil]. https://www.lume.ufrgs.br/bitstream/handle/10183/223552/001127314.pdf?sequence=1.

Xing, B., Gao, W. J., Battle, K., Nelwamondo, F. V., & Marwala, T. (2012). e-RL: the Internet of things supported reverse logistics for remanufacture-to-order. International Conference in Swarm Intelligence: Advances in Swarm Intelligence, (pp. 519–526). Springer.

Xu, X., & Hua, Q. (2017). Industrial Big Data Analysis in Smart Factory: Current Status and Research Strategies. *IEEE Access: Practical Innovations, Open Solutions*, *5*, 17543–17551. doi:10.1109/ACCESS.2017.2741105

Yaman, O., Karakose, M., & Akin, E. (2017a). Improved Rail Surface Detection and Condition Monitoring Approach with FPGA in Railways. *International Conference on Advanced Technology & Sciences (ICAT'17)*, (pp. 108-111). IEEE.

Yaman, O., Karakose, M., & Akin, E. (2017b). A Fault Diagnosis Approach for Rail Surface Anomalies Using FPGA in Railways. *International Journal of Applied Mathematics Electronics and Computers*, *2017*(Special Issue), 42–46. doi:10.18100/ijamec.2017SpecialIssue30469

Yan, B., & Huang, G. (2009). Supply chain information transmission based on RFID and internet of things in Computing. In *Communication, Control, and Management*, (pp. 166–169). ISECS International Colloquium.

Yang, T., Laugesen, R. S., Mehta, P. G., & Meyn, S. P. (2016). Multivariable feedback particle filter. *Automatica*, *71*, 10–23. doi:10.1016/j.automatica.2016.04.019

Yang, T., Mehta, P. G., & Meyn, S. P. (2013). Feedback particle filter. *IEEE Transactions on Automatic Control, 58*(10), 2465–2480. doi:10.1109/TAC.2013.2258825

Yan, H., Wan, J., Zhang, C., Tang, S., Hua, Q., & Wang, Z. (2018). Industrial Big Data Analytics for Prediction of Remaining Useful Life Based on Deep Learning. *IEEE Access: Practical Innovations, Open Solutions*, *6*, 17190–17197. doi:10.1109/ACCESS.2018.2809681

Yavus, B., Armbrust, M., Das, T., & Condie, T. (2017). Working with Complex Data Format with Structured Streaming in Apache Spark 2.1. *Databricks.* https://databricks.com/blog/2017/02/23/working-complex-data-formats-structured-streaming-apache-spark-2-1.html

Ye, X., Song, W. S., Hong, S. H., Kim, Y. C., & Yoo, N. H. (2022). Toward Data Interoperability of Enterprise and Control Applications via the Industry 4.0 Asset Administration Shell. *IEEE Access : Practical Innovations, Open Solutions*, *10*, 35795–35803. doi:10.1109/ACCESS.2022.3163738

Yu, S., Wu, Z., Zhu, X., & Pecht, M. (2019). A Domain Adaptive Convolutional LSTM Model for Prognostic Remaining Useful Life Estimation under Variant Conditions. *Proceedings - 2019 Prognostics and System Health Management Conference, PHM-Paris 2019*, (pp. 130–137). 10.1109/PHM-Paris.2019.00030

Yuan, H., Li, F., & Wang, H. (2012). Using Evaluation and Leading Mechanism To Optimize Fault Diagnosis Based on Ant Algorithm. *Energy Proscenia*, *1*(6), 112–116.

Yue, G., Ping, G., & Lanxin, L. (2018). An End-to-End model base on CNN-LSTM for Industrial Fault Diagnosis and Prognosis. [IEEE.]. *Proceedings of IC-NIDC*, *2018*, 274–278.

Yuvaraj, S., & Sangeetha, M. (2016). Smart supply chain management using internet of things (IoT) and low power wireless communication systems. In *Wireless Communication, Signal Processing and Networking, International Conference*, (pp. 555-558). IEEE.

Zabiński, T., Maoczka, T., Kluska, J., Madera, M., & Sęp, J. (2019). Condition Monitoring in Industry 4.0 Production Systems - The Idea of Computational Intelligence Methods Application. *Procedia CIRP*, *79*, 63–67. doi:10.1016/j.procir.2019.02.012

Zaczyński. (n.d.). *YAML: The Missing Battery in Python.* Available: https://realpython.com/python-yaml/

Zawadzki, P., & Zywicki, K. (2016). Smart product design and production control for effective mass customization in the industry 4.0 concept. *Management of Production Engineering Review*, *7*, 105–112.

Zemmar, A., Lozano, A. M., & Nelson, B. J. (2020). The rise of robots in surgical environments during COVID-19. *Nature Machine Intelligence*, *2*, 566–572.

Zhang. (n.d.). Sato: Contextual Semantic Type Detection in Tables. *Proceedings of the VLDB Endowment, 13*(11). doi:10.14778/3407790.3407793

Zhang, L., & Schaeffer, H. (2020). Forward Stability of ResNet and Its Variants. *Journal of Mathematical Imaging and Vision*, *62*(3), 328–351. doi:10.100710851-019-00922-y

Zhang, Y. D., Satapathy, S. C., Zhang, X., & Wang, S.-H. (2021). COVID-19 Diagnosis via DenseNet and Optimization of Transfer Learning Setting. *Cognitive Computation.* doi:10.100712559-020-09776-8 PMID:33488837

Zhu, Chen, Zhao, Zhou, & Zhang. (2017). Emotion Recognition from Chinese Speech for Smart Affective Services Using a Combination of SVM and DBN. *New Advances in Identification Information & Knowledge on the Internet of Things.*

About the Contributors

* * *

Ismail Fidan serves as a Professor of the Department of Manufacturing and Engineering Technology and Director of iMakerSpace at Tennessee Tech University. His research and teaching interests are in additive manufacturing, electronics manufacturing, distance learning, and STEM education. Dr. Fidan is a member and active participant of SME, ASEE, ABET, ASME, IEEE and Tennessee Academy of Science. He is also the Associate Editor of IEEE Transactions on Components, Packaging, and Manufacturing Technology and International Journal of Rapid Manufacturing, and Associate Author of Wohlers Reports. Dr. Fidan is an ABET Program Evaluator for Manufacturing Engineering and Manufacturing Engineering Technology Programs.

Rinat Galiautdinov is a Principal Software Developer and Software Architect having the expertise in Information Technology and Computer Science. Mr. Galiautdinov is also an expert in Banking/Financial industry as well as in Neurobiological sphere. Mr. Rinat Galiautdinov works on the number of highly important researches (including such the spheres as: Drones, Financial Systems and Private Money, Brain interface and neuro-biology) as an independent researcher and projects holder and developer.

David J. Kuhn received the B.Sc. degree in electrical engineering and the M.Sc. degree in engineering management from Clarkson University, Potsdam, NY, USA, in 1989 and 2008, respectively. From 1993 to 1997, he worked with Emhart Powers, NY, USA, as a Project Engineer. He also worked with Monarch Machine Tools as a Control Engineer from 1991 to 1993. From 1997 to 2009, he has worked with Corning Inc., New York, NY, USA, in numerous positions, including Senior Engineer Systems, Senior Engineer Controls, Project Engineer, and as a Supervisor in imaging systems and development. Since 2009, he has been the Manager of the System Integration and Innovation Department, Corning Inc. He has five U.S. patents in the area of glass molding apparatus design.

Victoria Martino is a student of computer science and engineering at the University of Tennessee at Chattanooga. She was a student researcher in Dr. Khalid Tantawi's lab.

Grethel Mulroy received the B.Sc. degree in mechanical engineering from the Worcester Polytechnic Institute, Massachusetts, MA, USA, in 1991, and the M.Sc. degree in engineering management from Clarkson University, Potsdam, NY, USA, in 2001. Since 1991, she held several positions at Xerox Corporation as the Director, Manager, Executive Technical Assistant to SVP, Project Engineer, and as a Manufacturing Process Engineer. Her experience in manufacturing and supply chain, as well as in commercial solutions development, included roles ranging from manufacturing process improvements, production management, development of messaging and communications, to creation and delivery of software solutions to extend Xerox's products and service offerings to its customers. She joined Corning Inc., New York, NY, USA, in 2018, and has been working as the Project Manager in Smart Manufacturing solutions. In this role, she leads collaborative, multiorganizational projects that apply digital technologies to Corning's manufacturing plants, and more broadly across the technical community. She has the Lean Six Sigma Green Belt certification and is a holder of three U.S. patents.

Yasmin Musa obtained her B.Sc. degree in Biochemistry from the University of Alabama in Huntsville in 2012, and the level 2 Mechatronics certification from Motlow State Community College in 2018.

Ahad Nasab is the head of the Department of Engineering Management and Technology at the University of Tennessee at Chattanooga since 2018.

Kamalendu Pal is with the Department of Computer Science, School of Science and Technology, City, University of London. Kamalendu received his BSc (Hons) degree in Physics from Calcutta University, India, Postgraduate Diploma in Computer Science from Pune, India, MSc degree in Software Systems Technology from the University of Sheffield, Postgraduate Diploma in Artificial Intelligence from Kingston University, MPhil degree in Computer Science from the University College London, and MBA degree from the University of Hull, United Kingdom. He has published over ninety international research articles (including book chapters) widely in the scientific community with research papers in the ACM SIGMIS Database, Expert Systems with Applications, Decision Support Systems, and conferences. His research interests include knowledge-based systems, decision support systems, teaching and learning practice, blockchain technology, software engineering, service-oriented computing, sensor network simulation, and supply chain management. He is on the editorial board of an international computer science journal and is a member of the British Computer Society, the Institution of Engineering and Technology, and the IEEE Computer Society.

Dajiah Platt is a chemical engineer, she obtained her bachelor's degree in Chemical Engineering from the University of Tennessee at Chattanooga in 2021.

Rohit Rastogi received his B.E. degree in Computer Science and Engineering from C.C.S.Univ. Meerut in 2003, the M.E. degree in Computer Science from NITTTR-Chandigarh (National Institute of Technical Teachers Training and Research-affiliated to MHRD, Govt. of India), Punjab Univ. Chandigarh in 2010. Currently he is pursuing his Ph.D. In computer science from Dayalbagh Educational Institute, Agra under renowned professor of Electrical Engineering Dr. D.K. Chaturvedi in area of spiritual consciousness. Dr. Santosh Satya of IIT-Delhi and dr. Navneet Arora of IIT-Roorkee have happily consented him to co supervise. He is also working presently with Dr. Piyush Trivedi of DSVV Hardwar, India in center of Scientific spirituality. He is a Associate Professor of CSE Dept. in ABES Engineering. College, Ghaziabad (U.P.-India), affiliated to Dr. A.P. J. Abdul Kalam Technical Univ. Lucknow (earlier Uttar Pradesh Tech. University).Also, He is preparing some interesting algorithms on Swarm Intelligence approaches like PSO, ACO and BCO etc.Rohit Rastogi is involved actively with Vichaar Krnati Abhiyaan and strongly believe that transformation starts within self.

Zurana Mehrin Ruhi is currently enrolled as a master's student in the DSAI program at the University of Saarland. Her primary interest lies in the intersection of Machine Learning and Computer Vision.

Parul Singhal received her B.Tech degree from AKTU Univ. Presently She is M.Tech. Second Year student of CSE in ABESEC, Ghaziabad. She is a Teaching Assistant in the CSE department of ABES Engineering. Ghaziabad, India. She is working presently on data mining (DM) and machine learning (ML). She is also working on Yagyopathy. She has keen interest in Google surfing. Her hobbies is playing badminton and reading books. She is young, talented and dynamic.

Khalid Tantawi is an Assistant Professor in Mechatronics at the University of Tennessee at Chattanooga. He holds a PhD and MSc. in Electrical Engineering from the University of Alabama in Huntsville, and a double MSc. in Aerospace Engineering from the Institut Superieur de l'Aeronautique et de l'Espace and University of Pisa. He was the elected chair of the Engineering section of the Tennessee Academy of Science in 2017 and for 2022, and was an academic auditor for the Tennessee Board of Regents, and a member of the Tennessee Textbook Advisory Panel for the Tennessee Department of Education, and the European Commission's Erasmus Mundus Association. He founded the Society of Manufacturing Engineers chapter at UTC, and the SkillsUSA, and American Chemical Society chapters at Motlow State Community College. With more than 30 journal and conference publications, Dr. Tantawi was also an invited speaker at several conferences, and chaired many conference sessions. He reviewed and judged many textbooks, student competitions, scientific papers, and grant proposals, and was awarded several grants in mechatronics and advanced manufacturing that amount to more than $750,000. He is an active member of the Society of Manufacturing Engineers, IEEE, the Industrial Engineering and Operations Management Society, and the Tennessee Academy of Science.

Omar Tantawi is an associate professor of Mechatronics at Motlow State Community College, he is a Siemens-certified Mechatronics Instructor. His primary research interests are in advanced manufacturing, industrial robotics, and vision systems.

Anwar Tantawy obtained his PhD in Telecommunications Science from the National Institute for Scientific Research, Montreal, Canada specializing in Artificial Intelligence - Automatic Speech Recognition.

Index

T

U

V

X

Y

Recommended Reference Books

IGI Global's reference books are available in three unique pricing formats: Print Only, E-Book Only, or Print + E-Book.

Order direct through IGI Global's Online Bookstore at **www.igi-global.com** or through your preferred provider.

ISBN: 9781799884552
EISBN: 9781799884576
© 2022; 394 pp.
List Price: US$ **270**

ISBN: 9781799897101
EISBN: 9781799897125
© 2022; 335 pp.
List Price: US$ **250**

ISBN: 9781668441398
EISBN: 9781668441411
© 2022; 291 pp.
List Price: US$ **270**

ISBN: 9781668441534
EISBN: 9781668441558
© 2023; 335 pp.
List Price: US$ **270**

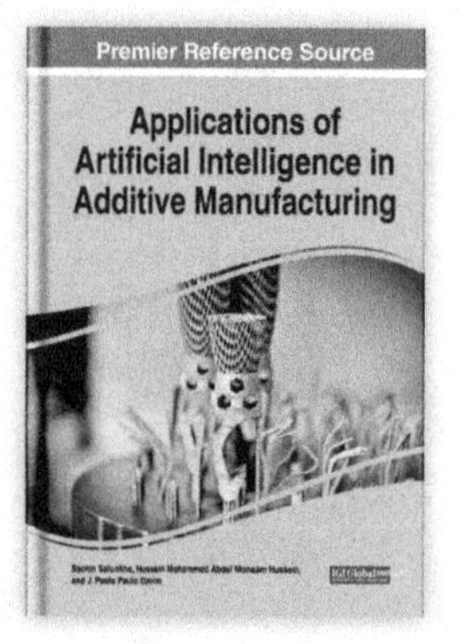

ISBN: 9781799885160
EISBN: 9781799885184
© 2022; 240 pp.
List Price: US$ **270**

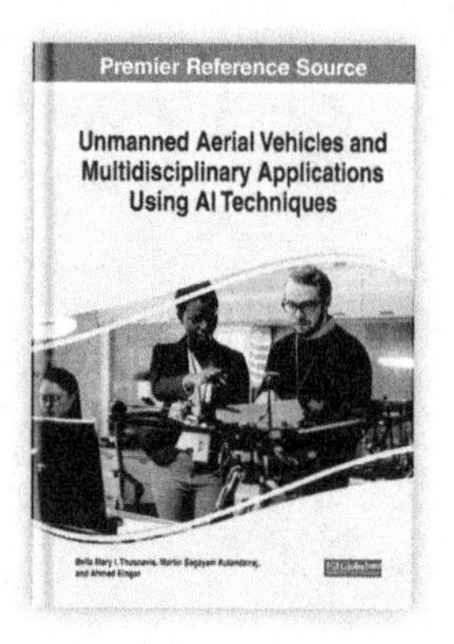

ISBN: 9781799887638
EISBN: 9781799887652
© 2022; 306 pp.
List Price: US$ **270**

Do you want to stay current on the latest research trends, product announcements, news, and special offers?
Join IGI Global's mailing list to receive customized recommendations, exclusive discounts, and more.
Sign up at: **www.igi-global.com/newsletters.**

Publisher of Timely, Peer-Reviewed Inclusive Research Since 1988

www.igi-global.com Sign up at www.igi-global.com/newsletters facebook.com/igiglobal twitter.com/igiglobal linkedin.com/igiglobal

Ensure Quality Research is Introduced to the Academic Community

Become an Evaluator for IGI Global Authored Book Projects

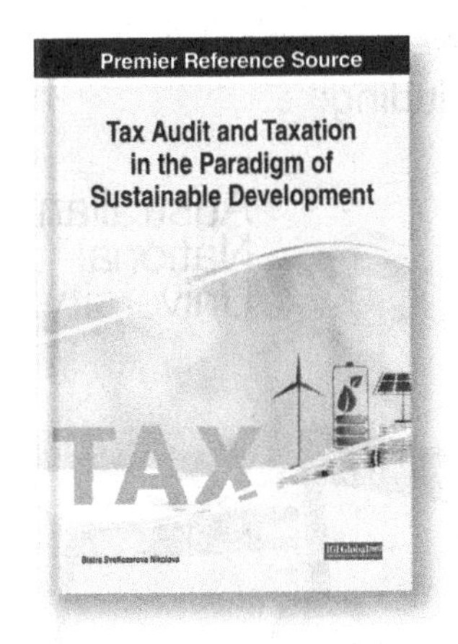

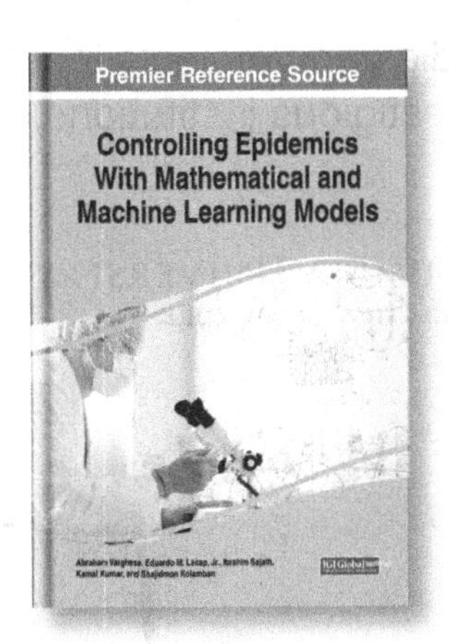

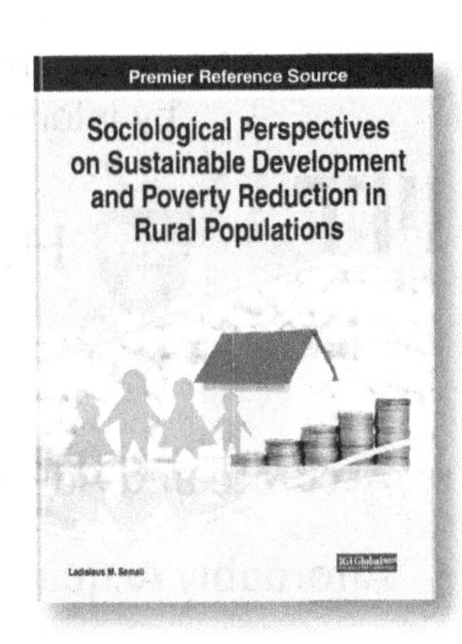

The overall success of an authored book project is dependent on quality and timely manuscript evaluations.

Applications and Inquiries may be sent to:
development@igi-global.com

Applicants must have a doctorate (or equivalent degree) as well as publishing, research, and reviewing experience. Authored Book Evaluators are appointed for one-year terms and are expected to complete at least three evaluations per term. Upon successful completion of this term, evaluators can be considered for an additional term.

If you have a colleague that may be interested in this opportunity, we encourage you to share this information with them.

Easily Identify, Acquire, and Utilize Published Peer-Reviewed Findings in Support of Your Current Research

IGI Global OnDemand

Purchase Individual IGI Global OnDemand Book Chapters and Journal Articles

For More Information:
www.igi-global.com/e-resources/ondemand/

Browse through 150,000+ Articles and Chapters!

Find specific research related to your current studies and projects that have been contributed by international researchers from prestigious institutions, including:

- Accurate and Advanced Search
- Affordably Acquire Research
- Instantly Access Your Content
- Benefit from the InfoSci Platform Features

"*It really provides* **an excellent entry into the research literature of the field.** *It presents a manageable number of* **highly relevant sources** *on topics of interest to a wide range of researchers. The sources are* **scholarly, but also accessible** *to 'practitioners'.*"

- Ms. Lisa Stimatz, MLS, University of North Carolina at Chapel Hill, USA

Interested in Additional Savings?

Subscribe to

IGI Global OnDemand

Learn More

Acquire content from over 128,000+ research-focused book chapters and 33,000+ scholarly journal articles for as low as US$ 5 per article/chapter (original retail price for an article/chapter: US$ 37.50).

7,300+ E-BOOKS. ADVANCED RESEARCH. INCLUSIVE & AFFORDABLE.

IGI Global e-Book Collection

- Flexible Purchasing Options (Perpetual, Subscription, EBA, etc.)
- Multi-Year Agreements with No Price Increases Guaranteed
- No Additional Charge for Multi-User Licensing
- No Maintenance, Hosting, or Archiving Fees
- Continually Enhanced & Innovated Accessibility Compliance Features (WCAG)

Handbook of Research on Digital Transformation, Industry Use Cases, and the Impact of Disruptive Technologies
ISBN: 9781799877127
EISBN: 9781799877141

Handbook of Research on New Investigations in Artificial Life, AI, and Machine Learning
ISBN: 9781799886860
EISBN: 9781799886877

Handbook of Research on Future of Work and Education
ISBN: 9781799882756
EISBN: 9781799882770

Research Anthology on Physical and Intellectual Disabilities in an Inclusive Society (4 Vols.)
ISBN: 9781668435427
EISBN: 9781668435434

Innovative Economic, Social, and Environmental Practices for Progressing Future Sustainability
ISBN: 9781799895909
EISBN: 9781799895923

Applied Guide for Event Study Research in Supply Chain Management
ISBN: 9781799889694
EISBN: 9781799889717

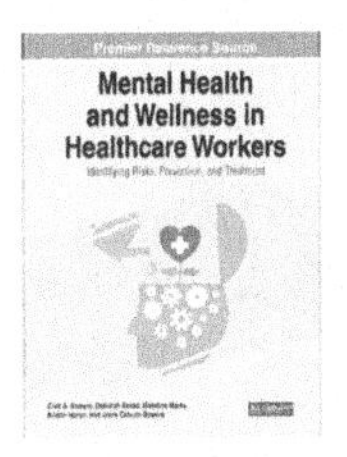

Mental Health and Wellness in Healthcare Workers
ISBN: 9781799888130
EISBN: 9781799888147

Clean Technologies and Sustainable Development in Civil Engineering
ISBN: 9781799898108
EISBN: 9781799898122

Request More Information, or Recommend the IGI Global e-Book Collection to Your Institution's Librarian

For More Information or to Request a Free Trial, Contact IGI Global's e-Collections Team: eresources@igi-global.com | 1-866-342-6657 ext. 100 | 717-533-8845 ext. 100

Printed in the United States
by Baker & Taylor Publisher Services